Water Demand Management

Water Demand Management

David Butler and Fayyaz Ali Memon

Department of Civil & Environmental Engineering
Imperial College London

Published by   IWA Publishing
                  Unit 104–105, Export Building
                  1 Clove Crescent
                  London E14 2BA, UK
                  Telephone: +44 (0)20 7654 5500
                  Fax: +44 (0)20 7654 5555
                  Email: publications@iwap.co.uk
                  Web: www.iwaponline.com

First published 2006
© 2006 IWA Publishing

Disclaimer

The information provided and the opinions given in this publication are not necessarily those of IWA or of the authors, and should not be acted upon without independent consideration and professional advice. IWA and the authors will not accept responsibility for any loss or damage suffered by any person acting or refraining from acting upon any material contained in this publication.

British Library Cataloguing in Publication Data
A CIP catalogue record for this book is available from the British Library

Library of Congress Cataloging- in-Publication Data
A catalog record for this book is available from the Library of Congress

ISBN: 9781843390787 (hardback)
ISBN: 9781789065381 (paperback)
ISBN: 9781780402550 (eBook)

Contents

8 Demand management in developing countries 180

Kalanithy Vairavamoorthy and
M.A. Mohamed Mansoor

9 Drivers and barriers for water conservation and reuse in the UK
Susan Roaf

Preface

A common characteristic of water demand in urban areas worldwide is its inexorable rise over many years, and projections of continuing growth over coming decades. The chief influencing factors are population growth and migration, together with changes in lifestyle, demographic structure and the possible effects of climate change. The detailed implications of climate change are not yet clear, and anyway will depend on global location, but must at least increase the uncertainty in security of supply. This is compounded by rapid development, creeping urbanization and, in some places, rising standards of living.

Meeting this increasing demand from existing resources is self-evidently an uphill struggle, particularly in water stressed/scarce regions, in the developed and developing world alike. There are typically two potential responses; either 'supply-side', meeting demand with new resources or 'demand-side', managing consumptive demand itself to postpone or avoid the need to develop new resources. There is considerable pressure from the general public, regulatory agencies, and some governments to minimise the impacts of new supply projects (e.g. building new reservoirs or inter-regional transfer schemes), implying the emphasis should be shifted towards managing water demand by best utilising the water that is already available.

In the UK, considerable effort is now being concentrated on addressing future water needs, and demand management is seen as a key element in the government's sustainable development policy, which concentrates on managing demand for water by controlling leakage and maximising its efficient use. The need to develop, investigate and implement environmentally sustainable, technically feasible, economically viable and socially acceptable options has never been more urgent.

This document on *water demand management* is written within the context just outlined and has been produced by the academic/government/industry network – *WATERSAVE*. This and other similar networks have been sponsored over three years by the UK's Engineering Physical Science and Engineering Research Council to foster collaboration and technology transfer in important industry sectors.

The book now in your hands was produced as one of a series of key deliverables of the network; the others being six national workshops, an international conference and a web-site. Further details of these other elements can be found at http://www.watersave.uk.net.

The concept of the book was to assemble a comprehensive picture of demand management topics ranging from technical to social and legal aspects, through expert critical literature reviews on the subject. We believe the depth and breadth of coverage to be a unique contribution to the field.

Finally, writing a chapter for a book of this type is no mean feat, and we would like to acknowledge the contribution and dedication of our team of authors. Throughout the book you will find a range of styles and content and approach, but we suggest the whole is greater than the sum of the parts. Indeed, this variety reflects the diversity of approaches needed to tackle the important goal of safely and wisely managing out water into the 21st century.

David Butler
Fayyaz A. Memon

Contributors

Dr John Blanksby

Research Fellow (Pennine Water Group)
Dept. of Civil and Structural Engineering,
Sir Frederick Mappin Building, Mappin Street, University of Sheffield,
Sheffield, S1 3JD, UK.
Tel: +44 (0) 114 222 5768; Fax: +44 (0) 114 222 5700;
Email: j.blanksby@sheffield.ac.uk

Professor David Butler

Head, Urban Water Research Group
Dept. of Civil & Environmental Engineering, South Kensington Campus
Imperial College London, SW7 2AZ, UK.
Tel: +44 (0) 207594 6099; Fax: +44 (0) 207594 6124
Email: d.butler@imperial.ac.uk
http://ewre-www.cv.imperial.ac.uk

Dr Andrew M. Dixon

Interdisciplinary Projects Officer for the Environment Division
Dept. of Animal and Plant Sciences,University of Sheffield,
Western Bank Sheffield S10 2TN, UK.
Tel: +44 (0) 114 2220063
Email: a.m.dixon@shef.ac.uk

Malcolm Farley

Principal Consultant
Malcolm Farley Associates
The Firs, Station Road, Alvescot, Bampton, Oxfordshire, OX18 2PS, UK.
Tel: +44 (0) 1993 841357; Fax: +44 (0) 1993 842924
Email: mfarley@alvescot.demon.co.uk

Alan Fewkes

Senior Lecturer
School of the Built Environment, The Nottingham Trent University,
Nottingham, NG1 4BU, UK.
Tel.: +44 (0) 115 941 8418; Fax: + 44 (0) 115 848 6438
Email: alan.fewkes@ntu.ac.uk

Mary Gearey

Research Scholar
School of Water Sciences, Cranfield University, Beds, MK43 0AL, UK.
Tel: +44 (0) 1234 750111 ext. 3336; Fax: +44 (0) 1234 751671
Email: m.gearey@cranfield.ac.uk

Nick Grant

Water Consultant
Elemental Solutions, Hereford, HR2 8SE, UK.
Tel: + 44 (0)1981 540728; Fax: + 44 (0)1981 541044
Email: nick@elementalsolutions.co.uk   http://elementalsolutions.co.uk

Paul Herrington

Part-time tutor in environmental economics in the Dept. of Geography at the
University of Leicester, and Consultant, Herrington & Pey Enterprises
The Old School, Tilton, Leicester, LE7 9LF, UK.
Tel. & fax: +44 (0) 116 259 7361   Email: paulrherrington@aol.com

Dr David Howarth

Strategic Environmental Planning Manager
Environment Agency Southern Region,
Guildbourne House, Chatsworth Road Worthing,
West Sussex, BN11 1LD, UK.
Tel: +44 (0) 1903 832391; Fax: +44 (0) 1903 832211
Email: david.howarth@environment-agency.gov.uk

Dr Paul Jeffrey

Principal Research Fellow
School of Water Sciences,
Cranfield University, Cranfield, Beds, MK43 0AL, UK.
Tel: +44 (0) 1234 754814; Fax: +44 (0) 1234 751671
Email: P.J.Jeffrey@Cranfield.ac.uk

Dr Christos K. Makropoulos

Research Fellow and Assistant Editor Urban Water Journal
Urban Water Research Group
Dept. of Civil & Environmental Engineering, South Kensington Campus
Imperial College London, SW7 2AZ, UK.
Tel: +44 (0) 207594 6020; Fax: +44 (0) 207594 6124
Email: c.makropoulos@imperial.ac.uk
http://ewre-www.cv.imperial.ac.uk; http://www.urbanwater.net

M.A.Mohamed Mansoor

Research Scholar
Water, Engineering and Development Centre (WEDC)
Department of Civil and Building Engineering,
Loughborough University, Leicestershire LE11 3TU, UK.
Tel: +44 (0) 1509 222809; Fax: +44 (0) 1509 211079
Email: m.a.mohamed-mansoor@lboro.ac.uk

Marcelle McManus

Sustainable City Team, CREATE Centre,
Bristol City Council, Bristol, BS1 6XN, UK.
Tel: +44 (0) 117 9224472
Email: marcelle_mcmanus@bristol-city.gov.uk

Dr Fayyaz Ali Memon

WaND Project Manager & WATERSAVE Network Co-ordinator
Urban Water Research Group
Dept. of Civil and Environmental Engineering, South Kensington Campus
Imperial College London, SW7 2AZ, UK.
Tel: +44 (0) 207 594 6020; Fax: +44 (0) 207 594 6124
Email: f.a.memon@imperial.ac.uk;
http://www.watersave.uk.net; http://www.wand.uk.net

Professor Susan Roaf

School of Architecture,
Oxford Brookes University, Gipsy Lane, Oxford, OX3 0BP, UK.
Tel: +44 (0) 1865 484075; +44 (0) 1865 483298
Email: sroaf@brookes.ac.uk

Stuart Trow

Director
Trow Consulting
The Vineries, Chester le Street, Durham, DH3 3ND, UK.
Tel & Fax: +44 (0) 1913 882296
Email: stuarttrow@aol.com

Dr Kalanithy Vairavamoorthy

Senior Lecturer
Water, Engineering and Development Centre (WEDC)
Department of Civil and Building Engineering,
Loughborough University, Leicestershire LE11 3TU, UK.
Tel: (+44 (0) 1509 222622; Fax: +44 (0) 1509 211079
Email: K.Vairavamoorthy@lboro.ac.uk
http://www.lboro.ac.uk/wedc/

Dr William S. Warner

Environmental Consultant
Oslo, Norway.
Email: williamswarner@hotmail.com

1

Water consumption trends and demand forecasting techniques

Fayyaz Ali Memon and David Butler

1.1 INTRODUCTION

This chapter gives an account of domestic water consumption patterns, factors driving change in consumption trends and demand-forecasting techniques currently in use. A major part of the paper is within the context of research that has taken place in the UK. However, wider references are made to present a broader picture and show similarities and contrasts in consumption tends.

1.2 THE BIG PICTURE

Will the available freshwater resources be sufficient to meet future demand if current water consumption trends remain unchanged? This is an important question, but understandably, not one that can be answered simply, since it requires a thorough assessment of the impact of some complex factors such as the pace of population growth, emerging socio-economic trends and the extent of climate change. The total requirement of water for domestic use in the world is about $200km^3$/year, which is some 0.5% of the average total runoff (Stephenson, 2003). Theoretically, it is possible to meet existing and future domestic water demands, but the problems associated with its distribution in time and space and affordability are some of the factors widening the gap between the demand and supply in many parts of the world.

Table 1.1 Water stress in terms of relative water demand (RWD)

Level of water stress	RWD
Low	<0.1
Moderate	0.1 to 0.2
Medium-high	0.2 to 0.4
High	> 0.4

The UN (1997) has classified the level of water stress in terms of relative water demand (i.e. the ratio of total use to total water from available resources) as shown in Table 1.1. Vörösmarty *et al.* (2000) made an attempt to predict the influence of population growth and climate change and assess the level of water stress for various regions of the world using a water balance model in combination with two global climate circulation models (CGCM1 and HadCM2). The water stress in 1985 and predictions for the year 2025 for each continent are shown in Table 1.2.

The results indicate that there is a considerable increase in the relative stress for all regions, with the increase mainly due to population growth rather than climate change. Comparison of the 2025 predictions with the water stress levels shown in Table 1.1 apparently paints a rosy picture. Yet, the global scale masks the water shortage and drought-like situations which emerge when investigations are made at smaller (country and regional) scale. For example, in 1993-94 in England and Wales, about half of the regions (covered by the Environment Agency) used more than 80% of the 1 in 50 year drought supply and one used more than 90% (DoE, 1996). This problem is exacerbated by an apparent change in climate that may be altering the reliability of water resource stock replenishment (Mitchell, 1999). There is an indication that the temporal distribution of precipitation is also changing with wetter winters and drier summers (Wigley and Jones, 1987). Therefore, there is no guarantee that sufficient quantities of water will be available, when needed, in the areas where long-term average annual precipitation is 'normal'. The 1995 drought caused severe problems in the UK, resulting in an additional cost of approximately £47 million to satisfy the water demand for that year. For the UK, as whole, 1997 represented the seventh year since 1989 that drought orders (legally enforced restrictions on non-essential use of water) were applied (Mitchell, 1999). Therefore, a better understanding of water consumption trends and the implementation of appropriately designed demand management studies is clearly essential.

In the UK, there are four main uses for water: domestic (public water supplies), power generation, industrial and agricultural. Figure 1.1 shows total water abstracted annually for these uses from 1971 to 1991 (in England and Wales). The figure shows that over the years there has been a substantial increase in public water supplies, a gradual reduction in water abstracted for industry and power generation and a marginal increase in agricultural sector. Despite the increase in public water supplies, the total quantity of water abstracted reduced by 16% in 1991 compared to base figures in 1971. Of the aforementioned four water uses, only the consumption aspects related to domestic supplies will be discussed in this chapter.

Table 1.2 Observed and predicted water stress in 1985 and 2025 (Vörösmarty *et al.*, 2000)

Area	Population (millions)		Available water (km^3/yr)		Water Stress in 1985	Change in water stress, relative to 1985, in 2025 (%)		
	1985	2025	1985	2025		Climate	Population	Combined
Africa	543	1440	4520	4100	0.032	10	73	92
Asia	2930	4800	13700	13300	0.129	2.3	60	66
Australia	22	33	714	692	0.025	2.0	30	44
Europe	667	682	2770	2790	0.154	-1.9	30	31
North America	395	601	5890	5870	0.105	-4.4	23	28
South America	267	454	11700	10400	0.009	12	93	121
Globe	4830	8010	39300	37100	0.078	4.0	50	61

1.3 PER CAPITA WATER CONSUMPTION

The primary aim of providing water to households is to meet the basic water requirements of the residents. Gleick (1996) proposed a minimum of 50 litres/person.day (lpd) as the basic water requirement for meeting the four basic human needs: drinking water for survival, water for human hygiene, water for sanitation services and modest household needs for preparing food.Unfortunately, over 50 sovereign states in the world are unable to meet this basic water requirement for domestic use. About 70% of these countries can only provide just less than 30 lpd, including countries such as Nigeria where there is substantial oil production. Per capita domestic consumption of water varies from country to country, largely depending on economic well-being, traditional habits regarding sanitation, availability of freshwater resources and political willingness to improve water efficiency. As an example, Figure 1.2 shows the domestic per capita consumption in a variety of countries. At the upper extreme is the USA with an average consumption above 300 lpd. At the lower end are countries such as Gambia and Nigeria where consumption is in the range of 4-30 lpd.

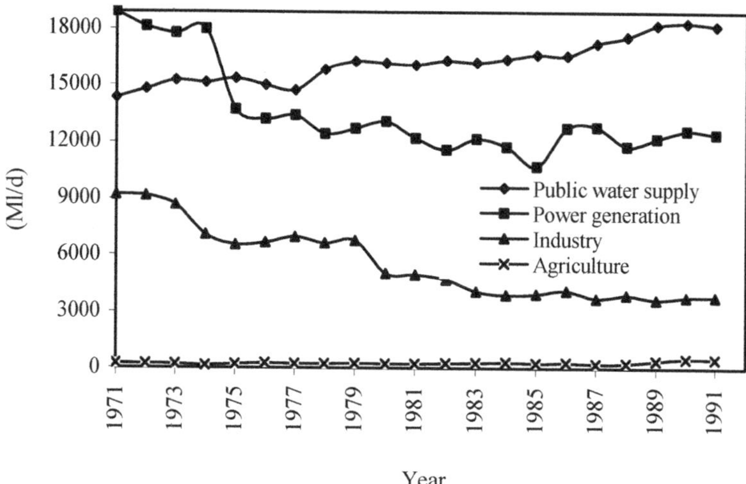

Figure 1.1 Licensed non-tidal water abstractions in England & Wales (Herrington, 1996).

The trend in water consumption changes from country to country, depending on several factors including climate, availability of resources, technological advancement, water price structure, incentives and legislative provisions. For example, an analysis of 27 years of data on per capita consumption in OECD countries showed that per capita average consumption in Japan has been reasonably constant since 1990. In England and Wales and Korea, water consumption has increased over the years. In Germany, a falling per capita consumption can be observed (Herrington, 1999).

In the UK, per capita consumption is about 150 lpd and rising, but this average figure masks a considerable variation between individuals. Figure 1.3 shows a log-normal plot, with elongated tail, reflecting high consumption by a small section of the sample population.

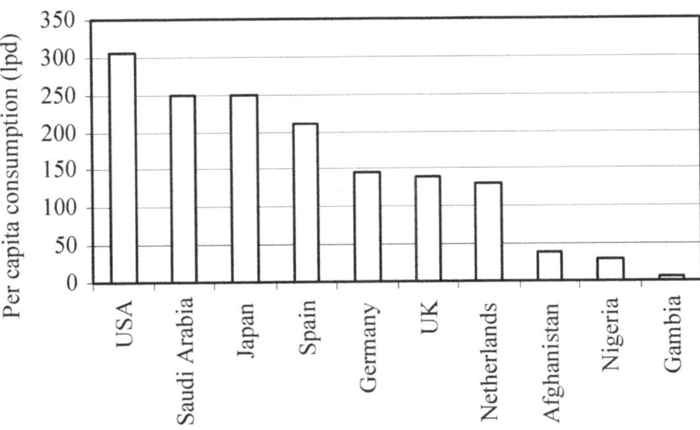

Figure 1.2 Per capita domestic water consumption in some countries.

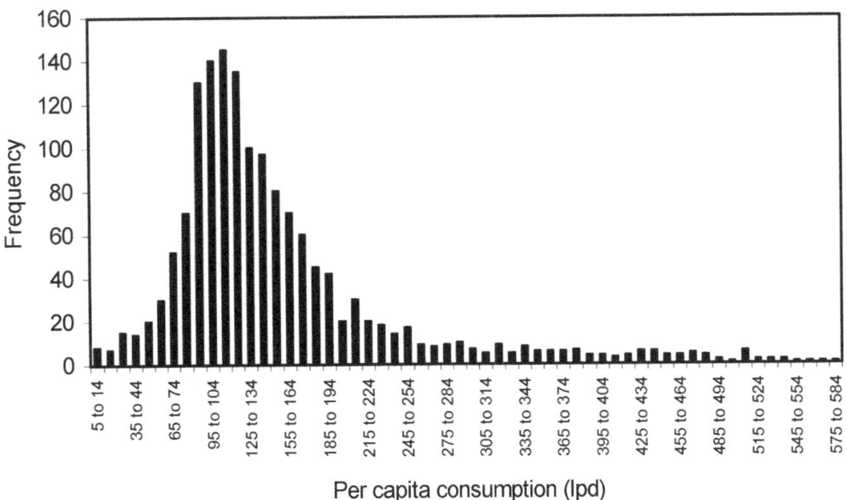

Figure 1.3 A typical frequency distribution profile for per capita consumption in the UK (Edwards and Martin, 1995).

1.4 FACTORS INFLUENCING CONSUMPTION

The level of domestic demand varies considerably from household to household depending on the socio-economic factors and household characteristics. Several studies, conducted to investigate the influence of such factors, have revealed that per capita consumption changes with, for example, household size, type of property, ages of household residents and time of the year. The results of these studies are briefly reviewed.

Occupancy (i.e. number of people living in a household) has a direct influence on per capita consumption. Although an increase in the number of inhabitants per household increases the total domestic water consumption, there is a general agreement that per capita consumption decreases with increased occupancy (Butler, 1991; Edwards and Martin, 1995). For example, in a single person household, per capita water consumption is 40% greater than in a 2-person household, 73% greater than in a 4-person household and over twice that in households of 5 or more people (POST, 2000). This trend is shown in Figure 1.4. From a future water demand perspective, this relationship between the occupancy and per capita consumption is very important since much of the projected growth in the number of households over the coming decades in developed economies is expected to be from single-person households and other small properties. According to the UK government projections, the number of new homes in England and Wales is expected to increase by 3.3 million between 1996 and 2016 and the trend is towards homes with smaller household size (EA, 2001).

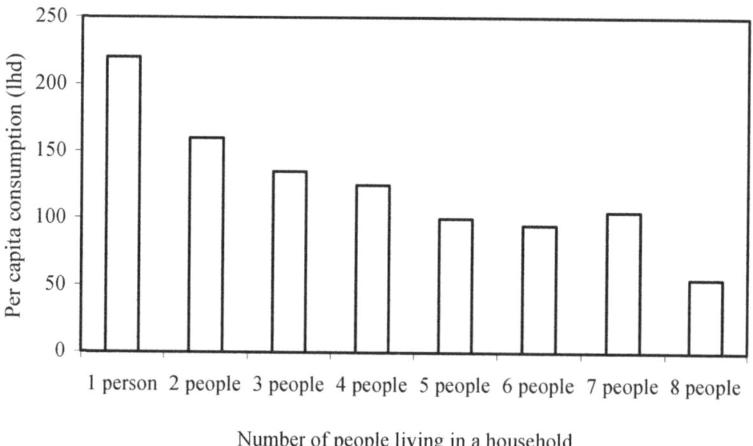

Figure 1.4 Impact of household size on per capita consumption (Edward and Martin, 1995).

The influence of household type (e.g. detached, semi detached house, flat) on average consumption per household was investigated by Russac *et al.* (1991). He found that the demand was highest in detached houses and lowest in flats. This was attributed to the relatively low per capita garden watering requirement for flats. Additionally, the high water demand in detached houses may be linked to the greater space available for appliances or socio-economic reasons. BSRIA (1998) and AWWA (1999) found a correlation with floor space.

Russac *et al.* (1991) also made an interesting observation with regard to consumers' age groups and water demand. Retired people in one-person dwellings consumed 200 litres per day on average as compared to 140 litres per day by an adult living in a single-person residence. This may be due to the fact that retired people stay at home for longer periods. Age-related diseases such as diabetes and, in men, prostate problems often result in the increased frequency of urination and hence use of the toilet (Green, 2003).

Seasonal variations are also reported to cause changes in the level of water demand. This demand variation is generally linked to garden watering. According to Herrington (1996), garden watering in houses in the South and East of England, using sprinklers, took place once every six days during May to August in an average year in the early 1990s. The estimated average volume for each irrigation session ranges between 1000-1200 litres. About 40% of households use hosepipes, on average, three times per week in hot dry weather consuming approximately 315 litres per use (Three Valleys, 1991). The UK based water consumption monitoring projects suggest that much of the peak demand is due to garden watering, with an estimated 4% of annual consumption taking place over an eight week period due to this activity (CC:DW, 2001). In long dry periods, garden watering could constitute up to 40% of the total consumption in some months.

Clearly, affluence is a key factor in influencing water consumption. This is most starkly illustrated in developing countries (see Table 1.3) where per capita water consumption can vary significantly, depending on economic well being of the community. This topic is returned to again in Chapter 8.

Table 1.3 Variation in domestic water use as a function of affluence (adapted from Stephenson, 2003)

Type of dwelling/supply source	Average consumption (lpd)
High-quality housing areas	225
Urban residential areas	180
Suburban low cost housing	95
Urban areas served by standpipes	60
Rural areas served by standpipes	40
Rural dwellings with distance to source >1km	20

1.5 CONSUMPTION BY MICRO-COMPONENT

To understand consumption patterns and trend more deeply, it is necessary to study the individual uses of water within the house (micro-components), whether for personal hygiene (e.g. water use in wash basins, WCs, shower and bath) or communal use (e.g. water use in washing machines, dishwashers, garden and car washing).

Significant information is now available and patterns of micro-component use have been studied with respect to their share of total household demand, volume per use, frequency of use, level of ownership and their peak use hours. The main findings of several key studies are reviewed here.

In the UK, a comprehensive study of domestic water consumption patterns (SoDCon) by Anglian Water has been ongoing since 1991. The study includes water consumption monitoring in about 3000 houses in the Anglian (eastern England) region with 100 houses exclusively monitored for every micro-component. Figure 1.5 shows the averaged values from 1993 to 1998 of the SoDCon data. Over a year, the greatest uses (60%) are for personal hygiene, followed by washing machines and dishwashers (21%), kitchen taps (15%) and outside taps (4%).

In the USA, domestic water consumption patterns for single-family houses were studied over two years in 12 different cities located in different climatic regions by AWWA (1999). The results showed that average water consumption was highest in toilets (27.6%), followed by clothes washing (21.7%), showers (16.8%), faucet (13.7%), baths (1.7 %), dishwasher (1.4%) and other domestic uses (2.2%).

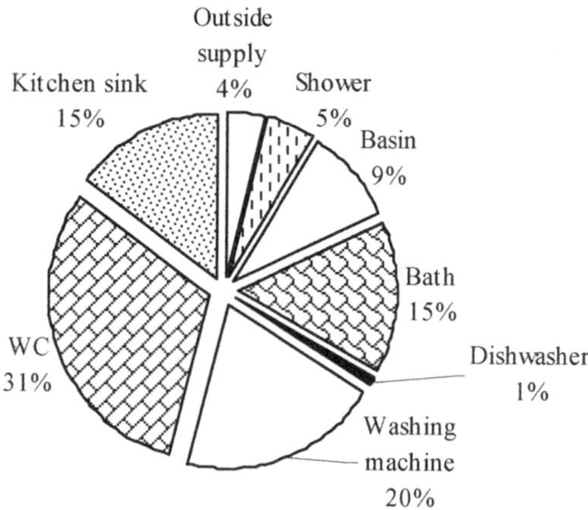

Figure 1.5 Water consumption share of different micro-components (POST, 2000).

In the Netherlands, the highest consumption is for toilets (37%), followed by bathrooms (26%), kitchen use (16%) and clothes washing (16%). In Sweden, the appliance share is slightly different. The major consumption is in bathrooms (32%) followed by kitchen use (23%), toilets (18%) and clothes washing (13%) (EAA, 2001).

Butler (1991) carried out a diary survey in 28 households in Southern England to investigate the pattern in which different appliances are used within a household throughout the day. The flow and the usage pattern of each micro-component varies with time. The peak frequency and time to peak discharge from several appliances are shown in Table 4.1, based on UK data.

Table 1.4 Approximate peak frequency and time to peak use for different appliances

Micro-component	Peak frequency (Uses/hour) Butler (1993)	Time of peak (Edwards and Martin, 1995)
Washing machine (WM)	0.03	10:45
Dishwasher (DW)	-	03:15
Bath	0.14	18:45
Shower	0.32	07:30
Toilet (WC)	1.2	08:00

In a separate study, Herrington (1996) observed that the frequency of bath/shower increased during 1976 to 1990, suggesting that a cultural change in personal washing has occurred and is continuing. A comparison of change in domestic water consumption pattern for Southeast of England over 25 years is shown in Figure 1.6. The per capita water consumption for the bath/shower has increased from 27% in 1976 to 33% in 2001 of the total consumption. Figure 1.6 also shows a marginal decrease in per capita consumption for toilet flushing, perhaps resulting due to introduction of low flush toilets.

Table 1.5 Average water consumption in domestic appliances (EEA, 2001)

Appliance	England & Wales	Finland	France	Germany
Toilet	9.5 l/flush	6 l/flush	9 l/flush	9 l/flush
Washing machine	80 l/cycle	74-117 l/cycle	75 l/cycle	72-90 l/cycle
Dishwasher	35 l/cycle	25 l/cycle	24 l/cycle	27-47 l/cycle
Shower	35 l/shower	60 l/shower	16 l/minute	30-50 l/shower
Bath	80 l/bath	150-200 l/bath	100 l/bath	120-150 l/bath

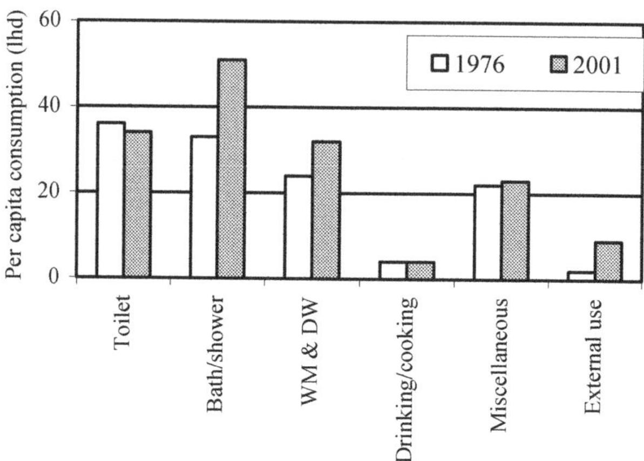

Figure 1.6 Change in domestic water consumption in South East England.

The average water consumption for each appliance per use for different countries is shown in Table 1.5.

1.6 WATER CONSUMPTION TRENDS AND SAVING POTENTIAL

The cumulative domestic water consumption profile, appearing as the result of the above-mentioned micro-component frequency and flow patterns varies significantly throughout the day (Figure 1.7). There are four distinct periods: a sharp morning peak around 08.00 am, moderate mid-day flow lasting up to 4.00 pm, an evening and relatively small late night peak and subdued low night flow until 4.00 am. The peak discharge from the WC, shower, bath and sink in the morning is on average simultaneous. Flows from the washbasin are rather more uniformly spread whilst washing machine use shows an increase after midday. Similar diurnal patterns are observed in the USA (AWWA, 1999).

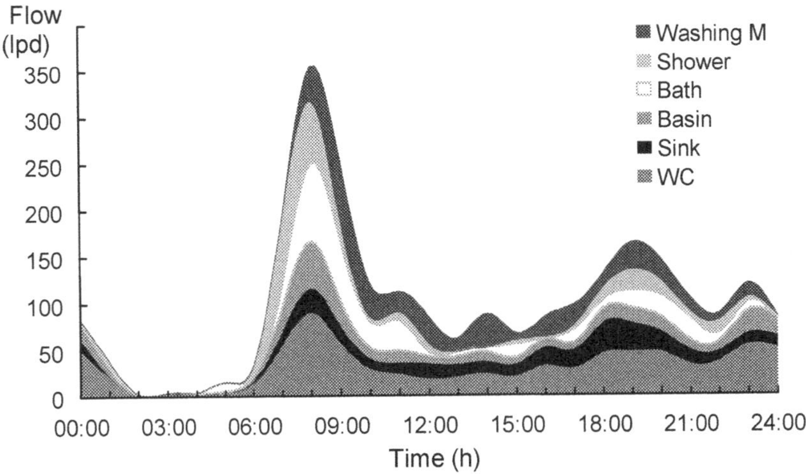

Figure 1.7 Appliances' daily wastewater discharge pattern (Butler and Davies, 2004).

The data in Table 1.5, refers to water for the appliances that have already been installed. However, new regulations and policy initiatives should ensure reduced consumption by future appliance installations. For example, the UK Water Supply (Water Fittings) Regulations of 1999 set 6 litres as the maximum volume for single flush toilets from the year 2001. Requirements for new WCs to operate at flush volumes lower than 6 litres were introduced in countries such as Singapore and Australia due to the lack of available freshwater supplies. For general use, Singapore has a maximum flush volume of 4.5 litres for newly installed WCs and Australia has a maximum flush volume of 4 litres. Australia has the lowest flush volume requirement for WCs that can be directly connected to sewers system. This requirement has reduced WCs' water consumption from 55 (in mid 1950s) to 18 litres per person per day in early 1990s (Cummings, 2001). The use of low-flush WCs is becoming increasingly popular and several low flush toilet retrofitting projects have been undertaken in different parts of the world. The estimated economic and financial gain from these projects has been reported as considerable (Green, 2003).

Technological advancements and policies designed to encourage efficient use of water have produced significant reductions in water consumption by different appliances. Figure 1.8, for example, shows the influence of water conservation measures and technological improvements giving a gradual reduction over time in the water used by washing machines. A focused discussion on the impact of various water saving devices and associated implications influencing water consumption trends is provided in Chapter 4.

BSRIA (1998) has investigated the impact of low water consuming devices and made a scenario-based assessment of potential savings in water and cost in various types of buildings (including households) in eight regions of the Environment Agency in England and Wales. The main statistics obtained from the study are shown in Table 1.6. The total water savings in England and Wales are reported to be in the range of 2.0 to 2.8% of current consumption. There lies a potential of a further 24% saving. The payback period is the lowest for factories (1.7 years) and the highest for households (29.6 years). This high payback period for households is surely one of the main barriers to be overcome. The trends for wider uptake of greywater recycling at domestic level are also not encouraging. The Environment Agency (EA, 2001) has anticipated that the uptake of greywater recycling systems at a domestic level is unlikely to exceed 10% even after 2016. The penetration of greywater recycling units is estimated to reach 18% by the end of 30 years (EA and UKWIR, 1996). The main barriers and potential drivers for the successful implementation of greywater recycling systems are identified in Chapters 3 and 9. Rainwater harvesting, as discussed in Chapter 2, appears to be a relatively more pragmatic and cost-effective option compared to greywater recycling.

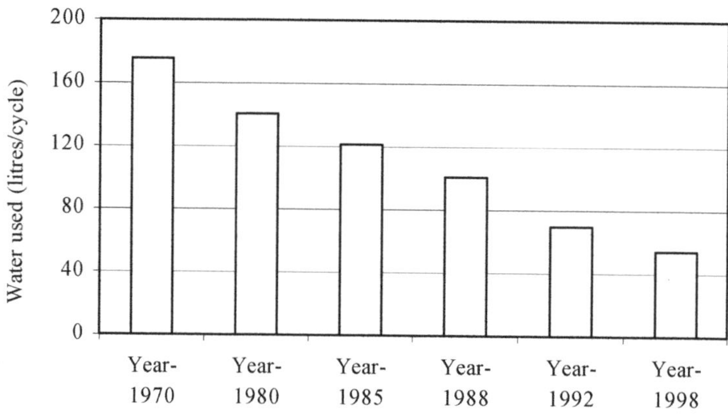

Figure 1.8 Reduction in water consumed by washing machines in last 30 years (EEA, 2001).

Table 1.6 Water savings made and potential for further savings in different types of buildings in England and Wales (BSRIA, 1998)[1]

Type of Building	Savings made already		Current saving potential	
	Water (Ml/d)	Financial £ k/year	Water (Ml/d)	Financial £ k/year
Households	55	2627	335	15958
Factories[2]	49.6	28545	37.2	21408
Hospitals	0.6	305	2.6	1271
Hotels/Motels	30.2	15989	11.6	53296
Leisure centers	1.0	524	2.7	1399
Nursing Homes	0.4	221	8.3	4865
Offices	24.9	13778	24.9	13778
Schools	43.7	23264	124.3	66219
Universities	30.5	15459	14.1	7135
Retail Stores	3.6	2447	4.7	3389

[1]As suggested by BSRIA, a thorough understanding of the assumptions and limitations is necessary before referring to the values reported in the table. Financial savings were calculated using water tariff schedules from the water companies operating in the UK.

[2]Does not include process water

1.7 DEMAND FORECASTING TECHNIQUES

In order to meet future water supply needs and assess environmental and financial sustainability of various demand management options, an accurate prediction of water demand is essential. According to Herrington (1987), forecasting water demand helps to serve the following purposes:

- Strategic planning;
- Investment appraisal;
- Operations planning;
- Appraisal of demand-management policies and innovations;
- Demand management in "crisis" periods;
- Calculation of future price trends as efficiency signals; and
- Some supply forecasting (via wastewater reclamation).

Water demand forecasting based on rules of thumb or naive extrapolation is now recognized as being inappropriate, since estimates obtained in this way have been shown to deviate significantly from the observed demand and also are of no help in appraising water conservation initiatives (Herrington, 1987). Owing to increases in population, constraints on freshwater supply and the rising costs (both economic and environmental) involved in the development of new water resources, there is a need to develop methods capable of demonstrating a high correlation with actual demand. Such methods should ideally incorporate the influence of:

- Spatial and temporal variability;
- Properly-appraised water conservation policies such as: metering, water saving devices, water recycling measures and the rate at which these measures would be adopted by consumers in the future
- Characteristics associated with various appliances used (e.g. ownership, frequency and volume of water consumed per use)
- Lessons learnt from the forecasting techniques used in the past; and
- Past water consumption trends.

Additionally, the forecasting methods should reflect a scientific basis, acceptability to the regulator and feasibility with regard to cost and data collection and validation requirements.

Selection of the forecasting technique also depends on the nature of demand that needs to be predicted. Demand varies in time. Peak daily consumption has been observed to be 1.8 times the average hourly flow and peak seasonal demand are about 1.4 to 1.8 times the daily demand (Green, 2003). Predictions of hourly and peak daily demands are helpful in managing/providing the water distribution network. For strategic and planning decisions, information on micro to macro scale (including seasonal) variations in demand is often desirable.

Forecasting techniques can be broadly divided into two groups:

- Techniques that build conceptually and require a relatively limited amount of data to produce future projections in water demand. These techniques are generally used for long-term forecasts.
- Techniques that require extensive data collection. The data is then used to formulate often complex statistical relationships and infer the rules that will indicate the level of demand. These methods are normally for short-term forecast.

Owing to a number of resource constraints and unavailability of sufficiently large sets of data on consumption patterns, it is hard to develop such forecasting techniques, which represent all the above-mentioned aspects. However, UKWIR (1997) made an attempt to develop forecasting methods in order to meet the needs of UK water companies. These methods have considered three different components of water consumption viz: unmeasured and measured household and measured non-household demand. Forecasting methods related to households will be reviewed here. These methods were designed to make long-term predictions.

1.7.1 Unmeasured household demand

UKWIR (1997) recommends two related forecasting methods using micro-component analysis and micro-component *group* analysis.

In micro-component analysis, the information on ownership level, frequency of use and volume of water consumed for each appliance (or household water using-activity) is used to calculate demand. The per capita consumption is estimated by adding up water contributions from each appliance (or activity) in unmeasured households using the following relationship.

$$pcc = \sum_i (O_i \cdot F_i \cdot V_i) + pcr \qquad (1.1)$$

Where

pcc	- per capita consumption
O_i	- proportion of household using appliance (or undertaking activity) i
F_i	- average frequency of use of appliance (or activity) i per capita among the proportion of household
V_i	- volume of water consumed by appliance (or undertaking activity) i per use
pcr	- per capita residual (miscellaneous) demand

The *pcr*, in effect, is the difference between the estimated cumulative demand from all appliances (activities) in the period under consideration and actual total demand observed during this period. For the purposes of forecasting, *pcr* is usually kept to a constant value or it can be projected by assuming it is a fraction of total *pcc*. This method does not require historical records of time series data covering several years. It is flexible enough to accommodate the change in appliance usage patterns (e.g. individual actions, water efficient devices). However, depending on the level of accuracy required in predictions, the method does require a considerable amount of resources to gather the data on appliance characteristics (e.g. ownership, frequency of use and volume of water per use).

In micro-component group analysis, the residential units showing distinct similarities in terms of appliance penetration rate, usage frequency etc. are classified into a single group. The group classification criteria are decided considering the factors having a direct or indirect influence on water consumption patterns. Some of the criteria suggested are: socio-economic (indicating the purchasing power of a particular household), type of house (e.g. flat, detached, semi-detached, terraced) and composition of household (e.g. retired, single adult, families with more than two children). The ACORN (A Classification Of Residential Neighborhoods) classification has also been widely used in the UK water industry to choose groups based on demographic attributes (Mitchell, 1999). In group analysis, the estimation procedure builds on the micro-component method as given below:

$$pcc_g = \sum_i \left(O_{i,g} \cdot F_{i,g} \cdot V_{i,g} \right) + pcr_g \qquad (1.2)$$

where

g - number of groups identified

The non-measured household (NMHH) demand can thus be estimated as:

$$NMHH = \sum_g \left(pcc_g \cdot pop_g \right) \qquad (1.3)$$

pop_g - population in group g

Group identification is the key feature of this method. This helps in examining the implications of targeted policies aimed at water conservation measures. The significance of a group-based approach increases in the case where a sharp increase in metering penetration (and hence a shift in consumer base) is expected.

1.7.2 Measured household demand

Metering household water supply can have some impact on water consumption trends. In the UK, national metering trials conducted between 1989 to 1992 found an average reduction in household consumption of 10.8% for 11 small-scale sites. The average demand effect of metering in the Anglian region is estimated at around 15 to 20% and for peak demand it is 25 to 30% (EA and UKWIR, 1996). However, installation of meters to measure household consumption is rather sluggish. According to Ofwat (2001), approximately 22% of households in England and Wales are metered. Thus sufficient amounts of historical data on water consumption in metered households are not yet available for establishing consumption patterns. However, due to a growing emphasis on water conservation in certain regions, a significant increase in the proportion of metered houses is anticipated, as discussed in Chapter 11.

UKWIR (1997) has reported on various approaches to forecasting this component of the water demand. Among these, the micro-component group analysis method was recommended, the method being broadly similar to that for unmeasured household demand. For measured households, it was suggested that the forecast is made by carrying out the micro-component analysis separately for each identified group first, and then projecting the size of the respective group. The emphasis in this method is on the significance of identification of different groups, since the composition of metered households is changing substantially. Other group identification criteria, in addition to the criteria mentioned for unmeasured demand, may include new homes and water tariff structure.

Confidence in the performance of these methods, particularly the methods using group analysis, is yet to be established. In group analysis, the data gathered from a small section is assumed to be representative of the entire water consumer base falling within that particular group. This assumption may not work well in its entirety. For example, Russac *et al.* (1991) found that similar demand occurred in different ACORN groups, and that markedly different demands occurred in the same ACORN group.

1.7.3 Scenario-based forecasting

The Environment Agency (EA, 2001) has developed a rather different scenario-based approach to forecast public water supply demands in the years 2010 and 2025 in 125 resource zones of water companies operating in England and Wales. The forecasting procedure is essentially the application of the micro-component method proposed by UKWIR (1997) but involves an extensive consideration of the influence of socio-economic changes, which could emerge in the future due to changes in social values and systems of governance in the UK. Four distinct scenarios were developed; following work undertaken by the UK government 'Foresight Programme':

Scenario Alpha (Provincial Enterprise): Under this scenario, society's interest in environmental issues and social equity is low due to low economic growth and lack of investment.
Scenario Beta (World Market): This scenario assumes a high level of economic growth but little consideration is given to social equity. Concern for the environment is low particularly in less well off sections of the community.
Scenario Gama (Global sustainability): Consideration of sustained economic growth and social equity driven by global institutions is the main feature of this scenario. The scenario assumes considerable investment in environmental research, which would produce clean technologies that help in resource conservation
Scenario Delta (Local stewardship): In this scenario, leadership at local level takes collective action to resolve environmental problems.

The application of these scenarios was demonstrated for the household sector using a set of drivers perceived to influence domestic demand. These are:

- Water policy drivers (metering and water regulations, pricing): These will affect the volume of water used in the household and restrict the ownership of high consumption appliances.
- Technology drivers (water efficient appliances, grey and rainwater reuse): Owing to investment in research and regulatory requirements, innovative technologies will render savings in water consumption.
- Behavioural drivers (type and pattern of personal washing, garden watering): This will change the pattern of the frequency of appliance use and the extent of ownership.
- Economic drivers (affordability): These influence the extent of ownership and affordability of low/high consumption devices.

The forecasting methodology is given in detail in EA (2001), but a summary of the key assumptions made to reflect the influence of four scenarios on water consumption by key micro-components is shown in Table 1.7.

Using the above-mentioned assumptions, the average measured and unmeasured per capita consumption and total household demand for water was calculated for 2025 and the results are shown in Figure 1.9. The figure shows the relative increase or decrease in forecast water demands for 2025 with respect to the corresponding data for the year 1997-1998.

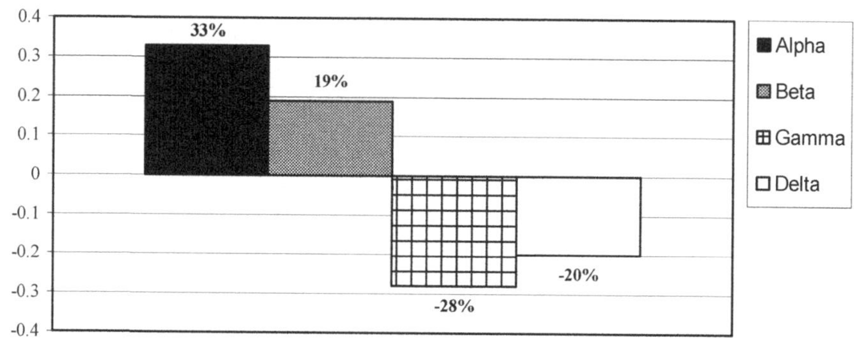

Figure 1.9 Percent change in total household demand predicted by each scenario for year 2025 as compared with 1997-1998.

Table 1.7 Key assumptions in the scenario based forecasting method (EA, 2001)

Factor		Assumptions
WC	Ownership	Alpha and delta: Rate of replacement of old WC with less water consuming device is 1 in 40 years, owing to slower technological development and lack of national regulation Beta: Replacement rate is in 1 in 30 as a result of effective implementation of existing regulation and uptake of available technology Gamma: Replacement rate is 1 in 20 due to increased affluence, awareness to environmental issues stringent regulations and efforts to develop efficient technologies
	Volume	Six devices differing in level of water consumption were considered, but in each scenario volume of particular device under consideration and frequency of use was maintained. However, the ownership for each cistern type varied according to specific scenario assumptions throughout the forecast period
Power shower	Ownership	Alpha: Power shower ownership is assumed not to exceed 50% by the year 2025 Beta: Due to high affluence 59% household will own power showers by the year 2025 Gamma: The trend for power showers is assumed to be reversed due to enforcement of strict regulations on maximum flow rate and people's willingness to replace the power showers with the normal devices. The assumed rate of replacement is 1 in 20 years. Delta: The community at a local level considers power showers inefficient and strives to minimize their personal impact on environment. Power shower use is almost eliminated by the year 2025.
	Volume	Alpha & beta: In 2025 the volume of water per power shower event is assumed to reach a maximum of 150 liters (15 l/minute for 10 minutes) Gamma & delta: Stringent regulations will limit the use of power showers exceeding the flow rate of 6 litres/minute.
Washing Machine	Ownership	Alpha, Beta and Gamma: Ownership reaches to 94% and then remains constant for the remaining forecast period. Delta: Ownership declines by 4% within the period of 2015 to 2025, reflecting the emergence of community laundries.
	Volume	Alpha: water consumed in washing machines is reduced to 80 litres by the year 2010 and afterwards it remains constant. Beta: 50 litres by 2025, Gamma and Delta: 40 litres by 2025.
Dish washer	Ownership	Alpha: Increase in the ownership at a rate of 1.7% per year and then increase at a reduced rate of 1.5% a year until 2025. Beta: Owing to high level of affluence the ownership increases at 2% a year. Gamma and Delta: Increase in the ownership at a rate of 1.7% per year and then increase at a reduced rate of 1% a year until 2025.
	Volume	Alpha: 30 litres by 2010 and remains constant thereafter. Beta: 20 litres by 2025, Gamma and Delta: 15 litres by 2025
Water recycling measures		Alpha & beta: Very limited recycling Gamma & delta: Reduction in mains water due to availability of water conservation technologies is assumed after 2010 for scenarios gamma and delta. It is assumed that 10% of the household will have greywater systems installed by the year 2025.

The EA (2001) water demand forecasting methodology provides demand projections in England and Wales for each resource zone of the Environment Agency. However, when analysing at a smaller scale, a considerable variation in domestic consumption has been reported in different parts of the same resource zone (Williamson, 1998). This variation is attributable to the change over time in demographic features, household size, social aspects (for example, change in the married and divorced fraction in the community), age groups and fertility and mortality rates. Williamson attempted to assess the impact of these factors on future water demand (at a ward level) using an official database for the metropolitan district of Kirkless, West Yorkshire. Population growth projections and variation in household composition were obtained using a static ageing technique within the micro-simulation framework, an established technique used in social and healthcare services. Although the crudeness of the forecasting assumptions to predict the water demand is acknowledged, it was demonstrated that the inclusion of spatial features in the forecasting process increases complexity, and differences in domestic water demand were visible in the spatial dimension. A detailed account of the micro-simulation approach adopted is available in Williamson (1998).

1.7.4 Statistical methods

In the USA, AWWA (1999) has developed a rather different forecasting approach. This builds on a number of complex statistical relationships developed using a part of the data collected over 2 years from metered single-family residential units in 12 cities in different climatic regions of the USA. The daily water consumption from each micro-component in a household was expressed in terms of several demand-influencing parameters such as household size and income, house floor area, degree of water conservation appliances and marginal price of water. The detailed statistical procedure adopted to derive the micro-component equations is given in full in AWWA (1999). Here, only the most relevant equations are presented. Equations 1.4-1.10 were proposed to model the water demand from each micro-component.

1. Toilet Water use Model (US gallons per household per day)

$$\hat{q}_{TOILET} = 14.483 \cdot (MPW)^{-0.225} \cdot (HS)^{0.509} \cdot (HSQFT)^{0.117}$$
$$\cdot\, e^{-0.091\,(PRE\,60s)-0.164\,(POST\,80s)-0.076\,(ULTRATIO)-0.539\,(ULTONLY)}$$

$$(1.4)$$

Where

MPW	- marginal price of water
HS	- Household size (average number of persons)
HSQFT	- Home square footage (average)
PRE60s	- fraction of houses built before 1960
POST80s	- fraction of houses built after 1980
ULTRATIO	- fraction of all toilets that are ultra-low-flow (ULF)
ULTONLY	- fraction of customers that are completely retrofitted with ULF toilets

2. Shower/Bath Water Use Model (US gallons per household per day)

$$\hat{q}_{SHOWER} = 3.251 \cdot \left(MPW\right)^{0.514} \cdot \left(HS\right)^{0.885} \cdot \left(INC\right)^{0.171} \cdot e^{0.349(RENT)-0.16(ULTSRATIO)}$$

(1.5)

where

INC	- household income ($, average)
RENT	- fraction of customers that rent
ULSRATIO	- fraction of all showerheads that are low-flow

3. Faucet Water Use Model (US gallons per household per day)

$$\hat{q}_{FAUCET} = 7.972 \cdot \left(HS\right)^{0.498} \cdot \left(HSQFT\right)^{0.077} \cdot e^{-0.254(RENT)+0.238(TRTMENT)}$$

(1.6)

where

TRTMENT	- fraction of customers with home water treatment systems

4. Dishwasher Water Use Model (US gallons per household per day)

$$\hat{q}_{DISHWASHER} = 0.409 \cdot \left(MPW\right)^{-0.5171} \cdot \left(HS\right)^{0.345} \cdot \left(INC\right)^{0.196} \tag{1.7}$$

5. Clotheswasher Water Use Model (gallons per household per day)

$$\hat{q}_{CLOTHESWASHER} = 2.293 \cdot \left(HS\right)^{0.852} \cdot \left(INC\right)^{0.162} \tag{1.8}$$

6. Leak Water Use Model (US gallons per household per day)

$$\hat{q}_{LEAKS} = 1.459 \cdot \left(MPW\right)^{-0.485} \cdot \left(MPS\right)^{-0.160} \cdot \left(HS\right)^{0.392} \cdot \left(HSQFT\right)^{0.214}$$
$$\cdot e^{-0.2641(RENT)+0.712(POOL)}$$

(1.9)

MPS	- marginal price of sewer ($/kgal)
POOL	- fraction of customers with swimming pools

7. Outdoor Water Use Model (US gallons per household per day)

$$\hat{q}_{OUTDOOR} = 0.046 \cdot \left(MPW\right)^{-0.485} \cdot \left(HSQFT\right)^{0.634} \cdot \left(LOTSIZE\right)^{0.237}$$
$$\cdot\, e^{1.116(SPRINKLER)+1.039(POOL)}$$

$$(1.10)$$

Table 1.8 Observed and predicted water consumption in US gallons per household per day (AWWA, 1999)

Micro-component	Boulder		Seattle		Waterloo	
	Observed	Predicted	Observed	Predicted	Observed	Predicted
Toilet	43.7	40.8	44.9	39.7	51.4	43.9
Clotheswasher	35	28.6	30.5	31.3	37.5	36
Shower/bath	32.4	28.4	34.3	26.4	28.5	33.3
Faucet	25.4	21.3	22.8	22.9	50.3	28.7
Leaks	5.5	5.9	9.3	5.4	17.0	6.6
Dishwasher	3.6	2.8	2.6	2.4	2.1	2.9
Other/unknown	2.6	3.0	2.8	2.5	3.8	2.9
Indoor	148.1	130.9	147.2	130.6	190.4	154.3
Outdoor	198.3	58	204.2	21.1	74.2	27.6

These equations can be used to formulate predictions of water use for each micro-component given assumptions about the demographic make-up of a particular water service area. Conceptually, one may derive a prediction of end usage over time from these equations as household and property characteristics change over time (e.g. study the effects of growth in household sizes, incomes, and home and lot sizes). In addition the toilet and shower models build in a mechanism to study the impact of particular water conservation programmes that seek to replace inefficient fixtures (AWWA, 1999).

The statistical relationships were applied using the input data collected from 13 different cities in order to assess the performance of the developed model. The results from three locations are shown in Table 1.8. For all micro-components (except outdoor water use), the difference between the observed and predicted values is small. The performance of the outdoor model is, however, poor. This may be due to the absence of various demand-influencing parameters such as weather (temperature and precipitation) and season (month of the year) from the outdoor use forecasting equation (Equation 1.10). In order to account for this, an extended version of the model was proposed. The modified version uses the output from the micro-component equations, monthly billing records and temperature and precipitation data to forecast the monthly total water demand. A marked improvement in predictions was observed when the refined model was tested (AWWA, 1999).

1.7.5 Forecasting techniques for network operations

Several other approaches have been developed to estimate total daily demand. These methods are intended for predicting the demand to address distribution network operational issues: optimisation of water head in the service reservoirs, achievement of required level of pressure in the water distribution network and reduction in pumping and thus the associated costs. Realisation of the implications of inaccurate estimations of instantaneous water demand on optimal operational management efficiency (of water distribution networks), has led to further research and the use of statistical and complex computational tools. A brief review of some of the demand prediction strategies is given here.

An *et al.* (1996) proposed an expert-system method based on the rough-set approach to automatically frame probabilistic rules for predicting the daily total demand. The method applies a rigorous statistical treatment and takes into account the uncertainty in the available time series data on various factors influencing the level of demand. These factors include day-to-day variation in minimum and maximum temperatures, rainfall, snowfall, average humidity and speed of wind and bright sunshine hours. An example of the most generic form of the rule is:

$$(53 < a_4 \leq 58) \wedge (22.98 < a_{14} \leq 28.45) \wedge (13.30 < a_{18} \leq 15.20) \xrightarrow{1} (124 < D \leq 134)$$

The rule means that: if today's average humidity (a_4) is between 53 and 58% and the day before yesterday's maximum temperature (a_{14}) is between 22.98 and 28.45 °C and the day before yesterday's bright sunshine hours (a_{18}) are between 13.30 and 15.20, then the water demand (D) is between 124 and 134 Ml with a certainty factor of 1. The rules framed using the proposed method have been reported to produce an average error of approximately 10%.

Although statistical methods, particularly the auto-regressive integrated moving average model (ARIMA), have been used in the past to forecast consumer demand by taking into account time series observed data on weather conditions and measured flows, the predictions often indicate a considerable estimation error. Recent developments show a shift towards more sophisticated approaches such as fuzzy logic and neural networks. These approaches tend to produce somewhat better results.

Lertpalangsunti *et al.* (1999), for example, have described the development of a software package: Intelligent Forecasters Construction Set (IFCS). The package provides a range of intelligent tools (e.g. artificial neural networks (NN), fuzzy logic (FL), knowledge-based and case-based reasoning (CBR)) which can be used singly and in combination to develop a particular application. The package was applied to water demand forecasting. A comparative study using the above-mentioned intelligent tools showed that multiple NNs (i.e. each NN predicting the separate feature, for example demand on weekdays and demand on weekends) approach produced the minimum error. Mukhopadhyay *et al.* (2001) also developed a neural network based model using a year long dataset collected on water consumption and associated social, economic and seasonal characteristics for 48 residential units in Kuwait.

The above-mentioned demand forecasting techniques require an extensive amount of data covering a wide band of variability to train the neurons. Thus the precise extent of confident forecasting would depend upon the degree of similarity of the input data sets compared to the data used for developing the model.

1.8 CONCLUSIONS

In the UK, per capita water consumption has shown a trend of steady growth over time. This appears to be the combined effect of improved affluence, level of service and change in traditional values. The diurnal pattern of consumption indicates the WC and washbasin as the most water-using and frequently used appliance, respectively. Per capita consumption increases with reduction in household size. This could have considerable implications for increased demand, since the number of households with single and two occupants is anticipated to increase significantly in future. There is a considerable scope for water savings if low flush toilets are installed in place of old high-water using WCs. A range of water efficient technologies for households is commercially available. The uptake of water conservation measures (installation of water efficient devices and greywater and rainwater recycling systems) is relatively slow compared to many developed countries. An apparent reason for this is the high cost and absence of subsidies from the government.

The importance of producing reliable water demand forecasts is now realised and there have been several attempts to devise water demand-forecasting strategies. The typical problem associated with most of these strategies is the scarcity of suitable historical data on water consumption trends, micro-components' characteristics, socio-economic influences and temporal and spatial factors responsible for altering the composition of existing consumer base. The use of techniques incorporating micro-component features seem to be the most promising approach to forecasting long term water demand, since it offers a flexible framework to accommodate the influence of emerging socio-economic changes.

1.9 ACKNOWLEDGEMENTS

The authors are grateful to David Howarth for facilitating access to the EA National Water Demand Management Centre Library. Comments from the members of the WATERSAVE Network, particularly Paul Herrington, are gratefully acknowledged. Thanks are also due for Helen Perish and Rob Westcott of the Environment Agency.

1.10 REFERENCES

An, A., Shan, N., Chan, C., Cercone, N. and Ziarko, W. (1996) Discovering rules for water demand prediction: An enhanced rough-set approach. *Engineering Applications in Artificial Intelligence* **9** (6), 645-653

AWWA (1999) *Residential End Uses of Water.* American Water Works Association and AWWA Research Foundation, USA.

BSRIA (1998). *Water Consumption and Conservation in Buildings: Potential for Water Conservation.* Building Services Research and Information Association Report No. 12586B/3.

Butler, D. (1991) A small-scale study of wastewater discharges from domestic appliances. *J.IWEM* 5, 178-185.

Butler, D. (1993) The influence of dwelling occupancy and day of the week on domestic appliance wastewater discharges. *Building and Environment* **28**(1), 73-79.

Butler, D. and Davies, J.W. (2004) *Urban Drainage*, 2nd Edn., SponPress, London

Butler, D., Friedler, E. and Gatt, K. (1995) Characterising the quantity and quality of domestic wastewater flows. *Water Science and Technology* **31** (7), 13-24.

CC:DW (2001) *Climate Change and Demand for Water.* Progress Report by the Environmental Change Institute, University of Oxford. http://www.eci.ox.ac.uk/.

Cummings, S. (2001) *Future directions for water closet and sanitation systems.* National Water Conservation Group Meeting on 14-9-2001, London.

DoE (1996) *Indicators of sustainable development for the United Kingdom.* Department of the Environment/Government Statistical Service, March 1996, HMSO.

EA (2001) *A scenario approach to water demand forecasting.* Environment Agency, Worthing.

EA and UKWIR (1996) *Economics of Demand Management – Practical Guidelines.* Environment Agency and UK Water Industry Research Ltd., (SO-8/96-B-AVQC).

Edwards, K. & Martin, L. (1995) A methodology for surveying domestic consumption. *J.CIWEM*, **9**, Oct., 477-488.

EEA (2001) *Sustainable Water Use in Europe (Part II): Demand Management.* European Environment Agency. Environmental Issue Report No. 19.

Gleick, P. H. (1996) Basic water requirements for human activities: Meeting basic needs. *Water International*, **21**, 83-92.

Green, C. (2003) *Handbook of Water Economics- Principles & Practice*. John Wiley & Sons Ltd.

Herrington, P. R. (1987), *Water Demand Forecasting in OECD Countries*. Organization for Economic Co-operation and Development (OECD), Environment Monograph No. 7. OECD Environment Directorate, Paris.

Herrington, P. R. (1996) *Climate Change and the Demand for Water*. HMSO: London

Herrington, P. R. (1999) *Household water pricing in OECD countries*. Organization for Economic Co-operation and Development. Environment Directorate document: ENV/EPOC/GEEI(98)12/FINAL [Please note that most of this text is reproduced with minimal amendments in the more accessible publication by the OECD (1999)]

Lertpalangsunti, N., Chan, C. W., Mason, R. and Tontiwachwuthikul, P. (1999) A toolset for construction of hybrid intelligent forecasting systems: application for water demand prediction. *Artificial Intelligence in Engineering* **13** (1), 21-42.

Mukhopadhyay, A., Akber, A. and Al-Awadi, E. (2001) Analysis of freshwater consumption patterns in the private residences of Kuwait. *Urban Water* **3** (1-2), 53-62.

Mitchell, G. (1999) Demand forecasting as a tool for sustainable water resource management. *International Journal of Sustainable Development and World Ecology* **6,** 231-241.

OECD (1999) *The Price of Water: Trends in OECD Countries*. Organisation for Economic Co-operation and Development: Paris.

Ofwat (2001) *Tariff Structure and Charges: 2001-2002 Report*. Office of Water Services.

Russac, D.A.V, Rushton, K.R. and Simpson, R.J. (1991) Insight into domestic demand from metering trial. *J.IWEM*, **5**, June, 342-51.

POST (2000) *Water efficiency in the home*. Parliamentary Office of Science and Technology Note 135, London.

Stephenson, D (2003). *Water Resources Management*. A. A. Balkema Publishers.

Three Valleys (1991). *Domestic Demand Study*. Three Valleys Water Services PLC

UKWIR (1997) *Forecasting Water Demand Components: Best Practice Manual*. UK Water Industry Research Report No. 97/WR/ 07/1.

UN (1997) *Comprehensive Assessment of the Freshwater Resources of the World (overview document)*. World Meteorological Organization, Geneva.

Vörösmarty, C. J., Green, P., Salisbury, J. and Lammers, R. B. (2000) Global water resources: Vulnerability from climate change and population growth. *Science* **289** (7), 284-288.

Wigley,T.M.L. and Jones P.D. (1987) England and Wales precipitation: a discussion and an update of recent changes in variability and an update to 1985. *Journal of Climatology*, **1**, 231-44.

Williamson P (1998) Estimating and projecting private household water demand for small areas. *Workshop on micro-simulation in the new millennium: challenges and innovations*, Cambridge 22-23 August.

2

The technology, design and utility of rainwater catchment systems

Alan Fewkes

2.1 INTRODUCTION

This chapter reviews the technology, design and utility of rainwater catchment systems applied to small scale systems using rainwater collected primarily from roofs for use by the building occupants. Initially, the application of systems in developing and developed countries to supply both potable and non-potable water is considered. The reasons underlying the renewed interest in rainwater catchment systems over the past twenty years are identified. Generally, these relate to economic, operational and environmental difficulties associated with centralised water systems. The second part of the review concentrates upon the different types of rainwater system, which can be used to supply non-potable water in developed countries. Systems can be categorised according to how they store and deliver rainwater within dwellings or in relation to their hydraulic properties. The main system components, namely, catchment area, treatment methods and storage tanks are discussed in relation to how they affect system performance. The different methods used to determine the capacity of rainwater stores are considered in detail. The capacity of the store is important both economically and operationally. The size of the store influences the volume of the water conserved, installation costs and the final quality of water supplied by the collector. Finally, the physical, chemical and microbiological quality of delivered rainwater is assessed. The acceptance of the system by the user is related to the aesthetic quality of the water expressed in terms of colour, odour and turbidity. The microbiological quality of the water determines the potential health risk the system represents to the user. The chemical and the physical parameters such as pH and dissolved solids will affect the selection of system components.

2.2 BACKGROUND AND APPLICATION OF RAINWATER CATCHMENT SYSTEMS

2.2.1 History

The concept of rainwater catchment systems is simple, consisting of the process of collecting, storing and using rainwater as a primary or supplementary water source. Systems are easily constructed and maintained with the ability to operate independently of central water supply systems. This review concentrates on small, scale catchment systems using rainwater collected primarily from roofs for use by the building's occupants. Examples of larger scale systems, which often use collection areas of many hectares in the form of roads and ground run-off catchments for the provision of water mainly for livestock and irrigation, are documented by Pacey and Cullis (1986) in their extensive review of rainwater harvesting methodologies.

Rainwater catchment systems have a long history, which is reviewed in detail by Gould and Nissen-Petersen (1999). Examples are given of systems dating back to 2000BC in the Negev desert in Israel, Africa and India. In the Mediterranean region there is archaeological evidence of a system used at the Palace of Knossos dating back to 1700BC. On Sardinia cisterns hewn from bedrock and plastered with hydraulic mortar to make them watertight have been found, which are thought to date back to 6 to 7 BC. In Western Europe there is documentary evidence recording the use of systems in Venice dating back beyond the sixteenth century. In fact examples of rainwater catchment systems can be found for most areas of the world.

During the twentieth century the use of rainwater collection systems has declined in many parts of the world due to the development of more sophisticated technologies. However, the concept is now receiving increased recognition as a source of water supply in many parts of the world. Rainwater catchment systems can be used to provide:

(1) The main source of potable water
(2) A supplementary source of potable water
(3) A supplementary source of non-potable water, for example, WC flushing, garden watering and vehicle washing

The main application in developing countries is for the provision of potable water. In developed countries examples of all three applications can be found but potable supplies are more common in rural areas and non-potable supplies in urban locations.

2.2.2 Application in Developing Countries

The application of rainwater collection systems in both rural and urban areas of developing countries is well documented. Typical examples of countries where systems are used include Bangladesh, Botswana, Caribbean Islands, India, Indonesia, Kenya, Malaysia, New Guinea, South Pacific Islands and Sri Lanka (Schiller, 1987). More recent developments in Africa, Thailand and the Philippines are reported by Gould (1993), Wirojanagud and Vanarothorn (1990) and Appan *et al.* (1989). The renewed interest and increased use of rainwater collection systems during the past twenty years is related to a number of factors. These include the failure of centralised piped systems due to operational and maintenance problems; problems with the contamination of ground and surface water supplies; increased demand on rural water supplies due to population growth; the widespread use of impervious roofing materials such as tiles and corrugated iron as opposed to traditional techniques such as grass and palm thatch; the development of low cost and effective tank designs. The application of demand management generally in developing countries is considered in detail in Chapter 8.

2.2.3 Application in Developed Countries

2.2.3.1 Potable Water Systems

The use of rainwater catchment systems for potable water supplies in developed countries occurs mainly in rural locations. In rural Australia it is estimated that one million people rely on rainwater as their primary source of supply (Perrens, 1982). In rural areas centralised piped supplies are usually uneconomical due to the low population densities. Similarly, the use of groundwater supplies is often not feasible due to economics, water quality or reliability of supply. Examples of rainwater catchment systems can also be found in urban areas due to the adoption during the last decade, by all levels of government in Australia, of the concept of ecologically sustainable design. One of the more radical innovations related to this initiative is water sensitive urban design, which recognises the role of rainwater catchment systems can have in substituting and/or supplementing reticulated urban water supply from centralised water supply facilities. A home has been constructed in south east Queensland, called 'The Healthy Home' to demonstrate the principles of ecologically sustainable development. In addition to various low energy innovations, rainwater is collected and used for all household uses including drinking (Gardner *et al.*, 1999). The Australian government has also published guidance on cistern sizing, materials and maintenance (Cunliffe, 1998).

A similar situation exists in the rural areas of the Canadian province of Nova Scotia. Rainwater systems have been in use for more than 50 years as an alternative drinking water source, where ground water supplies are inadequate or contaminated. The Nova Scotia Department of Health have published guidelines for system operation and construction (Scott & Waller, 1991). In an urban context the Research Division of Canada Mortgage and Housing Corporation have sponsored the design and construction of a self-sustaining and 'healthy' house in Toronto. Rainwater is collected in cisterns, buffered with limestone gravel and passed through a multi-media gravel, sand and activated carbon filter. After ultra violet disinfection the water is distributed on demand for potable and non-potable uses (Townshend *et al.*, 1997).

The use of rainwater systems is widespread in many rural areas of the United States; Grove (1993) estimated 200,00 rainwater systems are in use to provide the domestic water requirements of small communities and individual households. Some states regulate construction while both state and national agencies publish guidance notes relating to materials and system construction (Latham & Schiller, 1987).

The island of Bermuda is an example of a country, which uses rainwater catchment systems extensively. Until the 1930s the country was entirely dependent upon rainwater systems. Rainwater still provides approximately 50 percent of its total water demand and nearly all of its supplies to domestic properties. Hotels and commercial establishments are supplied with fresh ground water and desalinated brackish groundwater and seawater.

A typical Bermuda rainwater catchment system has an average roof area of $139m^2$ and a cistern capacity of 68 cubic metres, which is capable of supplying 106 litres/person/day for an average household of 3.5 persons during extreme dry periods. The 1949 Public Health Act and subsequent regulations require that cistern storage is provided in new buildings and regulate cistern sizes, system design and maintenance. Domestic roofs are usually covered with local limestone block, sealed with latex paint. Rainwater is collected from the roof using wedge shaped limestone 'glides', which direct rainwater to vertical leaders that discharge into cisterns under the houses. Electric pumps in conjunction with pneumatic vessels raise water from the cisterns into the plumbing systems (Waller, 1982).

2.2.3.2 Non-potable water systems

The utilisation of rainwater catchment systems to supply non-potable water to buildings during the last 15 – 20 years has become popular in urban areas of many developed countries. The increased awareness and application of these systems has arisen because a number of problems linked to centralised systems of water supply and disposal have been identified (Pratt, 1999 and Geiger, 1995), these include:

(1) Increasing water demand, which cannot be satisfied without the development of new resources.

(2) Available resources are not located in areas of high demand, which can result in water distribution over large distances.

(3) Over abstraction from ground sources resulting in low flow regimes in rivers.

(4) Urban expansion and highway developments increase surface run-off volumes resulting in increased risk of flooding and adverse changes to the receiving water's quality and aquatic eco-system.

The traditional solution to these problems has been the development of new water supplies, distribution networks and flood alleviation schemes, which are not environmentally friendly and require considerable capital expenditure. An alternative and sustainable strategy is the use of decentralised techniques (Konig, 1999). For example:

(1) The use of planted or green roofs, which result in partial water retention and reduced peak run-off flows into the storm sewer network

(2) On site infiltration of rainwater run-off, which recharge the ground water and eliminate storm sewer connections.

(3) Rainwater catchment systems, which reduce the utilisation of potable water. The benefits include conservation of water resources, relief of demand on public water supplies, potential attenuation of peak run-off into the sewer network and the reduction of combined sewer overflow emissions.

The potential of rainwater systems became apparent in Germany during the period 1970 – 75 when they experienced several problems with the management of their water supply and sewerage systems. Two main problems were experienced. In Germany a high proportion of water for potable use is abstracted from ground water sources. Increased abstraction in many towns resulted in the lowering of the water table with adverse environmental effects. The ground sources were also becoming polluted representing a potential health risk. Secondly in Germany a combined sewerage system predominates. During periods of heavy rainfall sewerage systems frequently reached their maximum capacity. Rainwater systems were considered as a possible solution to both of these problems (Sayers, 1999).

Initially to promote the use of rainwater systems many city councils offered financial incentives. The current trend is to divide charges for urban drainage into two components. One related to consumption dependent waste water discharges and the other linked to the impervious surface area of the property. A permanent financial incentive therefore exists to disconnect roofs from sewers (Herrmann & Schmida, 1999). The combined costs of water supply and disposal in Germany vary between regions but on average are 45% higher than in the UK (Sayers, 1999). The payback period for German systems varies dependent upon location but has been estimated to be between 12 – 19 years (Wessels, 1994).

The effect of rainwater systems upon combined sewer overflow emissions has been investigated by, Vaes & Berlamont (1999). Generally, more storage in rainwater tanks is required to obtain the same overflow frequency compared to centralised retention tanks. Herrmann & Hasse (1997) report specific storage volumes of $500m^3$/ha is required for rainwater tanks compared to $20 – 100m^3$/ha for centralised retention tanks. However, rainwater tanks compete economically with centralised retention tanks because of differences in construction costs. The specific construction costs of private rainwater a tank is approximately 250 Euro/m^3 compared to 600 to 3400 Euro/m^3 for centralised tanks depending upon the method of construction

In Europe, Germany is leading the way with the installation of rainwater collection systems. It is estimated that during the last ten years 100,000 systems have been installed in a range of building types including both low and high-rise housing and commercial buildings (Herman & Schmida, 1999). For example, Konig (2000a) describes an innovative scheme applied to high rise dwellings in Berlin. To satisfy the non-potable demand in the dwellings additional rainwater is collected from parking spaces, paths and surrounding streets to supplement the roof water. The rainwater is collected in a $160m^3$ capacity cistern before treatment in several simple stages before being used for WC flushing and garden watering. Further case studies relating to a wide range of building types in Europe but in particular Germany are described by Konig (2001).

The number of rainwater catchment systems in the UK is increasing; Hassel (2001) estimates there are approximately 1000 systems installed. Many of these are pilot or test schemes where data on water quality and system performance is being collected (Brewer *et al.*, 2001; Ratcliffe, 2002; Day, 2002: Chilton *et al.*, 1999). Schemes range from single houses to larger buildings such as shopping centres, schools, motorway services and commercial buildings. The largest scheme undertaken in the UK is probably the Millennium Dome, which was originally constructed as an exhibition hall. The dome is 50m high with a diameter of 320m. The roof, with a total area of 100,000m^2, is constructed of plastic coated glass fibre fabric suspended from twelve steel trellis masts 100m high. When used as an exhibition hall the rainwater satisfied 20% of the daily 500m^3 WC flushing demand (Lodge, 2000).

In the UK the financial benefits of rainwater systems remain to a large extent unproven. Brewer *et al.* (2001) monitored the performance of rainwater catchment systems in two office buildings and an ecological housing development, but it was only possible to conclude the operating costs of a rainwater system can balance the mains water cost dependent on the amount of rainfall. In another case study (Chilton *et al.*, 1999) the economics of a rainwater system installed in a supermarket were more favourable. A payback period of 12 years was reported, which was estimated could be reduced to 4 years if a more economical tank capacity of 10m^3 was installed. The economics of water demand techniques, including rainwater systems, are explored in Chapter 9.

Further research is required to confirm the sustainability benefits of rainwater collection systems. Rainwater systems require materials for their construction, such as tanks and pipe work, which will contain embodied energy and directly use resources. The level of such consumption compared to mains water use has not been reported (Leggett *et al.*, 2001). Rainwater systems also consume electricity during operation for pumping. However, the amount of electricity consumed in rainwater systems is comparable to that required to deliver mains water. Mikkelsen *et al.*, (1999) estimate energy consumption related with pumping rainwater from the storage tank is in the range 0.3 – 0.5 kWh/m^3 compared with the 0.39 kWh/m^3 used for the production and distribution of mains water. The subject of life cycle assessment in relation to water demand management is introduced in chapter 6.

In Asia, Japan is promoting the utilisation of rainwater in urban areas with subsidies available in many areas to encourage householders to adopt rainwater for non-potable uses (Murase, 1998). Larger scale applications also exist such as the dome stadiums in Tokyo, Nagoya and Fukuoka, which are used for baseball games, concerts and exhibitions. The systems are used to supply water for WC flushing but also have important secondary functions such as improved flood control, reduced river pollution, reduced groundwater abstraction and associated subsidence problems. The Nagoya stadium is the largest with a catchment area of 35,000m^2 and a total storage capacity of 28000m^3 (1300m^3 capacity for run-off control) (Zarizen *et al.*, 1999).

2.3 RAINWATER CATCHMENT SYSTEM CATEGORIES AND COMPONENTS

2.3.1 System Categories

The basic configuration of a rainwater catchment system is illustrated in Figure 2.1. For non-potable applications only filtration prior to entry into the tank is recommended (Konig, 2000b). For potable applications additional treatment such as chlorine or UV disinfection is required. This review concentrates upon the different system categories and components used in systems for supplying non-potable water in developed countries. Gould (1999) describes in detail the components and materials used to construct systems for supplying potable water in developing countries. Systems for potable supplies in developed countries are considered by Gardner *et al.* (1999) and Townshend *et al.* (1997).

Leggett *et al.* (2001) identify three basic types of system:

(1) Directly pumped system (Figure 2.1). Rainwater is pumped directly from the storage tank to appliances within the building. Mains top-up water is supplied to the storage tank.

(2) Indirectly pumped system (Figure 2.2). Rainwater is pumped from the rainwater storage tank into a feed cistern located in the building, which supplies water to appliances by gravity. Mains top-up water is supplied to the feed cistern.

(3) Gravity feed system (Figure 2.3). The rainwater storage tank is located in the roof void and supplies appliances by gravity. The main advantage of this system is a pump and electrical supply is not required. Konig (2000b) recommends these systems should only be used for industrial sheds and agricultural barns. The main concerns are the high structural loads, potential damage due to leakage and fluctuations in the temperature of stored water.

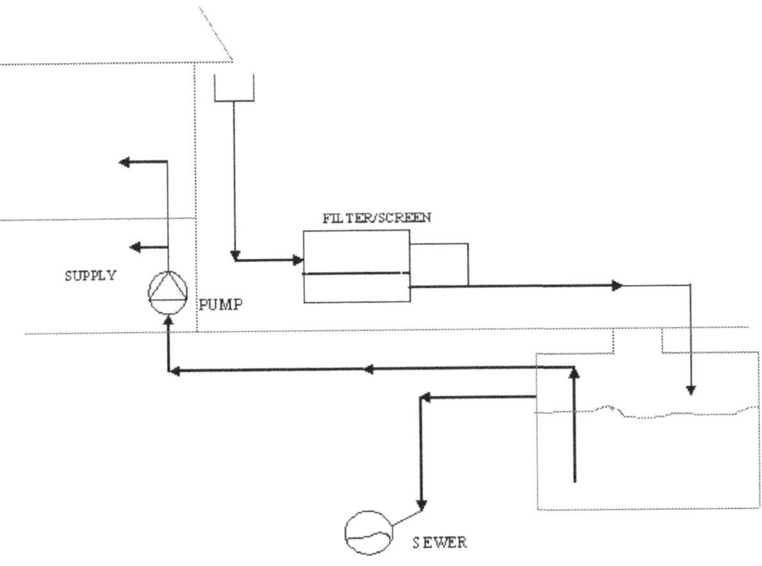

Figure 2.1 Schematic of directly pumped rainwater catchment system (Leggett *et al.*, 2001).

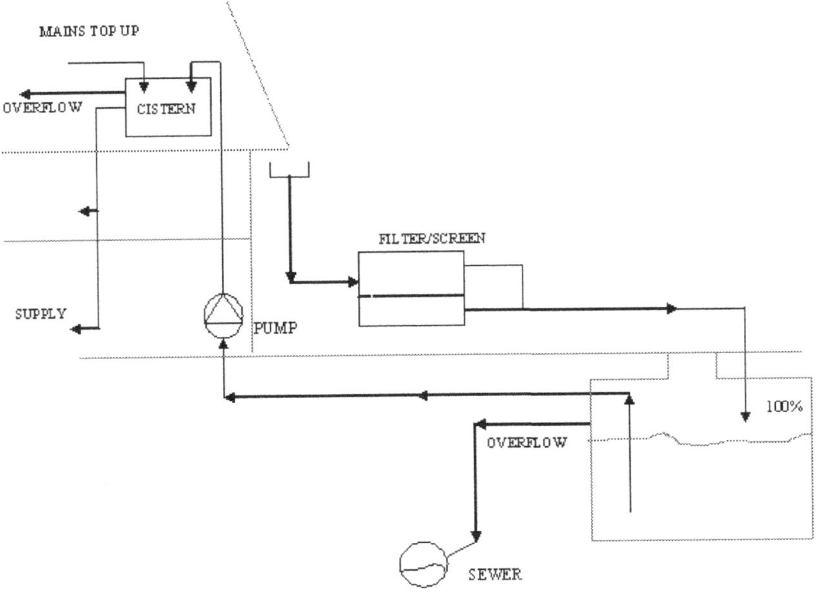

Figure 2.2 Schematic of indirectly pumped rainwater catchment system (Leggett *et al.*, 2001).

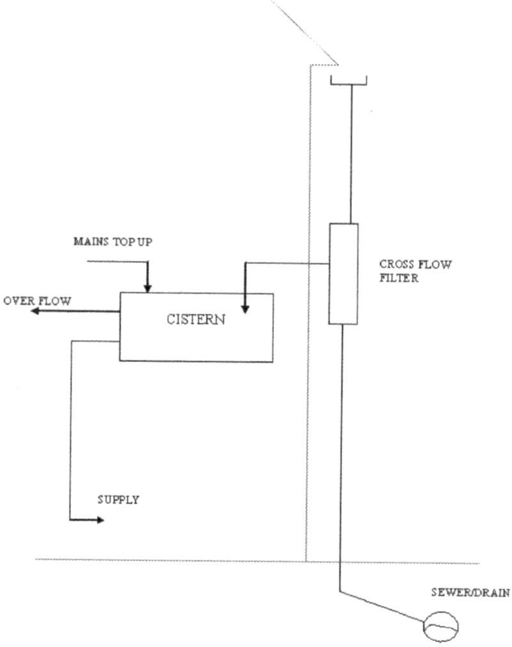

Figure 2.3 Schematic of gravity feed rainwater catchment system (Leggett *et al.*, 2001).

Hermann *et al.* (1999) offers a different classification of systems based upon their hydraulic properties:

(1)   The total flow type (Figure.2.1). The total runoff flows into the storage tank via a filter. The design of the filter should be such, that if it blocks the supply pipework upstream of the filter does not surcharge. The rainwater overflows into the sewer from the storage tank when it is full

(2)   The diverter type (Figure 2.4). A proprietary cross-flow filter is used to divert rainwater into the storage tank, a proportion of the rainwater will bypass the storage tank. The volume of rainwater that bypasses the store is dependent upon the rainwater flow rate in the collection pipe upstream of the cross-flow filter

(3)   The retention and throttle type (Figure 2.5). Additional storage is provided in the storage tank (retention volume), which is emptied at a low flow rate via a throttle valve into the sewer. The peak flow into the sewer is reduced decreasing the risk of sewage surcharge and possible flooding during periods of heavy rainfall.

(4)   The infiltration type (Figure 2.6). The overflow from the storage tank is allowed to infiltrate into the surrounding ground to recharge the local water table.

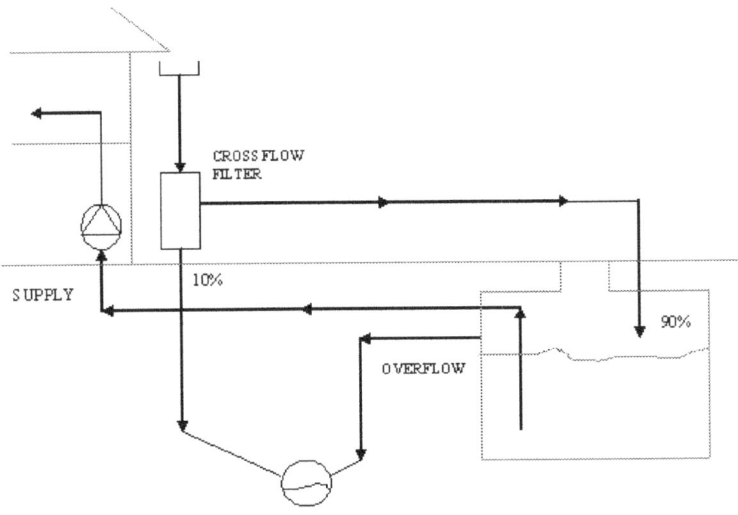

Figure 2.4 Schematic of diverter type rainwater catchment system (Herrmann & Schmida, 1999).

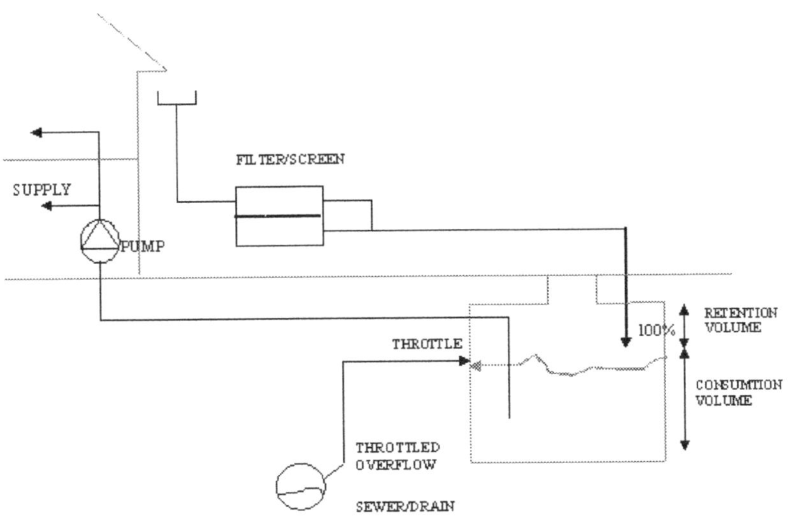

Figure 2.5 Schematic of throttle type rainwater catchment system (Herrmann & Schmida, 1999).

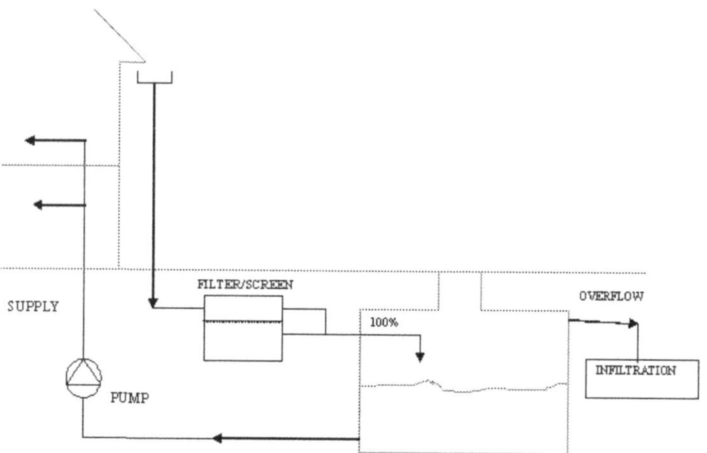

Figure 2.6 Schematic of infiltration type rainwater catchment system (Herrmann & Schmida, 1999).

2.3.2 System Components

2.3.2.1 Catchment Area

The most common catchment area for rainwater systems is the roof. The preferred surfaces are those, which are chemically inert such as slates. Metal roof coverings are acceptable but the slightly acidic nature of rainwater can produce some dissolution of metal ions from the surface. Water collected from roofs with bituminous surfaces may be discoloured and have a distinct odour (Konig, 2001). Rainwater collected from paved areas surrounding the building can be used but is likely to be more heavily polluted and require extra treatment. Permeable paving systems are an exemption because they provide some filtration in their sub-base and allow biodegradation of oils (Pratt, 1999). The quality of rainwater collected via permeable paving is particularly suited to irrigation.

Green or planted roofs can also be used as a catchment area for rainwater systems. However, this type of roofing system can retain in excess of 50 percent of the incident rainfall but is capable of filtering the rainwater depending upon the composition of it's substrata (Konig, 1999).

Generally, rainwater losses during collection can be quantified using a run-off coefficient that represents the proportion of rainwater collected from the roof being considered compared with an ideal roof for which no losses occur.

Approximate values of runoff coefficients for some different roof types and configurations are given in Table 2.1 (Fewkes & Warm, 2000)

Table 2.1 Approximate runoff coefficients for different roofs and system types

Roof/system type	Runoff coefficient
Pitched roof covered with tiles or slates (Total flow type)	0.9 – 1.00
Pitched roof covered with tiles or slates (Diverter type)	0.75 – 0.95
Flat roof covered with impervious membrane	0 – 0.5
Flat green roof (incorporating plants and growing medium)	0 – 0.5

2.3.2.2 First flush diverters

During periods of no rainfall roofs become polluted with atmospheric particulates and bird droppings. The first flush of rainwater from the roof is usually more polluted than subsequent runoff (Fewkes, 1996; Forster 1991). Various designs for first flush diverters exist (Michaelides, 1989) but Konig (2000b) suggests this initial separation of rainwater is unnecessary.

2.3.2.3 Treatment

For non-potable applications rainwater usually only needs filtration prior entry into the storage tank. Leggett *et al.* (2001) identify a range of different filter types including screen, cross flow, cartridge, slow sand, rapid sand, membrane and activated carbon filters. Current practice in Germany (Konig, 2001) does not recommend fine filters because they are susceptible to blockage and frequent maintenance. Cross flow filters or screen filters with a permeability of between 0.2 – 1.00mm are recommended. German regulations also require the inside of the filter feed line must be uniform throughout, i.e. a tennis ball, which enters the filter from the roof must be able to exit through the overflow. This ensures the system cannot surcharge if the filter becomes blocked.

2.3.2.4 Storage

The storage tank is an essential component of rainwater catchment systems, which can be constructed from a variety of materials such as plastic, concrete or steel. The preferred location is below ground sheltered from daylight, which minimises algal growth in the collected water (Konig, 2001). The volume of the storage tank is important because it significantly affects the systems initial capital costs and the volume of potable water conserved. In addition to the initial filtration further treatment occurs within the storage tank via flotation and settlement. Flotation occurs within the tank as particles less dense than water such as pollen float to the surface. The storage tank should be designed to overflow at least twice a year to facilitate the removal of these particles. Similarly particles more dense than water settle to the bottom of the tank. Within this settled layer a beneficial aerated biofilm can develop adding an additional stage of biological treatment (Sayers, 1999).

The rainwater just below the surface layer is the cleanest. Floating suction lines are often used to extract rainwater from just below the surface. Similarly various configurations of feeding rainwater into the storage tank are used to prevent the sediment at the bottom of the tank from being disturbed (Konig, 2001).

Alternative forms of storage are also available. Rainwater can be stored in the sub-base material of permeable paving systems or within the voids of geo-cellular modular units (Courier, 2002).

2.4 STORAGE CAPACITY OF RAINWATER COLLECTORS

2.4.1 Background

A storage tank or cistern is required to collect rainwater runoff because rainfall events occur more erratically than system demand. The capacity of the rainwater store is important both economically and operationally. For example, the size of the store will influence:

(1) The volume of water conserved.
(2) The installation costs of the system
(3) The length of time rainwater is retained, which affects the final quality of water supplied by the collector.
(4) The frequency of system overflow, which affects the rate of removal of surface pollutants.
(5) The volume of water overflowing into the surface drain or soak away.

A rainwater collector or store is a reservoir, which receives stochastic inflows (rainwater) over time and is sized to satisfy the demand on the system (WC flushing and garden irrigation for example). The basic problem is to analyse both inflow and outflow data characteristics to satisfy a certain level of system reliability. The sizing of storage reservoirs has been reviewed by McMahon & Mein (1978), who identify two broad categories of sizing model; Moran and critical period methods.

2.4.2 Moran related methods

The Moran related methods are a development of Moran's (1959) theory of storage. Moran derived an integral equation relating inflow to reservoir capacity and releases such that the probable state of the reservoir contents at any time could be defined. Using this approach, solutions were only possible for idealised conditions. Practical applications using this method were developed by Moran by considering time and water volumes both as discrete variables. The reservoir capacity, release and inflow can then be related to each other by a system of simultaneous equations. This method was subsequently modified by Gould (1961) to a general procedure of direct practical use to engineers. The method has not been widely applied to rainwater collectors although Piggot *et al.* (1982) uses the Gould matrix technique to assess the probability of failure for a range of roof areas, tank sizes and demands for the Brisbane region of Australia

2.4.3 Critical period methods

2.4.3.1 Mass Curve

Critical period methods use sequences of flows where demand exceeds supply to determine the storage capacity. The best known and earliest example of this approach to reservoir sizing is the mass curve method (Rippl, 1883). A simple storage system (Figure 2.7) will perform adequately provided the relationship;

$$S \geq Max \left\{ \int_{t_1}^{t_2} [D_t - Q_t] dt \right\} \tag{2.1}$$

is satisfied. Where:

S - Storage capacity.
D_t - Demand during time interval, t
Q_t - Inflow during time interval, t
T - Time and $t_1 < t_2$

In the mass curve method developed by Rippl the variable input (rainfall) is subtracted from a constant demand output. The maximum cumulative difference is the storage required for a system, which is 100% reliable or efficient. The mass curve method has been applied to the sizing of rainwater stores by Grover (1971) and more recently Ngigi (1999). Simplified mass curves have also been used in an attempt to reduce the amount of calculation; Watt (1978) used average monthly data while Keller (1982) used two thirds of average monthly values. However, the method lends itself well to computerisation and such simplified approaches would usually only be employed at the preliminary design stage.

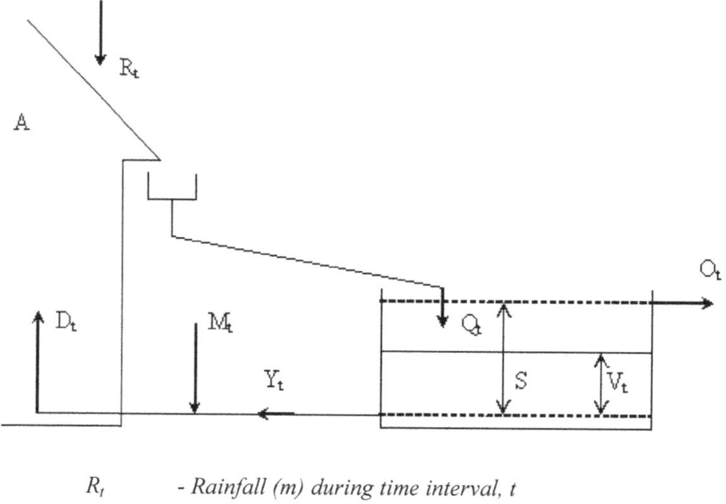

R_t *- Rainfall (m) during time interval, t*
Q_t *- Rainwater runoff (m^3) during time interval, t*
M_t *- Mains suppply make – up (m^3) during time interval, t*
O_t *- Overflow from store (m^3) during time interval, t*
V_t *- Volume in store (m^3) during time interval, t*
Y_t *- Yield from store (m^3) during time interval, t*
D_t *- Demand (m^3) during time interval, t*
S *- Store Capacity (m^3)*
A *- Roof Area (m^2)*

Note: Duration of time interval, t can be hourly, daily or monthly

Figure 2.7 Rainwater collection system configuration.

The main limitation of this method is that it is not possible to compute a storage size for a given probability of failure. To overcome this limitation a number of statistical methods have been developed, which can be used in conjunction with the mass curve method to determine the reliability of supply or probability of system failure. Ree *et al.* (1971) describe one such approach, which is based on a method originally proposed by Waitt (1945). The approach is based on an analysis of the frequency of occurrence of minimum rainfall amounts for periods between 2 to 84 months in a 75 year record of monthly rainfall levels. By applying standard statistical techniques, the minimum rainfall for a given probability can be determined for various time periods. Rainfall values are then selected for all periods, which have a given probability of occurrence and a mass curve calculation applied to the resultant series. This method therefore attempts to assign a given level of reliability or probability of failure to the tank size selected

2.4.3.2 Behavioural Analysis

Within the general category of critical period methods McMahon & Mein (1978) also include behavioural or simulation analysis, which simulates the operation of a reservoir with time. The changes in storage content of a finite reservoir (Figure 2.7) are calculated using a mass storage equation of the general form:

$$V_t = V_{t-1} + Q_t - D_t \tag{2.2}$$

Subject to $0 \leq V_t \leq S$

Where:
V_t - Water in storage at end of time interval, t

The water in storage at the end of a prescribed time interval is therefore equal to the volume of water remaining in the store from the previous interval plus any inflow and less any demand during the time period. Provided, that is, the computed volume in the store does not exceed the capacity of the store. Behavioural models therefore simulate the operation of the reservoir with respect to time by routing simulated mass flows through the algorithm that describes the operation of the reservoir. The operation of the system is usually simulated over a period of years. The input data, which is in time series form, is used to simulate the mass flows through the model and is based upon a time interval of an hour, day, month or year depending upon the application and degree of resolution required. The time series can be historical or generated data. It is possible to define the volume of the store such that it empties once for the period of the input time series. The storage is then the same as that found using the mass curve method. Various researchers have applied behavioural models to the sizing of rainwater stores, for example, Jenkins *et al.* (1978), Chu *et al.* (1999) and Walther & Thanasekaren (2001). Jenkins *et al.* (1978) identify two fundamental algorithms to describe the operation of a rainwater store (Figure 2.7).

The yield after spillage (YAS) operating rule is;

$$Y_t = min \begin{cases} D_t \\ V_{t-1} \end{cases} \tag{2.3}$$

$$V_t = min \begin{cases} V_{t-1} + Q_t - Y_t \\ S - Y_t \end{cases}$$

(2.4)

Where:

Y_t - Yield from store during time interval; t.

The YAS operating rule assigns the yield as either the volume of rainwater in storage from the preceding time interval or the demand in the current time interval, whichever is the smaller. The rainwater runoff in the current time interval is then added to the volume of rainwater in storage from the preceding time interval, with any excess spilling via the overflow and then the yield is subtracted.

The yield before spillage (YBS) operating rule is;

$$Y_t = min \begin{cases} D_t \\ V_{t-1} + Q_t \end{cases}$$

(2.5)

$$V_t = min \begin{cases} V_{t-1} + Q_t - Y_t \\ S \end{cases}$$

(2.6)

The YBS operating rule assigns the yield as either the volume of rainwater in storage from the preceding time interval plus the runoff in the current interval or the present demand, whichever is the smaller. The rainwater runoff in the current time interval is then added to the volume of the rainwater in storage from the preceding time interval before subtracting the yield and allowing any excess to spill via the overflow. Jenkins *et al.* (1978) used the YAS algorithm and a monthly time interval to investigate the performance of rainwater stores in North America. The reliability or performance of the rainwater store can be expressed using either a time or volume basis. In either case a reliability of 100% indicates complete security or provision of service water.

2.4.3.3 *The integration of different operating policies into behavioural models*

The performance or efficiency of a rainwater store is related to the rainwater catchment area, the rainfall level and the demand on the system. In situations where the catchment system is the only or main source of supply efficiency can be improved by careful compliance with rationing schedules by system users. The impact of rationing on system performance has been illustrated in a number of studies, particularly under seasonal or drought prone climatic conditions (Perrens, 1982 and Gieske *et al.*, 1995).

Perrens (1982) used a behavioural model, which incorporated rationing and an external source of supply. Rationing of demand occurred if the amount of water in storage at the beginning of the month was less than the monthly demand. The amount of water available was reduced to 75% of its normal level. The external stocking policy assumed water was available in standard volumes of $5m^3$ if the store could not meet the demand after rationing. Performance was based on failure rate defined as the number of occasions water had to be imported. For each roof area and demand level, failure rate was plotted against storage volume. A desired level of failure is chosen and the required storage for a given area and demand determined from the graph

2.4.3.4 Behavioural models and the time interval of the input data

The models reviewed use monthly rainfall data. Behavioural models using weekly or daily time series will give a more accurate prediction of system performance (Fewkes, 1995 and Heggen, 1993). The use of a large time interval, such as one month for the input rainfall series, results in a compact model and economic data set. However, the monthly time interval does not take into account rainfall fluctuations with a time scale of less than one month, which usually results in an inaccurate prediction of system performance

Latham (1983) approached this problem by using a behavioural model but defined the reservoir operating algorithm in a more general form;

$$Y_t = min \begin{cases} D_t \\ V_t + \theta Q_t \end{cases} \tag{2.7}$$

$$V_t = min \begin{cases} (V_{t-1} + Q_t - \theta Y_t) - (1-\theta)Y_t \\ S - (1-\theta)Y_t \end{cases} \tag{2.8}$$

Where θ is a parameter between 0 and 1. If $\theta = 0$, the algorithm is YAS and if $\theta = 1$, the algorithm is YBS. The poor resolution of the monthly interval model compared to the daily model is countered by the introduction of a parameter, referred to as the storage operating parameter. The short time scale fluctuations of the daily model are in effect replicated in the monthly model by the storage operating parameter (θ). A value of the parameter was selected by Latham (1983), such that the monthly model mimics the system performance predicted by corresponding model using a daily time interval. This approach provides a simple and versatile method of modelling the performance of a rainwater collector, which takes into account the temporal fluctuations in rainfall.

The model was subsequently used by Latham & Schiller (1987) to evaluate design methods developed by other researchers, namely, Jenkins *et al.* (1978), Perrens (1982), Rippl (1883) and Ree (1971). The rainwater storage capacities predicted by the models at a 98.3% time-based reliability level were compared with the capacities generated by Latham & Schiller's (1987) model using a 98.3% volume based reliability level. The models of Jenkins *et al.* (1978) and Rippl (1883) overestimated the storage requirements, Perrens (1982) model was in reasonable agreement and Ree's (1971) model underestimated the storage requirements compared to Latham & Schiller's (1987) model. There was no report of the extent to which comparisons between time-based and volume-based reliabilities distorted the comparison

2.4.3.5 The application of behavioural models to sizing rainwater collectors in the UK

The validity of using a YAS model to predict the performance of rainwater collectors and the sensitivity of a YAS model to the time interval of the input time series, related to the UK, has been investigated by Fewkes (1999a). A commercially available rainwater collection system was field tested and the measured performance of the system compared with the performance predicted by the YAS model using both hourly and daily time intervals. The performance of the rainwater collection system is described by its water saving efficiency (E_T) (Dixon *et al.*, 1999). Water saving efficiency is a measure of how much mains water is conserved in comparison to the overall demand and is given below

$$E_T = \frac{\sum_{t=1}^{T} Y_t}{\sum_{t=1}^{T} D_t} x100 \qquad\qquad (2.9)$$

Where:
$\qquad E_T \qquad$ - Water saving efficiency
$\qquad T \qquad$ - Total time under consideration

The results of this preliminary analysis indicated that a YAS model using either hourly or daily input series could be used to predict system performance.

In a more comprehensive study Fewkes & Butler (2000) report the results of a mapping exercise, which evaluated the accuracy of behavioural models for the sizing of rainwater collection systems using both different time intervals and reservoir operating algorithms applied to a range of reservoir capacities and collection areas. Small volume stores, which conserve a small but useful proportion of household water ($\leq 50\%$) were investigated. Previous researchers have concentrated on large volume stores with high reliability levels or water saving efficiencies $(\geq 95\%)$. For example, in Latham & Schiller (1987), storage capacities for the Ottawa (Lemieux) area ranged from $10 - 150\text{m}^3$ depending upon collection area and level of demand. The most appropriate time intervals (hour, day or month) for modelling the performance of the system using different combinations of store capacity and collection area were also investigated.

Different combinations of roof area, store capacity and demand were expressed in terms of two dimensionless ratios, namely the demand fraction and the storage fraction. The demand fraction is given by, D/AR, where D is the annual demand (m^3), A the roof area (m^2) and R the annual rainfall (m). The storage fraction is given by, S/AR, where S is the storage capacity (m^3).

The detailed analysis undertaken by Fewkes and Butler (2000) enables constraints to be proposed for the application of hourly, daily and monthly models expressed in terms of the storage fraction. It was recommended that hourly models should be used for sizing small stores with a storage fraction below or equal to 0.01. Daily models can be applied to systems with storage fractions within the range $0.01 - 0.125$. Monthly models were only recommended for use with storage fractions in excess of 0.125. Generally, daily models can be used to predict the performance of all stores except small stores with a storage fraction less than or equal to 0.01.

In the second part of the study values of storage operating parameter (θ) were also determined for a monthly model operating over a comprehensive range of demand and storage fractions. The system performance predicted by the monthly (θ) model was found to correlate well with corresponding values determined using a daily YAS model over a range of demand fractions likely to be encountered in practice. However, the results of the study by Fewkes & Butler (2000) were based on one geographic location within the UK and it was recognised that further research was required to confirm the results of their study are generally applicable to other areas of the UK.

In an extension of the previous study Fewkes (1999b) used the YAS algorithm and a daily time interval to model the performance of rainwater collectors using rainfall data from five different sites located within the Midland and Welsh regions of the UK. The performance curves for each location were closely grouped together, suggesting that system performance was relatively insensitive to daily fluctuations in rainfall at each site. The close grouping of the curves for each site also suggested collector performance could be adequately represented by a set of average curves, one for each demand fraction investigated. A set of generic performance curves was proposed that predicted the performance of rainwater collection systems within the geographic boundaries defined by the study. The investigation also explored further, the use of a monthly model in conjunction with the storage operating parameter (θ). An approximate method of sizing rainwater collection systems, within the geographic region studied was proposed using an overall average value of θ equal to 0.6.

In a more geographically extensive study Fewkes & Warm (2000) modelled the water saving efficiencies of rainwater collectors at 11 different UK locations. The correlation between the water saving efficiencies (E_T) predicted by a set of average curves for a range of demand fractions was good compared with the corresponding values defined by the curves for each of the individual sites. A set of generic performance curves was produced, which predict the performance of rainwater collection systems within the UK.

The curves provide a useful design aid for the accurate and therefore economic sizing of rainwater collection systems (Figure 2.8). The use of these curves, via a worked example, is given in Leggett *et al.* (2001). In the second stage of the research, a relationship for the water saving efficiency (E_T) in terms of the input ratio (AR/D) and storage period (S/d) was determined, where d is the daily demand (m^3/day). Initially, a series of exponential association type equations with two coefficients were fitted to the generic performance curves corresponding to storage periods (S/d) of 0.5, 1, 2, 4, 8, 16, 18 and 20 days. Finally, the relationship between the two coefficients in terms of the storage period (S/d) was determined, which enabled a generalised form of the initial exponential association type equation to be derived. The values of the water saving (E_T) predicted by the YAS model and the generalised form of the exponential association equation were compared and the errors calculated. The errors were within the limit +/- 5%, which was considered adequate for the initial estimation of storage volumes.

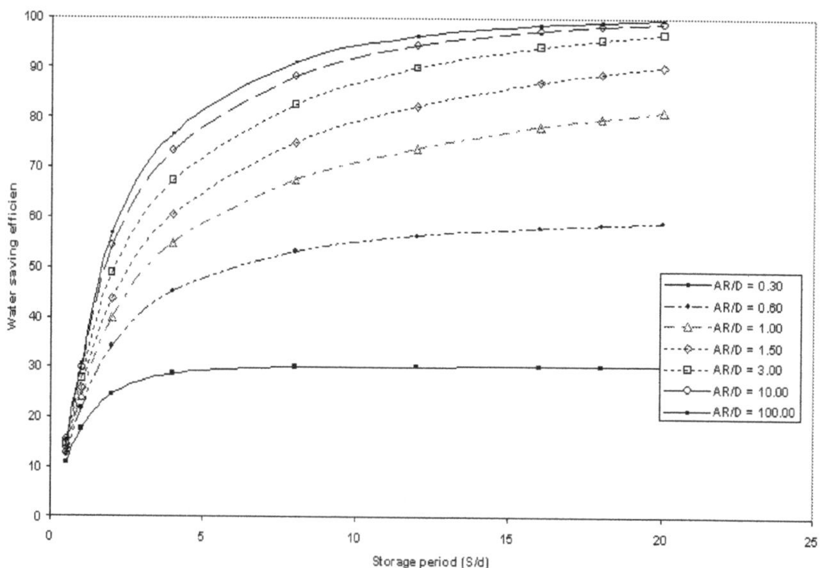

Figure 2.8 Average percentage of water conserved versus storage ratio (AR/D = 0.3 – 100).

2.4.4 Economic Considerations

An economic component can also be incorporated into the selection of rainwater cisterns. The volume of the store is optimised by considering the relationship between the cost of supplying any deficit in water required to meet the expected demand and the cost of providing the cistern. The optimum cistern volume is the one, which minimises the total cost. The basis of this approach is described in detail by Fok & Leung (1982). In another study Dominguez *et al.* (2001) apply this approach for determining the optimum cistern sizes for an urban zone of the city Queretaro, Mexico

2.4.5 Other Design Methods

In addition to the mass curve method and the behavioural models a range of other critical period methods have been developed such as : the semi-infinite reservoir method (Hazen, 1914); Hurst's procedure (Hurst, 1965); sequent peak algorithms (Thomas & Burden, 1963) and Alexander's method (Alexander, 1962). A comprehensive review of these methods is given by McMahon & Mein (1978) but are not considered here because they have not been widely adopted for the sizing of rainwater stores.

2.5 RAINWATER QUALITY

2.5.1 Introduction

The quality of rainwater delivered by a collection system is important for a number of reasons. The acceptance of the system by the user is related to the aesthetic quality of the water expressed in terms of colour, odour and turbidity. The microbiological quality of the water will determine the potential health risk the system represents to the user. The chemical and physical parameters such as pH and dissolved solids will affect the selection of system components and their durability. The presence of heavy metals and other chemical compounds will also be particularly relevant if the water is for a potable application.

The contamination of rainwater originates from four main sources:
(1) Pollution of rainfall during its passage through the atmosphere (wet deposition).
(2) Dry deposition of atmospheric pollutants onto the collection surface during dry periods.
(3) Chemical and/or physical reaction of rainwater with collection surface material and/or other system components e.g. storage tank
(4) Bird and animal faeces deposited onto the roof.

The first three potential sources of contamination affect the chemical and/or physical characteristics of the rainwater whilst the last will determine its microbiological quality.

2.5.2 Chemical and Physical Contaminants

The chemical and physical characteristics of rainwater collected from roofs have been investigated by a number of researchers. The values reported by Mottier *et al.* (1995), Forster (1991), Yaziz *et al.* (1989), Xanthopoulos & Hahn and Surendran and Wheatley (1998) are summarised in Table 2.2.

Table 2.2 Summary of run-off water quality data

	pH	BOD (mg/1)	COD (mg/1)	TOC (mg/1)	Turbidity	TS (mg/1)	SS (mg/1)
Roof runoff	5.2-8	7-24	44-120	6-13	10-56	60-379	3-281
Stored Runoff	6-8.2	3	6-151	-	1-23	34-421	0-19

The pollutant loads in the run-off from roofs are attributable mainly to the atmospheric pollutants, which accumulate on the roofs as dry deposition or are washed out from the atmosphere by the rain, wet deposition, (Mottier *et al.*, 1995; Thomas & Greene, 1993 and Zobrist *et al.*, 2000). However, the roofing material can influence the chemical and physical characteristics of the rainwater run-off. Roofs covered in copper or galvanised have been found to have higher concentrations of copper and zinc than rainwater collected from slate or concrete tiled roofs. Yaziz (1989) for example reports the zinc concentration to be five times higher when compared to a tiled roof but was below the WHO limit for drinking water. The study of copper and zinc run-off is considered in detail by He *et al.* (2001). The conductivity, turbidity and suspended solids concentrations have been found to be consistently higher in the run-off from concrete tiled roofs due partly to the weathering of the surface and their ability to trap particulate matter as a result to dry deposition. In contrast bitumen roofs with crushed stone or gravel can act as a pollutant sink. Due to the filter mechanisms and biochemical processes, which take place in the gravel layers the concentrations of most pollutants are reduced. (Klein and Bullerman, 1989; Mottier, 1995).

The pH of rainwater is usually slightly acidic, normally in the range of pH 5 to 6. Industrial activity is often associated with increased levels of atmospheric pollutants such as sulphur dioxide and nitrous oxides resulting in rainwater within the range of pH 4 to pH 5 with values sometimes dropping as low as pH

3 (Gould, 1999). However, the pH of the roof runoff is strongly influenced by the roof material and may differ by as much as 3.5 units (on the pH scale). For example, Forster (1991) found bitumen felt roofs to leave the pH almost unchanged. The largest increase was found with fibrous cement tiles, the pH increasing to an average value of 7.4 compared to the rain, which had a mean value of 3.9. Clay tiles, zinc sheet and gravel had average pH value of 5.5, 6.7 and 6.8 respectively.

The length of dry periods between rainfall events has a significant affect upon the collected rainfall quality due to the increased dry deposition of pollutants on to the roof surface. This results in the first part of the run off or first flush being more heavily polluted than the subsequent run off. For most contaminants, concentrations are in the range of the wet deposition after the first few mm of run off depth. (Zobrist *et al.*, 2000).

The influence of location upon the quality of roof runoff has been investigated by Thomas and Greene (1993). A water quality analysis of rainwater from different roof catchments in rural, urban and industrial areas was undertaken to determine its suitability for domestic use. Rainwater from the industrial area roof catchments had lead concentrations in the order of twice those recommended by the WHO drinking water guidelines and also had high levels of turbidity, suspended solids and zinc. The concentrations of nitrates from the rural area of roof catchments were higher and the pH slightly elevated. Lead was detected in the urban area roof catchments but concentrations were not as high as those recorded in the industrial area. The levels of heavy metals are important in rainwater systems, which supply potable water. Lead is particularly important because it is a cumulative poison. High lead levels can also be obtained from roofs with extensive lead flashings, (Waller & Inman, 1982).

The relationship between the chemical and physical properties of rainwater and the selection of system components and their durability are considered by Leggett *et al.* (2001) and Mustow *et al.* (1997). High levels of suspended solids can cause wear on system components such as pumps and valves. Low pH or acidic conditions can increase the likelihood of corrosion, as can high levels of total dissolved solids and chlorides. Dirty or contaminated surfaces are also more susceptible to localised corrosion. In contrast alkaline water reduces the effectiveness of chlorine disinfection.

2.5.3 Microbiological Contamination

The microbiological quality of collected rainwater, in particular the presence of pathogenic organisms is important because of the potential risk to the system user. Even in systems, that provide non-potable water, there is a risk associated with the accidental ingestion of rainwater. The degree of risk is related to the end use of the rainwater, this being greater for example where it is used for surface crop irrigation and an aerosol is produced, which can be inhaled compared to sub-surface irrigation

In the case of rainwater systems, that supply potable water various studies have been undertaken, that have indicated the collected runoff often fails to meet WHO limits for drinking water in relation to bacteriological quality. (Riddle & Speedy 1984, Fujioka & Chinn 1987, Lye 1987, Krishna 1989). In a more recent study by Appan (1997) the bacteriological water quality of rainwater collectors in Southeast Asian countries has been reviewed, which has confirmed the results of earlier studies. From samples collected in most locations, there were positive Total and Faecal Coliform counts. The faecal contamination was attributed to animal origins such as bird and rodent droppings because the Faecal/Faecal Streptococci ratio was less than unity. Studies in Germany (Lücke, 1999) confirm rainwater collection systems generally do not meet microbiological standards for drinking water. However, the collected rainwater usually complied with EC standards for recreational waters and was therefore argued to be suitable for non-potable uses such as WC flushing.

Four general pathogenic groups are identified by Lücke (1999), which can be transmitted by water:

(1) Pathogens adapted to humans e.g. *Salmonella, Typhi* and *Shigella.*
(2) Pathogens adapted to domesticated animals e.g. enterhaemorrhaging *Escherichia coli* (EHEC).
(3) Pathogens with a reservoir in birds or small mammals e.g. *Salmonella, Typhimurium, Campylobactor, Giardia* and *Cryptospiridium*
(4) Opportunistic pathogens able to survive in the water supply system e.g. *Pseudomonas aeruginosa*, certain Enterobacteriacea and *Legionella spp*

The first group of pathogens can only enter a cistern via sewage whilst the source of the second category originates from the faeces of livestock. Contamination of the cistern water by these pathogens is only likely to occur if polluted effluent enters a defective below ground level storage tank. The third group are the most likely source of pathogenic pollution. The faeces of birds or small mammals are commonly deposited on roof surfaces. The final category is opportunistic pathogens and may occur in biofilms, which are the result of stagnation in water systems.

The specific pathogens associated with bird faeces and opportunistic sources have been identified and isolated from rainwater by other researchers; *Salmonella* (Chareonsook 1986, Fujioka *et al.,* 1991), *Legionella spp* (Lye 1992), *Campylobacter* (Brodribb *et al.,* 1995), *Cryptosporidium* and *Giardia* (Crabtree *et al.,* 1996). However, there are few documented cases, that link the outbreak of diseases with rainwater. The best example of gastrointestinal illness directly linked to rainwater is reported by Koplan *et al.* (1978). This occurred in Trinidad on a Summer camp attended by children. The cause was thought to be *Salmonella arechevalata* within bird or animal excrement deposited on the camp roof and subsequently washed in to the rainwater collection tank.

In the WROCS (2000) study, microbial contamination was identified as the health risk of most concern for both grey water and rainwater recycling systems. The presence and growth of *Legionella spp* within a recycling system was considered to pose the highest risk. *Legionella spp* may be problematic because they are fairly resistant to conventional chlorination (Kuchta *et al.*, 1993). Aquatic biofilms are likely to exist within rainwater systems, which are ideal sites for their survival and proliferation. The peak growth temperature for these micro-organisms is about 37°C but multiplication of the bacteria will still occur in the range 25°C to 45°C (Leggett, 2001). Rainwater is likely to become contaminated with naturally occurring Legionella spp., however, there is no evidence it is more likely to occur in rainwater systems than potable water supply networks. Similarly because they require a minimum temperature of 25°C (Lücke, 1999) the risk of their development is likely to be small in rainwater collection systems, provided the storage tank is situated below ground level.

In some rainwater collection systems the rainwater may subsequently be heated for use in showers or washing machines. In these instances the growth of *Legionella spp* is more likely. Lye (1991 and 1992) has monitored the microbial quality of rainwater cisterns in Northern Kentucky, USA. The systems (an estimated 80,000) usually incorporate water heating to provide hot water for bathing purposes. *Legionella spp* and Legionella – like organisms have been isolated in the hot water pipework of these systems. A potential risk exists because the exposure route is through the inhalation of aerosols, which are likely when showers are used for personal hygiene.

A general increase in the bacterial activity of stored rainwater, as indicated by coliform counts, has been observed by other studies. For example, Ruskin and Callender (1988) have reported chlorination is only effective in controlling bacterial populations within rainwater systems for three to five days before re-growth. However, total coliforms do not necessarily indicate the presence of pathogenic organisms but do indicate a possible route for exposure

Generally, Leggett *et al.* (2001) consider the growth of pathogenic bacteria in rainwater storage tanks is unlikely. This view is supported by a study undertaken by Krampitz and Holländer (1999). Cistern collected rainwater was experimentally contaminated with the pathogenic bacteria, *Salmonella enteritidis, Yersina enterocolitica* and *Campylobacter jejuni.* The survival rates of the pathogens, which could enter a cistern through the faeces of birds or small animals was investigated using different temperatures and nutrient levels.

All of the pathogenic bacteria investigated were unable to grow at temperatures of 5°C, 15°C, 20°C or 37°C. *Campylobacter jejuni* was the most sensitive with rapidly decreasing colony counts at 15°C as compared with *Yersinia enterocolitica* or *Salmonella enteritidis*. Nutrient levels were investigated by the addition of bacteriological pepton or pigeon faeces. The rate of bacterial death was reduced but increases in colony counts at the temperatures investigated were not observed. It was concluded that even with increased nutrient levels and optimum water temperatures, which could arise occasionally in rainwater cisterns, the growth of enteropathogenic bacteria did not increase. The study also indicated the cleaning of rainwater cisterns may be counter productive. The fatality rate of *Salmonella enteritidis* increased in systems with biofilms and sediment layers. This finding is also supported by Lye (1991) who found heterotrophic bacterial levels were lower in rainwater cisterns which did not receive any type of maintenance or cleaning.

2.6 DISCUSSION

There is evidence of rainwater catchment systems being used in most parts of the world at some point in history. In the case of some locations such as Africa and India, systems can be traced back many thousands of years. Generally the use of rainwater collection systems declined in many developed and developing countries during the twentieth century due to the application of centralised water collection and distribution systems. However, the past twenty years has seen a revival of rainwater collection systems in rural and urban areas of both developed and developing countries. In developing countries the main application of rainwater systems is the provision of potable water. The renewed interest in these countries is attributable to a range of factors including operational problems associated with centralised supply systems and increases in demand due to population growth.

In rural areas of some developed countries such as Australia and Canada potable supplies from rainwater collectors have a long history because centralised supplies are often not possible on economic grounds or ground water supplies are either unreliable or of poor quality. However, the current trend in Australia and Canada is the spread of rainwater systems into urban areas to supply both potable and non-potable water supplies. The governments in these countries consider rainwater systems a viable and ecologically sustainable method of supplementing or in some cases substituting supplies from centralised water supply facilitates.

The use of rainwater catchment systems to supply non-potable water in urban areas of developed countries has become more widespread during the past 15 – 20 years, particularly in Germany. Problems have been encountered in meeting the increased demand from centralised sources. Urban expansion can also create problems due to the increased surface run-off and the associated risk of flooding.

The benefits of rainwater catchment systems in the urban context include conservation of water resources, relief of demand on public water supplies, attenuation of peak run-off into the sewer network and the reduction of combined sewer overflow emissions.

German experience indicates the economics of rainwater systems are favourable with payback periods of between 12 – 19 years. In the UK the financial benefits remain unproven. The sustainability benefits of rainwater collectors also require further research. The amount of embodied energy in the components of rainwater systems compared to centralised systems of supply has not been reported. However, the energy consumed by rainwater systems during operation is comparable to that used by centralised supply systems.

The basic configuration and construction of rainwater collection systems is relatively simple consisting primarily of a catchment area, storage tank and interconnecting pipe work. Systems can be classified as either direct, indirect or gravity feed depending upon how the rainwater is stored and distributed within the installation. Alternatively, systems can be classified according to their hydraulic properties, namely, total flow, diverter, retention and throttle, and infiltration types.

The materials from which the catchment area is constructed have a significant effect upon both the amount of run-off and its chemical, physical and possibly biological quality. The first flush of rainwater from the catchment area is more heavily polluted than the subsequent runoff. Current German practice suggests the diversion of the first flush is not required. For non-potable applications rainwater only requires treatment prior to entry into the storage tank using a cross flow or screen filter with an apperture size of between 0.2 to 1.00mm. Further treatment also occurs within the storage tank via flotation and settlement. Storage below ground is preferable to minimise the entry of daylight and maintain the collected water at a low temperature, which reduces microbiological activity such as algal growth.

The capacity of the store is critical in the design of rainwater collection systems because it affects both the economics and operation of the system. Two categories of reservoir sizing model have been identified and reviewed, namely Moran and critical period models. Moran related methods are based upon queuing theory but have not been widely applied to the sizing of rainwater systems. Critical period methods identify and use sequences of flows where demand exceeds supply to determine the storage capacity. This method has been adopted extensively for the sizing of rainwater systems but has one main limitation, it is not possible to compute a storage size for a given probability of failure. This limitation can be overcome using statistical methods but is more easily accommodated using behavioural analysis.

Behavioural or simulation analysis falls within the general category of critical period methods and simulates the operation of a reservoir with respect to time by routing simulated mass flows through an algorithm, which describes the operation of the reservoir. The operation of the system will usually be simulated over a period of years. Two fundamental algorithms have been identified, which define the operation of the rainwater, yield after storage (YAS) and yield before storage (YBS). The YAS algorithm is preferred because it has been found to give a conservative estimate of system performance.

The other problems associated with the modelling process relate to how the spatial and temporal fluctuations in rainfall can be incorporated into the model using rainfall time series from different geographic locations. Temporal fluctuations are related to the time interval of the input time series. The use of daily time series will give and accurate prediction of system performance. In the UK a set of generic performance curves using daily data have been produced to predict the performance of rainwater collection systems.

An alternative approach using monthly data is also reviewed. The poor resolution of monthly time interval models compared to daily models can be countered by the introduction of a parameter, referred to as the storage operating parameter (θ). The short time scale fluctuations of the daily model are effectively replicated in the monthly model by the storage operating parameter. Values of the parameter are selected so that the monthly model mimics the system performance predicted by the corresponding model using a daily time interval. This approach provides an alternative method of accurately incorporating temporal fluctuations into the modelling process. A value of the storage operating parameter equal to 0.6 has been found to be suitable for modelling the performance of systems in the UK.

The acceptance of a rainwater collection system by the user, the durability of the system components and the potential risk to health is directly related to the quality of rainwater supplied by the collector. The chemical and physical pollutants in the rainwater are due mainly to atmospheric pollutants, which contaminate the rainwater by two routes, dry and wet deposition. The amount of dry deposition is related to the time interval between rainfall events. The levels of pH, suspended solids, turbidity and metallic ions may also be affected by the type of roofing material used to collect the rainwater. Similarly the quality of the rainwater will also vary depending upon whether the system is located in a rural, urban or industrial environment.

The potential health risk to a user of a collection system is related to the microbiological quality of the rainwater. Pathogenic organisms can be introduced into a rainwater collector via bird and animal faeces deposited on the roof or via opportunistic pathogens, which can survive in water systems, for example *Legionella spp.* Pathogenic organisms present in bird or animal faeces such as *Salmonella, Campylobacter, Giardia* and *Cryptosporidium* are unlikely to grow in rainwater between the temperatures of 5 - 37°C even when nutrient levels are favourable. The presence and growth of *Legionella spp.*, represents the highest health risk. Growth occurs between the temperatures of 25 - 45°C with likely sites being aquatic biofilms, which are likely to exist in rainwater catchment systems.

The use of below ground storage will minimise the risk of *Legionella spp.*, because storage temperatures are unlikely to reach 25°C.

2.7 REFERENCES

Alexander, G.N. (1962) The use of Gamma distribution in estimating regulated output from storages. In *Civil Engineering Transactions, The Institution of Engineers, Australia* **4** (1), 29 – 34.

Appan, A. (1997) Roof water collection in some southeast Asia countries: status and water quality levels. *Journal of the Royal Society of Health* **117** (5), 319-323.

Appan, A. Villareal, C. and Wing, L. (1989) The planning, development and implementation of a typical rainwater cistern systems: A case study in the Province of Capiz. In *Proc. of the 4th Int. Conf. on Rainwater Cistern Systems*, Manilla, Philippines, 2-4 August.

Brewer, D. Brown, R. and Stanfield, G. (2001) *Rainwater and greywater in buildings: Project report and case studies.* Technical note TN7/2001, BSRIA, Bracknell, Berkshire.

Brodribb, R. Webster, P. and Farrell, D. (1995) Recurrent Campylobacter fetus species bacteraemia in a febrile neutropaenipatient linked to tank water. *Communicable Disease Intelligence* **19** (5), 312-313.

Chareonsook, O. (1986) The Entero-pathogenic bacteria and pH of rainwater from three types of container. *Communicable Disease Journal,* Thailand **12** (1), 50-58.

Chu, S. C., Liaw, S. C., Huang, S. K., Tsai, Y. L. and Kuo, J.J. (1999) The assessment of agricultural rainwater catchment systems in mudstone areas. In *Proc. of the 9th Int. Conf. on Rainwater Catchment Cistern Systems*, Petrolina, Brazil, 20-22 July.

Courier, S. (2002) GARASTOR – The benefits of a new storm water retention system for residential developments. In *Proc. Standing Conf. on Storm Water Source Control*, Coventry University, 13 February.

Crabtree, K., Ruskin, R., Shaw, S. and Rose, J. (1996) The detection of Cryptosporidium Oocysts and Giardia cysts in cistern water in the US Virgin Islands. *Water Research* **30** (1), 208-216.

Cunliffe, D. (1998) *Guidance on the use of rainwater tanks.* National Environmental Health Forum Mongraphs, Water Series 3, Public and Environmental Health Service, P O Box 6, Rundle Mall, SA 5000, Australia.

Day, M. (2002) The Millennium Green development rainwater project. In *Proceedings of Standing Conference on Storm Water Source Control*, Coventry University, 13 February.

Dixon, A., Butler, D. and Fewkes, A. (1999) Computer simulation of domestic water reuse system: investigating grey water and rainwater in combination. *Water Science & Technology* **38** (4), 25-32.

Dominguez, M.A., Schiller, E. and Serpokrylov, N. (2001) Sizing of rainwater storage tanks in urban zones. In *Proc. of the 10th Int. Conf. on Rainwater Cistern Systems,* Mannheim, Germany. 10-14 September.

Fewkes, A., and Butler, D. (2000) Simulating the performance of rainwater collection systems using behavioural models. *Building Services Engineering Research and Technology.* **21** (2) 99-106.

Fewkes, A., and Frampton, D. (1993) Optimising the capacity of rainwater storage cisterns. In *Proc. of the 6th Int. Conf. on Rainwater Cistern Systems*, Nairobi, Kenya, 1-6 August.

Fewkes, A., and Warm, P. (2000) Method of Modelling the performance of rainwater collection systems in the United Kingdom. *Building Services Engineering Research and Technology* **2**(4), 257-265.

Fewkes, A. (1996) The field testing of a rainwater collection and re-use system. In *Proceedings of Conference on Water recycling: Technical and social implications.* London, 2-6 March 1996

Fewkes, A. (1999a). Modelling the performance of rainwater collection systems: towards a generalised approach. *Urban Water* **1** (4) 323-333.

Fewkes, A. (1999b) The use of rainwater for WC flushing: the field testing of a collection system. *Building. & Environment.* **34** (6) 765-772.

Fok, Y. S., and Leung, P.S. (1982) Cost analysis of rainwater cistern systems. In *Proc. of the 1st Int. Conf. on Rainwater Cistern Systems*, Hawaii, Honolulu, June 15-18.

Foster, J. (1991) Roof runoff pollution. In *Second Junior Workshop: Hydrological and Pollution Aspects of stormwater infiltration*, Kastanlenbaum, Luzern, Switzerland, 4 – 7 April.

Fujioka, R., and Chinn, R. (1987) The microbiological quality of cistern waters in the Tantalus area of Honolulu, Hawaii. In *Proc. of the 3rd Int. Conf. on Rainwater Cistern Systems*, Khon Kaen University, Thailand, F3, 15-16 January, 1987.

Fujioka, R., Inserra, S., and Chinn, R. (1991) The bacterial content of cistern waters in Hawaii. In *Proc. of the 5th Int. Conf. on Rainwater Cistern Systems*, Keelung, Taiwan, 4-10 August.

Gardner, T., Coombes, P., and Marks, R. (1999) Use of rainwater in Australian urban environment. In *Proc. of the 10th Int. Conf. on Rainwater Catchment Systems*, Mannheim, Germany, 10-14 September.

Geiger, W.F. (1995) Integrated water management for urban areas. In *Proc. of the 7th Int. Conf. on Rainwater Cistern Systems*, Beijing, China, 21-25 June.

Gieske, A., Gould, J.E. and Sefe, F. (1995) Performance of an instrumented roof catchment system in Botswana. In *Proc. of the 7th Int. Conf. on Rainwater Catchment Systems*, Beijing, China, 21-25.

Gould, B.W. (1961) Statistical methods for estimating the design capacity of dams. *Journal of the Institution of Engineers, Australia* **33** (12), 405-416.

Gould, J. (1993) A review of the development, current status and future potential of rainwater catchment systems for household supply in Africa. In *Proc. of the 6th Int. Conf. on Rainwater Catchment Systems*, Nairobi, Kenya 1-6 August.

Gould, J. (1999) Is rainwater safe to drink? A review of recent findings. In *Proc. of the 9th Int. Conference on Rainwater Cistern Systems, Petrolina*, Brazil, 20-22 July.

Gould, J. and Nissen-Petersen, E. (1999) *Rainwater catchment systems for domestic supply: Design, construction and implementation*, Intermediate Technology Publications, London.

Grove, S. (1993) Rainwater harvesting in the United States – Learning lessons the world can use. *Raindrop* **8** (1), 1-10.

Hassel, C. (2001) Save rainwater save money. *Bldg. Eng.* **76** (2) 10-11.

Hazen, A. (1914) Storage to be provided in impounding reservoirs for municipal water supply. *Transactions of the American Society of Civil Engineers* **77**(1), 15-39.

He, W., Wallinder, O. and Leygraf, C. (2001) The Laboratory study of copper and zinc runoff during first flush and steady state conditions. *Corrosion Science* **43** 127-146.

Hegen, R. (1993) Value of daily data for rainwater catchment. In *Proc. of the 6th Int. Conf. on Rainwater Cistern Systems*, Nairobi, Kenya, 1-6 August.

Herrmann, T. and Schmida, U. (1999) Rainwater utilisation in Germany : efficiency, dimensioning, hydraulic and environmental aspects. *Urban Water* **1** (4), 307-16.

Hurst, H.E., Black, R.P. and Simaika, Y.M. (1965) *Long term storage*. Constable, London.

Jenkins, D., Pearson, F., Moore, E., Sun, J.K. and Valentine, R. (1978) *Feasibility of rainwater collection systems in California*. Contribution No. 173 (Californian Water Resources Centre), University of California.

Keller, K. (1982) *Rainwater harvesting for domestic water supplies in developing countries: A literature survey*. Working paper 20, Water and Sanitation for Health (WASH), USAID, Washington, DC.

Klein, B. and Bullerman, M. (1989) Qualitative aspects of rainwater use in the Federal Republic of Germany. In *Proc. of the 4th Int. Conf. on Rainwater Cistern Systems*, Manilla, Philippines, 2-4 August.

Konig, K.W. (1999) Rainwater in cities : A note on ecology and practice. In *Cities and the Environment: New approaches for Eco-Societies*, United Nations University Press, New York.

Konig, K.W. (2000a) Berlin's water harvest. *Water 21*, June, 31-32.

Konig, K.W. (2000b) Rainwater Utilisation. In *Technologies for urban water recycling*. Cranfield University.

Konig, K.W. (2001) *The rainwater technology handbook : Rainwater harvesting in building*. Wilo-Brain, Germany.

Koplan, J., Deen, R., Swanston, W., and Tota, B. (1978) Contaminated roof collected rainwater as a possible cause of an outbreak of Salmonellosis. *Journal of Hygiene* **8** (5), 303-309.

Krampitz, E., and Holländer, A. (1999) Longevity of pathogenic bacteria especially Salmonella in cistern water. *Zentralblatt für Hygiene und Umweltmedizin* **202** (6), 389-397.

Krishna, J. (1993) Water quality standards for rainwater cistern systems. In *Proc. of the 6th Int. Conf. on Rainwater Cistern Systems*, Nairobi, Kenya, 1-6 August.

Kuchta, J.M., Navratil, J.S., Shepherd, M.E., Wadowsky, R.M., Dowling, J.N. and States, S.J. (1993) Impact of chlorine and heat on the survival of Hartmamnella vermiformis and subsequent growth of Legionella pneumophila. *Applied Environmental Microbiology* **59**, 4096-4100.

Latham, B.G. (1983) Rainwater collection systems: the design of single purpose reservoirs. MSc thesis, University of Ottawa, Canada.

Latham, B.G., and Schiller, E.J. (1987) A comparison of commonly used hydro logic design methods for rainwater collectors. *Water Resources Development, 3* (3), 165-170.

Leggett, D. J., Brown, R., Brewer, D., Stanfield, G., and Holiday, E. (2001) *Rainwater and grey water use in buildings: Best practice guidance*. Report C539, CIRIA, London.

Lodge, B.N. (2000) A water recycling plant at the Millennium Dome. *ICE Proceedings*, Special Issue 1, 58-64.

Lücke, F.K. (1999) *Process water of potable quality: sense or nonsense*. Elemental Solutions, The Green Shop and Construction Resources, Gloucestershire, UK.

Lye, D. (1987) Bacterial levels in cistern water systems in Northern Kentucky, *Water Resources Bulletin*, **23**, 1063-68.

Lye, D. (1991) Microbial Levels in cistern systems : Acceptable or unacceptable. In *Proc. of the 5th Int. Conf. on Rainwater Cistern Systems*, Keeling, Taiwan, 4-10 August.

Lye, D. (1992) Legionella and Amoeba found in cistern systems. In *Proc. of the Regional Conference of the Int. Rainwater Catchment Systems Association*, Kyoto, Japan, 4-10 October.

McMahon, T. A., and Mein, R.G. (1978) *Reservoir capacity and yield: Developments in water science.* Amsterdam, Elsevier.

Michaelides, G. (1989) Investigation into the quality of roof harvested rainwater for domestic use in developing countries. In *Proc. of the 4th Int. Conf. on Rainwater Cistern Systems*, Manilla, Philippines, 2-4 August.

Moran, P.A.P. (1959) *The theory of storage.* Methuen, London.

Mottier, V., Bucheli, T., Kobler, D., Ochs, M., Zobrist, J., Ammann, A., Eugsler, J., Mueller, S., Schoenberger, R., Sigg, L., and Boller, M. (1995) Qualitative aspects of roof runoff. In *Eighth Junior Workshop: Urban rainwater resourcefully used.* Deventer, the Netherlands 22 – 25 September.

Murase, M. (1998) Rainwater utilisation and sustainable development in cities. *Wasser-Wirtschaft-Zeitschift for Wasser und Umwelt* **88** (7), 401-3.

Mustow, S., Grey, R., Smerdon, T., Pinney, C., and Waggett, R. (1997) *Water conservation : Implications of using recycled grey water and stored rainwater in the UK.* Final Report 13034/1, BSRIA, Berkshire.

Ngigi, S. N. (1999) Optimisation of rainwater catchment system design parameters in the arid and semi arid lands of Kenya. In *Proc. of the 9th Int. Conf. on Rainwater Catchment Cistern Systems*, Petrolina, Brazil, 20-22 July.

Pacey, A. and Cullis, A. (1986) *Rainwater Harvesting: The collection of rainfall and run-off in rural areas,* Intermediate Technology Publications, London

Perrens, S.J. (1982) Design strategy for domestic rainwater systems in Australia. In *Proc. of the 1st Int. Conference on Rainwater Cistern Systems*, Hawaii, Honolulu, 15-18 June.

Perrens, S.J. (1982) Effect of rationing on reliability of domestic rainwater systems. In *Proc. of the 1st Int. Conf. on Rainwater Cistern Systems*, Hawaii, Honolulu, 15-18 June.

Piggot, T.L.; Schiefelbein, I.J. and Duram, M.T. (1982). Application of the Gould matrix technique to roof water storage In *Proc. of the 1st Int. Conf. on Rainwater Cistern Systems*, Hawaii, Honolulu, 15-18 June.

Pratt, C.J. (1999) Use of permeable, reservoir pavement constructions for storm water treatment and storage for re-use. *Water Science & Technology* **39** (5), 145-151.

Ratcliffe, J, (2002) Telford rainwater harvesting project. In *Proc. Standing Conf. on Storm Water Source Control,* Coventry University, 13 February.

Ree, W.O., Wimberley, O., Guinn, W.R. and Lauritzen, C.W. (1971). *Rainwater harvesting system design.* Report A R S 41 – 184 Agricultural Research Service, USDA, Oklahoma.

Riddle, J., and Speedy, R. (1994) Rainwater Cistern Systems: The Park Experience. In *Proc. of 2nd Int. Conf. on Rainwater Cistern Systems*, US Virgin Islands, 15-18 June.

Rippl, W. (1883) The Capacity of Storage reservoirs for Water Supply. In *Proc. of the Institute of Civil Engineers* **71**, 270-278.

Ruskin, R.H., and Callender, A. (1988) *Maintenance of cistern water quality and quantity in the Virgin Islands.* Tech. Report No. 30, Caribbean Research Institute, St Thomas, USA Virgin Islands.

Sayers, D. (1999) *Rainwater recycling in Germany.* National Water Demand Management Centre, Worthing, West Sussex.

Schiller, E. (1987) Rainwater collection systems: A literature review. In *Proc. of the 3rd Int. Conf. on Rainwater Cistern Systems*, Khon Kaen University, Thailand.

Scott, R.S., and Waller, D.H. (1991) Development of Guidelines for rainwater cistern systems in Nova Scotia. In *Proc. of the 5th Int. Conf. on Rainwater Catchment Systems*, Keelung, Taiwan, 4-10 August.

Surrendran, S., and Wheatley, A., (1998) Grey water reclamation for non-potable re-use. *J. CIWEM* **12**(6), 406-13.

Thomas, H.A., and Burden, R.P. (1963) *Operations Research in Water Quality Management*. Havard Water Resources Group.

Thomas, P.R., and Greene, G.R. (1993) Rainwater quality from different roof catchments. *Water Science & Technology* **28** (3), 291-299.

Townshend, A.R., Jowett, E.C., LeCrow, R.A., Waller, D.H., Paloheimo, R., Ives, C., Russel, P., and Liefhebber, M., (1997). Potable water treatment and reuse of domestic waste water in the CMHC Toronto 'Healthy House'. *American Society for Testing and Materials* **1324**, 176-187.

Waitt, F.M.F. (1945) Studies of droughts in the Sydney catchment areas. *Journal of the Institution of Engineers, Australia*, **17** (4), 90-97.

Waller, D., and Inman, D. (1982) Rainwater as an alternative source in Nova Scotia. *In Proc. of the Int. Conf. on Rainwater Cistern Systems*, Hawaii, Honolulu, 15-18 June.

Waller, D.H. (1982) Rainwater as a water supply source in Bermuda. In *Proc. of the 1st Int. Conf. on Rainwater Cistern Systems*, Hawaii, Honolulu, 15-18 June.

Walther, D., and Thanasekaren, K. (2001) Cost effective dimensioning of artificial rainwater harvesting storage systems. In *Proc. of the 10th Int. Conf. on Rainwater Catchment Systems*, Mannheim, Germany, 10-14 September.

Watt, S.B. (1978) *Ferrocement water tanks and their construction*. Intermediate technology publications, London.

Wirojanagud, P., and Vanvarothorn, V. (1990) Jars and tanks for rainwater storage in rural Thailand. *Waterlines* **8** (3), 29-32.

WROCS (2001) *Final Report of the Water Recycling Opportunities for City Sustainability (WROCS) project*. Cranfield University, Imperial College London and The Nottingham Trent University.

Xanthopoulos, C., and Hahn, H. (1994) Priority pollutants from urban storm water runoff into the environment. *Journal European Water Pollution Control* **4** (5), 32-41.

Yaziz, M.I., Gunting, N., Sapari, N., and Ghazali, A. (1989) Variations in rainwater quality from roof catchments. *Water Research.* **23** (6), 761-765.

Zaizen, M., Urakawa, T., Matsumoto, Y., and Takai, T. (1999) The collection of rainwater from dome stadiums in Japan. *Urban Water* **1**(4), 355-9.

Zobrist, J., Muller, S.R., Ammann, A., Bucheli, T.D., Mottier, V., Ochs, M., Schoenenberger, R., Eugster, J. and Boller, M. (2000). Quality of roof runoff for groundwater infiltration. *Water Research* **34** (5), 1455-1462.

3

Understanding greywater treatment

William S. Warner

3.1 INTRODUCTION

A private concern drawing public interest is the health risk associated with on-site greywater treatment. Treatment alternatives vary in complexity, efficiency and efficacy. And the spectrum of solutions continues to expand as the public becomes aware that greywater can be reused for various purposes (e.g. irrigation or flushing toilets) and thus reduce the demand for water. This benefit, however, is predicated on proper treatment of potential contaminants.

3.1.1 Storms of controversy

The threat of potential contamination focuses on two fronts: (1) compounds found in household chemicals (e.g., detergents, bleaches, dyes, softeners, etc.), and (2) pathogenic micro-organisms found in transmission routes (e.g., groundwater infiltration and aerosol inhalation). The former has precipitated a flood of information on greywater composition and sources of hazardous compounds. The latter has generated a blizzard of data on indicator bacteria and probability of infections. And in both cases, we find controversial information about the safety of reusing greywater.

An underlying cause for the storming controversy is the general atmosphere of greywater information. It is polarised. On the one hand, there is the *high* information, consisting of scientific papers that are written by (and primarily for) those with an understanding of wastewater treatment *processes*. On the other hand, there is the *low* information, consisting of leaflets, brochures and Internet sites that are often written by (and primarily for) those with an interest in wastewater treatment *products*. When the two fronts meet the atmosphere for a perfect storm arises.

Forecasting the future of on-site greywater treatment systems is as difficult as predicting their health risks. To begin with, the characteristics of greywater are varied, and quantitative information on potential contaminants is limited. Equally important, quantitative information related to the factors that influence survival and transmission of pathogens (e.g. growth/reduction of selected pathogenic bacteria and viruses) is inconsistent. Moreover, the dynamic relationships associated with microbial risk assessment (e.g. dose-response) are as complex as the relationships associated with the fate of chemical compounds (e.g. sorption, volatilisation and degradation). And to further cloud the atmosphere, alternative solutions, mainly small-scale and mostly new (for domestic greywater treatment), are wide-ranging: infiltration systems, biofilters, membrane bioreactors, disinfection products, aerobic/anaerobic fixed film processes, etc. With such variance, confusion about treatment efficiency and efficacy is understandable.

The following helps explain why controversies materialise, with an emphasis on the cloudy issues of greywater characterisation and health risks. Although the future of domestic greywater treatment looks promising, the horizon is anything but clear. The fog of confusion still looms.

3.1.2 Treatment principles

The principles of greywater treatment are not universal but context specific. Treatment depends on several variables, including (but not limited to):
- greywater characteristics
- flow rate
- ground conditions: soils, percolation, depth of groundwater, etc.
- climate
- regulations

Consequently it is difficulty to assess the success (or failure) of a particular system outside its site-specific application. Nevertheless, generalisations about treatment concepts and system designs can help mitigate the confusion, provided they are understood as generalisations.

In general terms, greywater has lower concentrations of micro-organisms and some nutrients (e.g. nitrogen and potassium) than combined wastewater. However, the concentration of phosphorus, heavy metals, and xenobiotic organic pollutants are around the same level (Ledin *et al.*, 2001). Although the concentration can be similar to domestic wastewater the composition is not. The COD:BOD ratio can be high as 4:1 (Jefferson *et al.*, 2001). And though greywater can have an organic strength similar to domestic wastewater, it is relatively low in solids, suggesting that a greater fraction of the organic load is dissolved (Jefferson *et al.*, 2001).

Issues about greywater characteristics will be discussed later, however, it is critical to realise from the outset that greywater varies considerably from its source. Hence, treatment concerns can often be source targeted, for example, grease in kitchen greywater and non-biodegradable fibres in laundry greywater. In addition, the source fluctuates in contributing basic constituents, such as:

- sodium hydroxide soaps, which inhibit water/nutrient transport to plants
- detergents, which precipitate calcium and magnesium ions and prevent solids from adhering to surfaces, and
- disinfectants (e.g. chlorine), which can kill beneficial organisms and reduce the efficiency of some treatment systems.

Since greywater exhibits differential composition, depending on its source, and the scale of most treatment schemes is relatively small scale, variations within individual households can have pronounced impact on the overall characteristics of the greywater to be treated (Jefferson *et al.*, 2001).

Nevertheless, generalised characteristics of greywater can be identified, which in turn establish design parameters and rule-of-thumb concerns for treatment technologies. For instance, greywater is nitrogen deficient but rich in carbon. Therefore, carbohydrates (i.e., sugars and starches) metabolise easy and are often treated first. Undigested grease, cellulose, and difficult compounds tend to be treated later. As a result clogging is a common problem. Clogging is compounded when non-biodegradable fibers enter a system. And new clothing fabrics with fibres smaller than 40 μm in diameter are clogging the problem further. Thus, as a general rule, filtering is vital, and a variety of filter dimensions is recommended (Del Porto and Steinfeld, 1999).

3.1.3 Treatment technology

Treatment options are numerous, ranging from simple systems that require little maintenance to complex units that need monitoring. Like any technology, every system has its advantages and disadvantages, which explains, in part, the controversy associated with treatment efficiency and efficacy.

Jefferson *et al.* (2001) summarize greywater technologies from small-scale single house applications to systems for large-scale reuse. They note that although several companies market systems that recycle greywater for toilet flushing, the products tend to be based on the same generic two-stage process: a coarse filtration stage for removal of large particles and either a disinfecting process (typically using either chlorine or bromide) or UV irradiation. Generally, these units rely upon short residence time. The high organic load and turbidity may lead to problems during disinfection for two reasons: (1) the disinfectant might not diffuse far enough into flocculent particles > 40 µm, and (2) organic matter imparts a disinfectant demand (Jefferson *et al.,* 2001). The authors note that regular maintenance is critical, and all systems contain a potable water back-up device to ensure a continual supply of water for toilet flushing. Although 30% water saving has been achieved with many of these systems under trial (Diaper *et al.,* 2000), the economic benefits are directly related to occupancy level. For single occupancy households, the payback period for a system with a representative annual water saving of around 20% is more than 50 years for the highest UK water charge (Sayer, 1998).

Physical systems such as depth filters and membranes have been developed because they produce a higher quality effluent than the single house systems described above (Jefferson *et al.,* 2001). Depth filters – typically a 250-litre drum containing coarse gravel at the base and sand on the top – require occasional backwashing to regenerate the media. Water quality from a membrane is generally high but the process suffers operational and economic constraints (Stephenson *et al.,* 2000). A major concern is the fouling of the membrane surface, which can be mitigated by cleaning. Fouling can decline flux by up to 90% after just one hour of operation (Gander *et al.,* 2000). Residence time has been identified as having a major effect on system performance: when stored for extended periods, greywater can become anaerobic and generate organic compounds that are less readily rejected by the membrane (Holden and Ward, 1999). Jefferson *et al.* (2001) omit the simplest physical treatment products. For good reason: those that do not properly filter (and sell for about US$ 500) are little more than expensive grease traps (Del Porto and Steinfeld, 1999).

Natural treatment systems for greywater often demonstrate the influence of a new design but the treatment concept(s) are essentially the same as nature-based systems used for treating sewage effluent. In fact, the processes involved in a natural system include many of those used in mechanical or in-plant treatment systems (sedimentation, filtration, gas transfer, adsorption, ion exchange, chemical precipitation, chemical oxidation and reduction, and biological conversion and degradation). However, they exploit processes unique to natural systems such as photosynthesis, photooxidation, and plant uptake. In natural systems the processes occur at *natural* rates and tend to occur simultaneously in a single *ecosystem reactor*; unlike mechanical systems in which processes occur sequentially in separate reactors or tanks at accelerated rates as a result of energy input (Metcalf and Eddy, 1991). Moreover, reuse of treated greywater is possible with all natural systems (Angelakis, 2001).

The heart of many natural treatment systems is soil infiltration, with the reuse objective being either to recharge groundwater or irrigate plants. The relatively low concentrations of nitrogen and phosphorous indicate that the risk of eutrophication of receiving waters is less than for combined wastewater. But Ledin *et al.* (2001) caution that the concentrations of these nutrients will impact microbial activity of the soil and will thereby contribute to the microbiological removal of organic matter from the infiltrating water. And although the pH of greywater is generally similar to combined wastewater (7.5-8), greywater from laundry tends to be alkaline (pH range 9.3-10), which could influence the fate of chemical compounds, as well as the biological activity of the soil (Ledin *et al.*, 2001). The expected quality of effluent varies depending on the principal type of infiltration: slow rate, rapid infiltration or overland-flow (Angelakis, 2001).

Some form of pre-treatment precedes all natural treatment systems. One of the most common natural treatment systems is the conventional septic tank, with the effluent flowing to an absorption system (leach lines, mounds, or drainfield). Since greywater contains a lower concentration of solid contaminants than domestic sewage the tank will require less frequent emptying. Septic tanks can be combined with other treatment options to improve the quality of effluent, such as an anaerobic filter at the tank's outlet to reduce potential solids that might clog drainage or irrigation pipes.

Irrigation beds can recycle greywater effectively provided sodium-based soaps and chlorine bleach are avoided (Del Porto and Steinfeld, 1999). These engineered gardens typically grow selected plants in an evapotranspiration bed lined with impervious membrane, often enclosed in a greenhouse. For larger volumes, aquatic systems can treat effluent from several homes. These are aerated tanks (sometimes combined with wetlands), where microbes, seed plants, and often fish sequentially treat the wastewater within a greenhouse.

Ponds and artificial wetlands (also called constructed wetlands) provide high treatment with minimal management. Wetlands reduce pathogens, remove nutrients and mineralize organic matter. The principal types are free-water surface or subsurface water flow. The latter requires less land and avoids odour and mosquito problems, but the media is often costly and may clog. Some wetlands are simple single-stage systems; others are more complex multi-stage designs. Wetland performance is predicated, in part, on design, which must consider, amongst other factors, soils, greywater composition, volume, rate... and climate. Constructed wetlands perform best in warmer climates, however, Scandinavian studies have shown that they can be effective in cold climates when greywater is pretreated with an aerobic biofilter (Jenssen and Vråle, 2004). Although pre-treatment is required for constructed wetlands, the optimal design, as well as the specific area demand for a wetland, remain matters of discussion (Müllegger *et al.*, 2004).

With treatment designs constantly changing it is not surprising that system performance is variable. Arrays of technologies exploit different (physical, chemical and biological) processes. For some systems the primary objective is to treat effluent rather than save potable water; for others it is to save water by reusing it. For others still it is a combination, which implies a multiple quality concept where different levels of treatment are used for disposal and/or reuse (Tchobanoglous, 1999). Since system objectives differ – and the efficiency and efficacy of technologies vary as much are as greywater itself – confusion and controversy about greywater treatment is understandable.

3.2 CONFUSION AND CONTROVERSY

The general public has greater access to specialised fields of science and technology now more than ever. This is especially true with the advent of the Internet and proliferating web-sites, which enable all equal access to information once devoted to scientists. With a few keystrokes the layman can type in *greywater* and generate a plethora of hits. Point and click, and the links cascade information on treatment products, health risks, etc. Piping information is basically good; however, there are two downsides.

First, the layman often does not understand the science of treatment processes. How treatment works is one thing, why is quite another. The difference is critical if the system fails, because the consequences to public health and private cost can be far-reaching. Unlike conventional sewage systems, on-site greywater treatment places the burden of responsibility on the private individual rather than a public utility. Wastewater treatment is a long-term commitment. And since most treatment schemes do not have backup systems – and failure has dire consequences – common sense suggests homeowners should prepare themselves for potential problems. Unfortunately, many do not. They commit themselves to a treatment system with a vague notion of how it works, but with little if any understanding of why it might fail. Ignorance is not always bliss, especially when it comes to on-site wastewater treatment. Infectious diseases, costly repairs and malodorous environments are all examples of ignorance – or misinformation.

The second downside is misinformation. This bothers professionals more than laymen because they are more aware of the potential risks involved with greywater treatment. Few laymen are aware that myths leak in the information pipeline, polluting the flow of scientific knowledge. Misunderstanding flourishes like a virus (intestinal or digital) in the right environment, such as the Internet. Unfortunately, product manufacturers, as well as proselytising ecological engineers, often cultivate misunderstanding (albeit unwittingly) with their unbridled enthusiasm for new treatment designs.

3.2.1 Language and logic problems

Language is a general problem for the layman's understanding of science. It also poses a problem for scientists because "it is normal in scientific thought for terms to change meaning over time" (Dow 2002). Currently, even a simple term like *demand* can have at least four different meanings when applied to water Merrett (2003):

(1) Water use: A generic way of referring to the quantitative flow per unit time of water arriving at a user's property. This flow may be used beneficially or it may be wasted in various ways.

(2) Water consumption: After the water use process is complete the remaining fraction is wastewater and irrigation returns. These flows are reused for other purposes, recycled to surface water, or recharged to groundwater.

(3) Water need: A social, cultural and health-related concept, it refers to desirable or recommended levels of use for drinking, cooking, cleaning, irrigation, etc. The difference between use and need is that the former represents an actual flow to users whereas the latter represents a desired or recommended flow.

(4) Economic demand for water: The relationship, at a given time within a defined market, between price per unit of water and the quantity users are estimated to be willing to pay for it.

Merrett (2003) concludes that when the concept of *demand* is deployed in a catchall manner the confusion that arises may have far-reaching impact (e.g. it obscures the water needs for more than half the world's population).

Vague terms that have a social cachet exacerbate the problem. Both laymen and professionals vaguely understand clichés with cachet (e.g. sustainable wastewater management). The rub comes when the critical word (in this case, *sustainable*) is not clearly understood. Although sustainability criteria for wastewater are under development (Urban Water, 2001), the pivotal term remains elusive to most.

Another factor that leads to misunderstanding is faulty syllogism. For example, some laymen assume that since the presence of *Escherichia* (*E. coli*) bacteria indicates contaminated greywater, the absence of *E. coli* indicates the greywater is not contaminated. Or, since *E. coli* is found in human faeces, the presence of *E. coli* in greywater indicates faecal contamination from humans. The blunder in logic is as obvious as *All cows have four legs; this animal has four legs, therefore, it must be a cow.* When dealing with treatment processes, however, the logic of cause-effect is not always clear. Consider the following assumption: if a little treatment is good, more treatment is better. Generally speaking that may be true, but not always. Dixon *et al.* (1999) examined the impact of storage on wastewater quality. They found that storage for 24 h improved the quality of water, but storage for more than 48 h could be a serious problem as the dissolved oxygen was depleted.

If greywater characteristics, treatment and health risks can be logically mapped as relationships, many misunderstandings become self-evident. The rudimentary problem is that all three entities are complex. Presenting complex entities is one thing, where simple descriptive inventories often suffice, but showing relationships is quite another. One of the basic characteristics of mapping – whether it is cartographic, schematic, or a narrative presentation of relationships – is that the subject matter is generalised. Unfortunately, it is the generalisation of greywater (characterisation, treatment and health risks) that germinates the confusion and cultivates the controversy.

3.3 GREYWATER CHARACTERISATION

Many misunderstandings about greywater characterisation (and associated health risks) result from generalisations about greywater composition. The composition of greywater depends on its sources and installations from where the water is drawn: kitchen, bathroom or laundry. Composition will vary from source to source, where lifestyles, customs, installations and use of chemical household products all influence its characteristics. As a result, composition fluctuates significantly, depending on a multitude of complex factors, e.g. chemical and biological degradation of compounds and pathogens within the transportation network and storage time. Consequently, the vast majority of laymen's literature on greywater composition consists of two components: (1) a summary of its sources, often depicted in simplistic pie-diagrams, and (2) an inventory of commonly found compounds and pathogens, often listed in tables with little if any quantification.

A critical issue often overlooked, because it is difficult to generalise, is the influence of storage on greywater quality. Little has been reported, but what has indicates a decline in organic strength, with waste becoming more biodegradable (Jefferson *et al.*, 2001). During storage, biological growth may lead to increased concentrations of micro-organisms including faecal coliforms (Jefferson *et al.*, 2001). This may also cause new organic and inorganic compounds to be produced: metabolites from partly degraded chemicals present in the greywater (Ledin *et al.*, 2001). The presence of nutrients such as phosphate, ammonium/nitrate and organic matter will produce this microbial growth (Ledin *et al.*, 2001). And chemical reactions could also take place during storage, thus changing its chemical composition. Consequently, the characteristics of collected greywater may differ in residence time, which can range from minutes to days.

Another major concern is the comparison of real and synthetic greywater. The nature of synthetic greywater is very different to real greywater in terms of concentrations of macro and micro nutrients (Jefferson *et al.*, 2000b). Equally important, the biological degradation pathways are different (at a fundamental level) for the two greywaters.

As in all fields of science, gross generalisations may be helpful for the layman, but have limited value for scientists. Scientists require specifics, but the value of detailed information in project reports is also limited because, as noted, greywater characteristics differ considerably in terms of place and time. Estimated average values for a greywater component are sometimes available for a state, region, country (e.g., EPA, 2002), but these are typically based upon merging independent studies, which have used different sampling strategies. Moreover, the average estimates are often too crude to be of useful application.

In short, both generalisation and specification of greywater characteristics have limited value. Without mapping relationships between composition and treatment processes, the data are innocuous at best, misleading at worst.

What is needed is systematic mapping of characteristic relationships. Ericksson *et al.*'s (2002) *Characteristics of Grey Wastewater* is an excellent example of such an exercise. Combining available measurement data from greywater with a survey of chemical compounds and micro-organisms that theoretically could be expected to be present, estimates were applied to a method for environmental hazard identification.

3.3.1 Compounding compounds

Although Ericksson *et al.*'s (2002) report does not visually map the composition of greywater, in the way a cartographer would map the relationship between water sources (e.g. rivers and lakes), the narrative shows the relationship between compounds and source material. Consider xenobiotic organic compounds (XOCs), a heterogeneous group of compounds that *could be* present in greywater.[1] XOCs originate from chemical products commonly found in households: detergents, soaps, shampoos, etc. The amounts consumed in Europe, not surprisingly, vary from country to country. For example, softeners used in 1991 ranged from 2.5 kg per person in Italy to 9.2 kg per person in Belgium (Puchta *et al.*, 1993). But these data reflect consumption of household *products*, not the consumption of individual *compounds*. Eriksson *et al.* (2002) suggest an alternative way to map relevant compounds in greywater: namely, to combine XOCs potentially found in household chemicals with an environmental hazard identification.

The risk assessment was based on the classification of XOCs with respect to toxicity, bioaccumulation and biodegradation, according to a method commonly used for evaluating new chemicals to be introduced in the market (see, e.g. Van Leeuwen and Hermens 1995). Compiling the list of potential compounds was daunting. Based upon the information available in the declaration-of-contents of common Danish household products, they noted at least 900 different organic chemical substances and compound groups. Similar to the way a cartographer creates a classification legend for a map theme, Eriksson *et al.* (2002) classified the compounds into eight different groups, depending on how environmentally hazardous the compounds were. Out of the approximately 900 substances, 66 compounds were classified (or generalised) as priority pollutants. That is, they were prioritised in the first three groups with the highest environmental impact:

(1) Not biodegradable and potentially bioaccumulative.
(2) Biodegradable and potentially bioaccumulative.
(3) Biodegradable and not potentially bioaccumulative.

Unfortunately, less than a quarter of the substances could be evaluated based on the available information about toxicity, bioaccumulation and degradation. Moreover, Eriksson *et al.* (2002) suggest that the number of priority pollutants may increase dramatically if more information becomes available for the remaining 700 compounds.

3.3.2 Source sampling

Another relationship is the *source* of greywater (e.g. kitchen, laundry, or bathroom) and its microbial and chemical composition, including oxygen demand and nutrients. The content of BOD and COD indicate the risk of oxygen depletion due to degradation of organic matter during transport and storing. Eriksson *et al.* (2002)

---

1 The presence of substances in soil that are not naturally produced by biological species is of great public concern. Many of these so-called xenobiotic (from Greek xenos, "stranger," and bios, "life") chemicals have been found to be carcinogens or may accumulate in the environment with toxic effects on ecosystems.

found that, generally speaking, most COD derives from laundry and kitchen chemicals, such as dishwashing and laundry detergent. Likewise, nutrient concentrations vary with the source, with kitchen wastewater generating the most total nitrogen and laundry greywater the most total phosphorus. Based upon the relationship between chemical composition and wastewater sources, a two-fold conclusion is drawn: (1) different types of greywater can be suitable for different types of reuse, and (2) different pre-treatment requirements depend on both the type of greywater and the intended reuse.

Having said that, Eriksson *et al.* (2002) also summarise (i.e. generalise) data available in the literature, noting that there are differences in the quality of the data. Some references only report average or single values, while others have taken many more samples over a long time period. Sampling regimes, whether for chemical compounds or micro-organisms, are critical. Scientists (unlike laymen) are acutely aware that a grab sample might be very misleading since the concentration varies over the day, and is different on different days of the week.

Nevertheless, Eriksson *et al.* (2002) argue for more data to evaluate the potential of greywater for different reuses and infiltration. The current literature focuses on the content of oxygen consuming compounds (BOD and COD), nutrients and some micro-organisms. A few studies have included measurements of heavy metals, while information about the presence of levels of specific XOCs is totally missing. (Their list of XOCs that potentially could be present in greywater was based upon product information and knowledge about the presence of XOCs in domestic *wastewater*.) Equally important, the relationship between greywater reuse and health risks remains largely uncharted. Clearly a thorough characterisation of greywater and evaluation of possible sources of pollutants is required to establish proper treatment methods.

Terpstra (2001) maps functional relationships between greywater sources but focuses on water quality for potential reuse. The analysis is based upon intrinsic qualities of water, which parallel energy. Like energy, water is not permanently consumed. And like energy quality, water quality can be distinguished between high and low. Furthermore, in the same way that high quality energy can easily be converted into energy of lower quality – but the reverse process cannot take place spontaneously – the same holds true for water quality. As purity decreases so do the quality and number of potential applications. Terpstra (2001) classifies water quality into four classes, and models a theoretical relationship between source and water quality (both input and output). Based on indicators of water quality, efficiency and hygiene requirements, quality classes are chosen for the water supplied and discharged at different greywater sources.

3.3.3 Organising organisms

Pathogenic viruses, bacteria, protozoa and helminths can be introduced into greywater from several sources: from uncooked vegetables and raw meat in the kitchen, from diapers in the laundry, from hand washing and showering in the bathroom. Most of the concerns are associated with excreta, which transport pathogens from the bodies of infected persons to others via exposure in wastewater. Consequently, *E. coli* is commonly used as an indicator of faecal contamination, and by measuring its content (in greywater) health hazards can be deduced. But keep in mind, *E. coli* can be found in other animals, e.g. household pets, domestic livestock and wildlife. Consequently the mere presence of *E. coli* in greywater effluent does not necessarily imply faecal contamination from the influent. Likewise, its presence in treated greywater does not imply a failed or incomplete treatment. Moreover, its absence does not imply pathogen-free water.

Many laymen (and certainly all wastewater scientists) know that *E. coli* is a faecal indicator. But the term can be somewhat confusing because *indicator* is interpreted and used differently. It can indicate faecal contamination. It can infer – but only infer – that pathogens may be present. And it can be used to assess the efficacy of a process (e.g. food processing or wastewater treatment).

When dealing with wastewater treatment, indicator organisms are generally coliform bacteria, which includes *E. coli, Citrobacter, Enterobacter* and *Klebsiella* species. Because coliform bacteria are generally hardier than disease causing bacteria, their absence from water is an indication that the water is *bacteriologically* safe. Conversely, the presence of the coliform group is indicative that other kinds of micro-organisms capable of causing disease may also be present and therefore the water unsafe. Although experience shows that the absence of coliforms in 100 ml of drinking water suggests the prevention of bacterial waterborne-disease outbreaks, there are deficiencies in using this indicator (Cleeson and Gray (1997), cited in Gerba (2000)):

- regrowth in aquatic environments,
- suppression by high background bacterial growth,
- not indicative of a health threat,
- no relationship between enteric protozoa and viral concentration

Of equal concern is the growth or recovery of injured coliform bacteria in a distribution system, because this may give a false indication of faecal contamination (Gerba, 2000). Coliforms may colonise and grow in the biofilm found on distribution pipes, even in the presence of free chlorine. *E. coli* is 2400 times more resistant to free chlorine when attached to a surface than as free cells in water (LeChevallier *et al.,* 1988).

3.3.4 Actual versus potential micro-organisms

Since toilet waste is not included in greywater, faecal contamination should be minimal. Of course, some activities, like washing diapers may add minor amounts. But greywater may have heavy loads of easily degraded organic material, which can favour growth of enteric bacteria such as faecal indicators, and such growth in wastewater has been reported (Manville *et al.*, 2001). Consequently, focus on bacterial indicator numbers may lead to an over-estimation of faecal loads and thus risk (Ottoson and Stenström, 2003). Several studies noted by Ottoson and Stenström (2003) report high numbers of traditional faecal indicators in greywater, that from a "regulatory point of view" would indicate substantial faecal contamination (Table 3.1).

The heart of Ottoson and Stenström's (2003) investigation – which quantifies faecal contamination using faecal indicator bacteria and chemical biomarkers – suggests that if greywater should be reused, there is a need to differentiate between the actual faecal loads and potential re-growth of indicators if actual risks should be assessed. Their premise is that coliform indicators over-estimate the risks due to the potential bacterial re-growth within the system. Therefore, they chose to make a comparative validation of a greywater treatment system based on three different approaches. Evaluate faecal load based on:

(1) coprostanol concentrations and using epidemiological data to estimate the pathogen load,

(2) faecal enterococci numbers and a dose-response model derived from bathing water exposure,

(3) faecal enterococci as an index organism for *Salmonella*.

Their results showed that high and variable numbers of different groups of faecal indicator bacteria were found in the greywater, and the variability could not be explained by seasonality. More importantly, bacterial indicator densities overestimated the faecal load by 100-1000-fold when compared to chemical biomarkers. The different simulated exposure-scenarios (direct contact, irrigation of sports fields and groundwater recharge) gave unacceptably high rotavirus risks despite a low faecal load.

There are two other organisms used in risk assessment. *Index* micro-organisms are a group indicative of pathogen presence, and *model* micro-organisms are species indicative of pathogen behaviour. For example, *E. coli* may be used as an index for *Salmonella*, and F-RNA coliphages as models of human enteric viruses. Generally speaking, the distinction between the above terms – essentially pathogen presence versus pathogen behaviour – is clearer to microbiologists, epidemiologists and pathologists than to civil engineers, public health practitioners, and laymen.

Table 3.1 Reported number of indicator bacteria in greywater (log10/100 mL) (adapted from Ottoson and Stenström, 2003)

Wastewater origin	Total coliforms	Therm tolerant coliforms	E. coli	Faecal enterococci	Reference
Bath, hand basin			4.4	1.0-5.4	(Albrechtsen, 1998)
Kitchen sink, hand basin		5.0		4.6	(Gunther, 2000)
Kitchen sink		7.6	7.4	7.7	(Swedish EPA, 1995)
Laundry	3.4-5.5	2.0-3.0		1.4-3.4	(Christova *et al.*, 1996)
Shower, hand basin	2.7-7.4	2.2-3.5		1.9-3.4	(Christova *et al.*, 1996)
Greywater		5.2-7.0	3.2-5.1		(Lindgren, 1998)
Greywater		5.8	5.4	4.6	(Swedish EPA, 1995)
Greywater	7.9	5.8		2.4	(Casanova *et al.*, 2001)
Shower, bath	1.8-3.9	0-3.7		0-4.8	(Faechem *et al.*, 1983)
Laundry, wash	1.9-5.9	1.0-4.2		1.5-3.9	(Faechem *et al.*, 1983)
Laundry, rinse	2.3-5.2	0-5.4		0-6.1	(Faechem *et al.*, 1983)
Greywater	7.2-8.8				(Gerba *et al.*, 1995)

Whether micro-organisms are indicator, index or model, merely identifying the presence of potential pathogens has limited value unless the data are related to functional relationships. For instance, the major risk associated with the presence of micro-organisms in greywater is infection due to direct contact with the water during infiltration, where the application on/in the soil is the most critical moment (Ledin *et al.*, 2001). There is also potential risk of the contaminated soil used for gardening or agriculture or receiving water used as drinking water supplies. But to understand the cause-effect relationship requires an explanatory analysis. For example, if the greywater is reused for irrigation or infiltration, the *spatial* relationship between the size of the micro-organisms and infiltration material can be critical. The reason being this: the risk of contaminating groundwater is generally greater from smaller organisms (e.g. bacteria and virus), because larger organisms (e.g. parasitic protozoa and helminthes) are more likely to be removed by filtration.

Evaluating the impact of micro-organisms requires knowledge of the fate of the micro-organisms in the soil and water, including the relationship between residence time and transport mechanisms to which pollutants are subjected. Thus to understand the operational risk of potential pathogens one must understand the theoretical (time and space) relationships between micro-organisms and the water, sediments, and air. Combined with the main processes determining the fate of pollutants in soil and water (sorption, volatilisation and degradation (Connell, 1997)), the dynamic matrix of relationships is anything but simple and straightforward.

3.4 STANDARD ERRORS

Although Ottoson and Stenström (2003) clearly map the relationship between indicators and health risks, many others cloud the relationship. One reason is the careless use of logs. Unlike the implicit term *indicator* that infers quality, *log* is an explicitly straightforward expression of quantity. The problem is how it is used, as in "concentrations were reduced by 5 logs." That may be easy enough to understand, but the question remains, "Reduced from what?" The blunder is similar to presenting a value in percent without qualifying, percent of what? Without a context of reference, a reduction log value can be misleading. Admittedly, the quantitative value of a log reduction may look impressive, but its qualitative value is relative to both the original concentration (before treatment) and the desired goal (after treatment). Even if a concentration log value is extremely low, that does not mean it is innocuous. The quantity of a substance is meaningless unless one knows its quality (e.g. toxicity). Perhaps that is one reason guidelines and standards are rather difficult to establish.

Few countries, and regional authorities within countries, have/or are working on guidelines of (non-potable) reuse of treated greywater, and most are oriented around reclaimed water for irrigation (Salgot and Angelakis, 2001). In the USA, for example, California limits the level of *total coliforms* to 2.2 per 100 ml (Eriksson *et al.,* 2002). Florida specifies *faecal coliforms* and raises the bar to 0 per 100 ml (Crook and Surampalli, 1996), which stands in stark contrast to WHO's threshold of 1000 per 100 ml (World Health Organisation, 1989). Australia compounds the problem with four levels for *thermotolerant coliforms*, ranging from 10 to 150 per 100 ml, depending on the use (Gregory *et al.,* 1996). The only apparent standard regarding reuse of greywater is that coliforms are measured per 100 ml.

Standard. It is one of the most common terms in greywater treatment, yet it means different things to different people. Not just in terms of regulatory limits for wastewater professionals. The very word implies different nuisances to different people, especially when it applies to hygiene issues. So again we return to the sticky web of semantics. To some, *standard* means an accepted operational procedure, e.g. a hygiene standard for physicians is to wash their hands before surgery. To others it implies an attainable goal, e.g. to meet the sanitation standards prescribed by the World Health Organisation. To others still, it simply means meeting a basic minimal requirement, e.g. 100 coliforms per 100 ml. To clarify, for wastewater treatment *standards* are *regulations*, which are actual rules that have been passed and are enforceable by government agencies. Guidelines, however, are not enforceable, but they can be used in the development of a reuse program (Salgot and Angelakis, 2001).

Standards have been a matter of discussion among scientists, health and legislation officers and engineers, primarily because of the numerical expression of such standards and secondarily because of the parameters to be controlled (Salgot and Angelakis, 2001). For example, if a treatment system meets a reuse water standard based upon *E. coli* count, but ignores viruses, does that mean the treated water is safe to use? Of course not. Although most standards on greywater reuse do not specify threshold levels for viruses, the mere absence of limits does not imply the threat is absent. In fact, Ottosson's (2004) microbial risk assessment showed that treatment of greywater should be directed primarily at virus reduction.

Regarding numerical expression, standards are frequently based upon estimates. And sometimes the estimate is so rough it is rather meaningless for risk assessment. In fact, exposure factors for different waters are often so variable that a standard appears little more than guesswork. For example, the US EPA (cited in Gerba (2000)) merely suggests *default* values (whatever *default* means) for swimming water. The EPA claims a standard exposure frequency of 1-10 days per year, with 10-100 ml of water consumed per swimming event. A quick calculation suggests an annual intake of swimming water to be between 10 and 1000 ml, compared with 700 litres of potable water (another EPA estimate). When comparing the disease frequency within swimming and non-swimming groups, Cabelli *et al.* (1982) found that swimming in even marginally polluted bathing water was a significant route of transmission. Clearly, standards based upon default values are not well suited for risk assessment of greywater.

3.5 RISK ASSESSMENT

Risk is an inherent quality of human existence and a vital component in decision making. Everyone understands the generic meaning of the word, but the term *risk assessment* (like *standard*) has different meanings to different people. Most concur, however, that it is a concept applied to calculating (or estimating) risk and expressed in terms of probability.

Microbial risk assessment is important in setting standards for contaminants in the environment and for protecting public health. In the past risk assessment was largely qualitative and subjective; today quantitative risk assessment (QRA) estimates the risk of infection, illness, or mortality from microbial pathogens in a more objective, numerical fashion. Generally speaking, it consists of four basic steps (Gerba, 2000):

(1) Hazard identification – Defining the hazard and nature of harm. For example, identifying a contaminant (e.g., *Legionella*) and documenting its toxic effect on humans.

(2) Exposure assessment – Determining the concentration of a contaminating agent in the environment and estimating its rate of intake. For example, determining the concentration of *E. coli* in greywater and determining the average dose a person would come in contact with.

(3) Dose-response assessment – Quantifying the adverse effects arising from exposure. This assessment is usually expressed mathematically as a plot showing the response in the living organisms to increasing doses of the agent (e.g. rotoviruses).

(4) Risk characterisation – Estimating the potential impact (e.g. human illness or death) of a micro-organism or chemical based on the severity of its effects and the amount of exposure.

Once the risks are characterised, *risk management* evaluates regulatory options, taking into consideration social, political, and economic issues as well as engineering problems inherent in a proposed solution. To evaluate risk implies to rank or compare it with something. In other words, to show a relationship. But comparison with other risks does not itself establish the acceptability of a risk. As Gerba (2000) crystallises, "Voluntary risk is always more acceptable than involuntary risk." For example, the same people who willingly drive their cars every day – and thus face a 2:100 lifetime risk of death by automobile – are quite capable of refusing to accept the 1:10,000 involuntary risk of mortality from rotavirus diarrhoea.

Gerba (2000) also emphasises the *de minimis* principle, which means that there are some levels of risk so trivial that they are not worth bothering about. This concept is easy to understand but difficult to define.

> *Understandably, regulatory authorities are reluctant to be explicit about an acceptable risk... Some prefer the term* tolerable risk, *i.e. the level of risks we can accept given the economic costs, and social and scientific constraints. But it is generally agreed that a lifetime risk on the order of one in a million (or 10^{-6}) is trivial enough to be acceptable for the general public.* (Gerba, 2000).

The notion of a one-in-a million tolerable risk is obscure and theoretical. Nevertheless, its impact is clear and real, e.g. the amount of money spent on cleaning up waste sites to meet this 10^{-6} risk is often measured on the order of 10^7 US dollars. Gerba (2000) claims that levels of risk at the higher end of the range (10^{-4} rather than 10^{-6}) may be acceptable if just a few people are exposed rather than the entire population. His claim has subtle but profound implications for on-site greywater treatment. He notes, for example, that workers dealing with production of solvents can often tolerate higher levels of risk than the public at large; and these higher levels are justified because (amongst other reasons) employment is voluntary. Since most on-site greywater treatment is voluntary, one is tempted to apply the same logic and suggest higher levels of risk for those willing to accept them.

3.6 CONCLUSIONS

Simple answers to complex questions are rarely the best. Although greywater may appear simple to treat, the health issues are not. That might explain why regulations and guidelines vary as much as greywater characterisation and treatment technologies. The acquired complexity of standards and the inherent nature of language and logic are vulnerable to misunderstanding, especially when the public interprets information from different sources (e.g. scientists, engineers and product manufacturers). The pool of information on greywater is broad – shallow in some places, deep in others – which often confuses rather than clarifies complex issues. One reason being the variability of concern between interested parties: wastewater scientists tend to be interested in processes; engineers tend to focus on performance; manufacturers tend to promote products; and laymen tend to look for panaceas.

3.7 REFERENCES

Angelakis, A.N. (2001) Management of wastewater by natural treatment systems with emphasis on land-based systems. In: *Decentralised Sanitation and Reuse* edited by Lens, P., Zeeman, G., and Lettinga, G. IWA Publishing, London.

Cabelli, V.J., Dufour, A.P., McCabe, L.J., and Levin, M.A. (1982) Swimming associated gastroenteritis and water quality. *Am. J. Epidemiol.* 115, 606-16.

Casanova, L., Gerba, C., and Karpiscak, M. (2001) Chemical and microbial characterization of greywater. *Journal of Environmental Science and Health* 36 (4), 395-401.

Christova-Boal, D., Eden, R.E., and McFarlane, S. (1996) An investigation into greywater reuse for urban residential properties. *Desalination*, 106, 391-397.

Connele, D.W. (1997) *Basic Concepts in Environmental Chemistry*. CRC Press, Boca Raton, Fl.

Crook, J. and Surampalli, R.Y. 1996. Water reclamation and reuse criteria in the US. *Wat. Sci. & Tech.* 33 (10-11), 451-462.

Del Porto, D. and Steinfeld, C. (1999) *The Composting Toilet System Book*. The Center for Ecological Pollution Prevention. Concord, Mass. USA.

Diaper, C., Dixon, A., Butler, D., Fewkes, A., Parsons, S., Strathern, M., Stephenson, T., and Strutt, J. (2000) Small scale water recycling systems – risk assessment and modelling. In: *Ist World Congress of the IWA*, Paris, 3-7, July.

Dixon, A., Butler, D., Fewkes, A. and Robinson, M. (1999) Measurement and modelling of quality changes in stored untreated grey water. *Urban Wat.* **1** (4), 293-306.

Dow, S. (2002) *Economic Methodology: an Inquiry.* Oxford University Press. Oxford.

EPA (2002) *Onsite Wastewater Treatment Systems Manual.* EPA/625/R-00/008. Office of Research and Development, U.S. Environmental Protection Agency. Washington.

Eriksson, E., Auffarth, K., Henze, M. and Ledin, A. (2002) Characteristics of grey wastewater. *Urban Water,* **4** (1), 85-104.

Faechem, R.G., Bradley, D.J., Garelick, H. and Mara, D.D. (1983) *Sanitation and Disease: Health Aspects of Excreta and Wastewater Management.* Wiley. Washington.

Gander, M., Jefferson, B., and Judd, S. (2000).MBRs for use in small wastewater treatment plants. *Wat. Sci. & Tech.,* **41** (1), 205-211.

Gerba, C.P. (2000) Risk assessment. In *Environmental Microbiology,* (eds. Maier, R.M., Pepper, I.L., and Gerba, C.P.), Academic Press, New York, 557-570.

Gerba, C.P., Straub, T.M., Rose, J.B., Karpisack, M.M. and Foster, K.E. (1995) Water quality of greywater treatment system. *Wat. Sci. & Tech.* **31** (1), 109-116.

Gleeson, C. and Gray, N. (1997) *The Coliform Index and Waterborne Disease.* E and FN Spoon. London.

Gregory, J.D., Lugg, R., and Sanders, B. (1996) Revision of the national reclaimed water guidelines. *Desalination,* 106, 263-268.

Gunther, F. (2000) Wastewater treatment by greywater separation: outline for a biologically based greywater purification plant in Sweden. *Ecological Engineering* **15**, 139-146.

Holden, B., and Ward, M. (1999) An overview of domestic and commercial re-use of water In: *Proc. International Quality and Product Centre Conference on Water Recycling and Effluent Reuse.* Copthorne Effingham Park, London, U.K.

Jefferson, B., Judd, S., and Diaper, C. (2001) Treatment methods for grey water. In *Decentralised Sanitation and Reuse,* edited by Lens, P., Zeeman, G., and Lettinga, G., IWA Publishing, London.

Jenssen, P. and Vråle, L. (2004) Greywater treatment in combined bio-filter/constructed wetlands in cold climate. In *Proc. 2nd Int. Symp. on Ecological Sanitation.* Lübeck, Germany, 7-11, April.

LeChevallier, M.W., Cawthen, C.P., and Lee, R.G. (1988). Factors promoting survival of bacteria in chlorinated water supplies. *Applied and Environmental Microbiology* **54**, 649-654.

Ledin, A.Erisksson, E., and Henze, M. (2001) Aspect of groundwater recharge using greywater. In *Decentralised Sanitation and Reuse,* edited by Lens, P., Zeeman, G., and Lettinga, G.. IWA Publishing, London.

Lindgren, S., and Grette, S. (1998) *Water and sewerage system.* SABO Utveckling. Trycksak 13303/1998-06.500 (in Swedish).

Manville, D., Kleintop, E.J., Miller, B.J., Davis, E.M., Mathewson, J.J. and Downs, T.D. (2001) Significance of indicator bacteria in a regionalized wastewater treatment plant receiving waters. *International Journal of Environmental Pollution,* **15** (4), 461-466.

Merrett, S. (2003) Demand-side concepts in the management of water resources. Tech. paper. *WATERSAVE Network, Fifth Meeting,* London. June 17, 2003.

Metcalf and Eddy, Inc (1991) *Wastewater Engineering: Treatment, Disposal and Reuse*, 3rd edn, McGraw Hill, New York.

Müllegger, E., Langergraber, G., Jung, H., Starkl, M., Laber, J. (2004) Potentials for greywater treatment and reuse in rural areas. In: *Proc. 2nd Int. Symp. on Ecological Sanitation*. Lübeck, Germany, 7-11, April.

Ottosson, J. (2003) *Hygiene Aspects of Greywater and Greywater Reuse*. Department of Land and Water Resource Engineering. Royal Institute of Technology. ISSN 1650-8629.Stockholm

Ottosson, J. (2004) Faecal contamination of greywater – assessing the treatment required for hygienically safe reuse or discharge. In: *Proc. 2nd Int. Symp. on Ecological Sanitation*. Lübeck, Germany, 7-11, April.

Ottosson, J, and Stenström, T.A. (2003) Faecal contamination of greywater and associated microbial risks. *Water Research* **37**, 645-655.

Puchita, R., Krings, P. and Sandkuhler, P. (1993) A new generation of softeners. *Tenside Surface Detergents*, **30** (3), 186-191.

Salgot, M. and Angelakis, A. (2001) Guidelines and regulations on wastewater reuse. In: *Decentralised Sanitation and Reuse* edited by Lens, P., Zeeman, G., and Lettinga, G. IWA Publishing, London.

Sayer, D. (1998) *A study of domestic greywater recycling*. Interim Report, National Water Demand Management Centre, Environment Agency, Worthing, U.K.

Stephenson, T., Judd, S., Jefferson, B. and Brindle, K. (2000) *Membrane Bioreactors for Wastewater Treatment*. IWA Publishing. London, U.K.

Swedish EPA (1995) *What does household water contain? Nutrients and metals in urine, faces and dish-, laundry and shower waters*. Naturvårdsverket, Report 4424. Stockholm. (in Swedish).

Tchobanoglous, G. (1999) Wastewater: an undervalued water source for sustainable development. In *Proc. Environmental Technology for the 21st Century*, Heleco 99, Thessaloniki, 3-6 June, 1, 3-9.

Terpstra, P.M.J. (2001) Potentials of water reuse in houses and other buildings. In: *Decentralised Sanitation and Reuse,* edited by Lens, P., Zeeman, G., and Lettinga, G. IWA Publishing, London.

Urban Water (2001) Progress report. ISSN 1650-3791. Chalmers University of Technology.

Van Leeuwen, C. J. and Hermens, J. L. (eds.) (1995) *Risk Assessment of Chemicals: An Introduction*. Kluwer Academic Publishers. Dordrecht.

World Health Organization (1989) *Health Guidelines for the Use of Wastewater in Agriculture and Aquaculture*. Technical Report Series 778, ISSN 0512-3054. Geneva.

4

Water conservation products

Nick Grant

4.1 INTRODUCTION

Water efficient products offer the potential for significant water savings at point of use. However, evaluating the potential savings and acceptability to users is very complex. This chapter will consider design approaches such as optimising pipe sizes to reduce dead-legs, and actual products such as washing machines and WCs.

In looking at actual technologies the chapter explores a number of related questions:

- Does reduced water consumption mean reduced performance and hygiene standards?
- Are water efficiency measures cost effective?
- How does efficiency compare with recycling and harvesting?

The focus is on technology but there are many factors that influence the potential savings. These are explored in detail elsewhere (e.g. EA 2001) and in other chapters. They include:

1. Uptake
   a. Replacement period (e.g. washing machines or bathrooms)
   b. Fashion, trends
   c. Acceptability of technology
   d. Other drivers and barriers.
2. The proportion of total water uses by the component (e.g. an 80% reduction in water use for tooth brushing might be equivalent to a 5% reduction in WC consumption.
3. Rebound effects (longer showers, half flush used for facial tissues rather than bin).
4. Consumption trends, bigger baths, more frequent showers, cleaner cars, multi-head showers, economic recession or increasing affluence.

Whilst this chapter focuses only on hardware, the author does not wish to suggest a purely technical approach to water efficiency, as it is likely that greater savings can be had by changes in personal habit. It is also clear that technical improvements can be negated by unforeseen changes in lifestyle (1, 3 and 4 in the list above).

4.2 A FRAMEWORK FOR CONSIDERING WATER CONSERVATION PRODUCTS

Before evaluating 'water conservation products' it is worthwhile questioning what we mean by 'water conservation'.

The following definitions bring the answers to questions about barriers and drivers into focus. For each category the barriers and drivers are different. For example *Water Conservation* is often challenged on grounds of hygiene, cost, perception or the difficulty in 'educating the public'. Using the suggested categories below, this charge might be justified for 'water conservation' measures but can hardly be argued for 'water efficiency' or 'water sufficiency' measures. The breakdown also provides a checklist that can be applied to product or system design when looking for potential savings.

Some products or techniques will fit neatly into a single category whilst others will feature efficiency, sufficiency, substitution and even recycling or reuse. For example a clothes washer/dryer or vacuum WC.

4.2.1 Definitions

Some suggested definitions outlining generic approaches to water saving:

Water conservation; doing less with less

Particularly appropriate at times of drought or when camping but moderate application such as washing cars less often (but keeping lights and glass clean) may have some general applicability. Generally standards such as hygiene and aesthetics will usually suffer. Acceptability is very culturally dependent.

Examples:
- If it's yellow let it mellow.
- Don't water lawns.
- Don't wash the car (as often).
- Take shallow baths and short showers.

Water efficiency; doing the same (or more) with less

This approach to water saving should be the least sensitive to issues of human interface, i.e. no lifestyle changes should be required. Efficient fittings should do the same or even a better job with less water. Often efficiency improvements lead to other advantages such as reduced energy consumption, lower noise, better performance etc.

Examples:
- Hydraulically efficient WC pans and cisterns (i.e. not just reduced flush volume).
- Optimised pipe dead legs and insulation.
- Tap aerators and sprays.
- Shower head design.
- Efficient white goods.
- Garden design, drought tolerant plants and grasses, mulch.
- Fix leaks.
- Optimised bath shape.

Water sufficiency; enough is enough

This approach is one of optimisation. As with efficiency there should be no loss of effectiveness. Optimising water sufficiency usually involves technical and user input. For example baths can be optimised in shape and limited in size but users can still choose how deep to fill them. Similarly dual flush WCs promise savings dependent on correct use. Even habit dependent savings such as turning off taps when brushing teeth can be influenced by technical innovations, as we will see.

Examples:

- Hogs and Hippos and adjustment of WC flush volume to suit application.
- Dual flush.
- 'Ergonomics' – eco button on shower, water-brake taps.
- Flow regulation.
- Control – auto or manual, e.g. turn off tap when brushing teeth or timed taps.
- Careful garden watering.
- Bath sizing.

Water substitution; replace water with something else such as air

This is self-explanatory and covers technical solutions such as vacuum and compost toilets as well as simply using a broom rather than a hose to clean paths. Some alternatives to water may have a higher environmental impact (energy use of drying toilets, solvents for dry cleaning) than water.

Examples:
- Dry toilets.
- Waterless urinals.
- Vacuum drainage (uses some water but air used for transport).
- Clothes brush.
- Dry cleaning (not done for water saving).
- Hand-wipes (e.g. for remote toilets without water).
- Air for industrial cleaning processes.
- Broom rather than (or before) wash-down of floors.

Water reuse, recycling and harvesting; a potentially virtuous circle

Definitions vary but in this chapter *reuse* refers to direct reuse with minimal treatment whilst *recycling* refers to a process of treatment prior to reuse. Direct reuse is generally a low cost option but requires a good availability and match between resource and sink, both of quantity and quality, to avoid the need for treatment and storage. Recycling introduces the need for extra energy and possibly chemicals for treatment. Other impacts should be considered if the aim of the scheme is environmental improvement. Municipal scale strategies such as groundwater recharge are not discussed here.

Examples:
- Direct *reuse*:
  - Rainwater harvest
  - Water butts
  - Greywater irrigation (direct)

- o Process water reuse
- o Shared bath water
- *Recycling* – treatment, storage reuse:
  - o Greywater recycling
  - o Blackwater recycling

4.2.2 The potential for technical solutions for demand management

The following tables are offered as a preliminary summary of the potential savings from current and future technologies. It is intended as a starting point for discussion rather than prophesy.

The first table looks at frequency of use of micro-components. The last column shows the provisional values used in this chapter. A proper analysis of usage and trends is beyond the scope of this chapter (see Chapter 1) but is crucial to understanding.

Table 4.1 Assumptions for initial analysis of potential savings

Use	%	SODCON (1994) (lpd) [1]	Assumed vol/use, litres [2]	Frequency of appliance use (per person) [3]	Compare, frequency, EA 2001 [4]	Assumed uses/p/day- for BATNEEC and future [5]
WC	35	52.5	10	5.25	4.12	4.12
Bath	15	22.5	80	0.28	0.34	0.34
Shower	5	7.5	15	0.5	0.6	0.6
Kitchen sink	15	22.5	10	2.25	?	2.25
WHB	8	12	6	2	0	2
Wash m/c	12	18	100	0.18	0.157	0.157
Dishwasher	4	6	28	0.21	?	0.214
Outside	6	9	9	1	0	0
Totals	100%	150 litres				

Notes for Table 4.1

1. % from Anglian Water Survey of Domestic Consumption (SODCON) data 1994. lpd calculated for 4 person house.
2. Assumptions to generate frequency data.
3. Frequency calculated from previous data and assumptions.
4. Frequency values interpreted from the Environment Agency (EA, 2001).
5. Estimate of frequency for a single scenario comparison of water saving for improved technologies in table 2. These figures are not definitive and only for use in the context of this chapter.

Table 4.2 Estimate of potential water saving with Best Available Technologies Not Entailing Excessive Costs (BATNEEC) and estimated future technologies.

Use	BATNEEC vol/use [1]	BATNEEC Vol/day [2]	Reduction (% of SODCON figure) [3]	Technical potential: volume/use [4]	Vol/day [5]	% reduction from SODCON
WC	4	16.48	69	3	12.36	76
Bath	70	23.8	-6	50	17	24
Shower	20	12	-60	6	3.6	52
Kitchen sink	3	6.75	70	2.8	6.3	72
WHB	3	6	50	2.5	5	58
Wash m/c	45	7.07	61	35	5.5	69
Dishwasher	18	3.85	36	14	2.996	50
Outside	0	0	100	0	0	100
Totals		76	49		52.8	65

Notes for Table 4.2
1. Available technologies, 2001.
2. Calculated from assumed frequencies in Table 4.1.
3. Reduction in water use compared with SODCON data.
4. Estimate of technical limit using current technologies.
5. Calculated from frequency and [4]

Again, a scenario-based analysis is beyond the scope of this chapter but the broad issues must be considered in that they should influence technical design solutions. The figures should be seen as technical potential and assume water-aware users.

4.3 AVAILABLE TECHNOLOGIES; ANALYSIS BY CATEGORY

The Environment Agency *Conserving Water in Buildings* fact cards (EA, 2001) detail a wide range of currently available water saving technologies and techniques and represent the output from a desktop study of manufacturer's products carried out from February to April 2001.

The numbering and 'product' grouping of the cards will be repeated here:
1. Domestic Appliances.
2. Garden appliances/water efficient gardening.
3. Grey-water.
4. Rainwater.
5. Taps.
6. Supply restrictor valves.
7. Urinals, waterless, controls and washroom controls.
8. Waterless and vacuum toilets.

9. Water efficient WCs and displacement techniques/retrofits.
10. Showers and baths.
11. General management.

4.3.1 Domestic Appliances

Discussion

The energy and water efficiency of dishwashers and washing machines has improved significantly over the last 10 years. (However, tests published by the Consumers' Association in 'Which?' magazine suggest a discrepancy between the energy label water use and that measured in simulated use during their tests.)

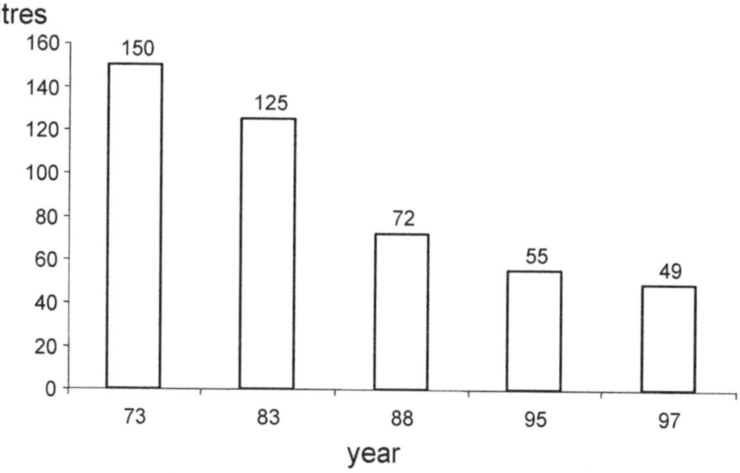

Figure 4.1 The trend in water use for Bosch washing machines (5kg hot wash).

Obviously water and energy use per cycle depends on the program. The UK Water Regulations and EU Directive 95/12/EC specifies a maximum water consumption of 27 litres per kilogram of wash load for horizontal axis clothes washing machines. Thus a 5kg load would use 135 litres compared with an increasingly common 50 litres of less.

Figure 4.1 shows the trend for reduced water use by Bosch washing machines, which suggests that the new Water Regulations may have been a useful conservation measure twenty years ago.

Water use and cleaning performance

Concern is often expressed about compromised hygiene and cleaning performance due to water efficiency measures so it is interesting to analyse available data to check this assumption. Figure 4.2 shows the wash performance and water use of machines tested by the Consumers' Association (Which?) on a 40°C wash. It shows that wash performance does not seem to be dependant on or guaranteed by high water use. The most water (and energy) efficient machine was as good or better at washing than all the machines except one. Figure 4.3 shows the Which? 'Total Score' (cleaning performance (22%), running costs (24%), spin efficiency (7%), water use (11%), time, rinse and balance (14%) drying (22%)) against water use. Again this would need to be repeated with more data for newer machines but the results seem to challenge at least two assumptions, namely that 'less water equals less performance' and that 'it is not worth buying a water efficient machine as there is no payback'. The graphs and other results suggest that the more water efficient machines offer other advantages and that water and energy efficiency are merely a spin off of good design rather than something achieved at the cost of performance and price. Thus efficiency is a bonus and should not be assumed to require a payback calculation.

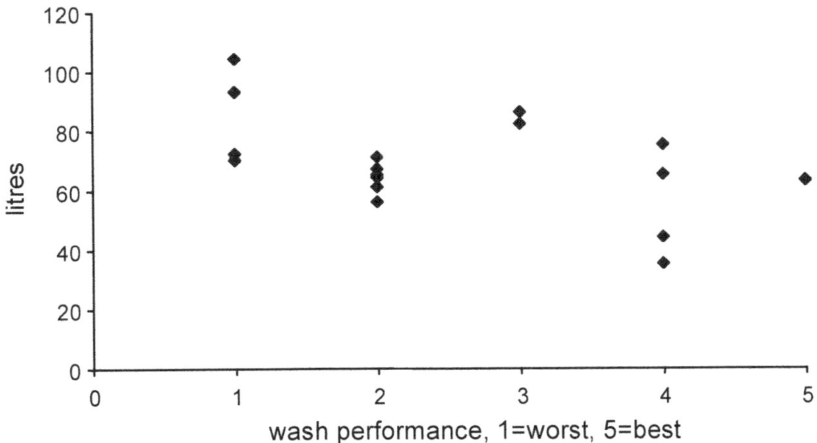

Figure 4.2 Wash performance and water use for the 19 washing machines (for cotton at 40°C) tested by Which? in 1997 (more recent data only rated water use from best to worst).

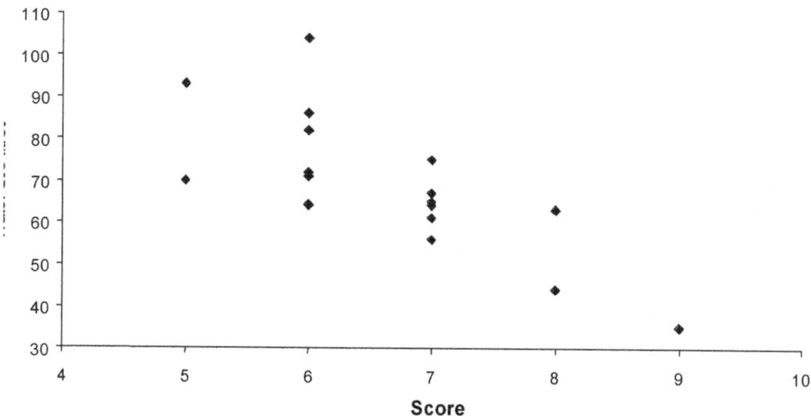

Figure 4.3. Water use against the total test score for the 19 washing machines tested by Which? in 1997 (more recent data only rated water use from best to worst).

Interestingly since this original analysis new test results (EA, 2003) have revealed that the previous situation of more efficient machines costing more is no longer the case. Thus having argued against the need to justify a payback for more efficient machines it seems that there is now no premium anyway (See Figure 4.4). The argument still stands in principle and will apply in other situations.

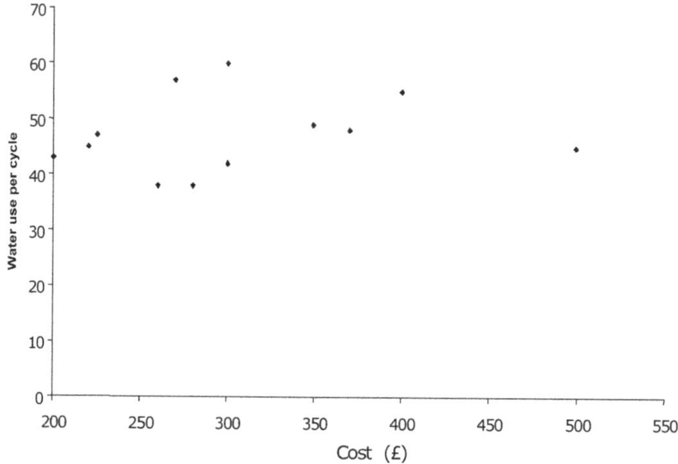

Figure 4.4 Water use against machine purchase price from Which? test data 2003 showing that water efficient machines no longer cost more.

Water efficiency of dishwashers and washing machines is strongly influenced by use, as part-loads are much less efficient than full ones. Half load buttons and fuzzy logic are only a partial solution and may encourage people to use part loads. Some half-load dishwasher programs have been found to use the same amount of water as full ones.

BATNEEC (best available technology not entailing excessive cost)

Specifications constantly change and, as already mentioned, water use depends on the program and how the machine is tested. Washing machines using around 40 litres for a 5kg, 40°C cotton wash and dishwashers using around 14 litres per cycle (12 place settings) are widely available. Manufacturers claim that dishwashers are more environmentally friendly than hand washing as less water can be used but the jury is still out when considering a real world life cycle assessment.

Future

It looks as if minimum water use will level out at around 30–40 litres per 5kg load and this is probably limited by rinse performance. Technical innovations that may reduce water use further include ultrasonic agitation, easy rinse detergents with controlled dosing and more sophisticated control systems. Payoffs with other environmental and health issues are possible, for example low temperature detergents that save energy but use enzymes, zeolytes rather than phosphates, etc. A machine using 40 litres per wash used every day for a family of four should use about 14.6 m^3 of water a year and about 365 kW.h of electricity. At a water and sewage charge of £1.50/m^3 and electricity price of 7p/kW.h this equates to £22 for water and sewage and £26 for electricity or about 13p/wash. On this basis water saving is unlikely to be driven by running costs alone.

Trends that may offset savings include the demand for larger machines and faster wash cycles.

4.3.2 Garden appliances/water efficient gardening

Discussion

Whilst garden water use only accounts for a small percentage of annual water use, it peaks at a time of highest water stress. Gardening and water use in the garden is a complex and controversial subject. All we will say here is that in the UK it is possible to garden using no mains water. Technical fixes such as drip irrigation and sprinkler timers are probably unnecessary complications for all but the driest parts of the UK and can lead to water wastage if incorrectly used or maintained.

Fashion is likely to be the greatest driver for water use in the garden, which could head towards a waterless, low maintenance future or large green lawns with leaking ponds and water features.

BATNEEC

Mains-waterless gardens are possible now with good design and appropriate planting. Water butts can provide a surprising amount of the water needs for all but the largest garden.

Future

More of the same perhaps with direct greywater reuse if extra water is required in summer. In dry climates and perhaps drier parts and seasons of the UK, the potential for direct greywater irrigation is considerable.

4.3.3 Grey-water and blackwater reuse

Discussion

Commercial and experimental systems have been developed to treat and store light grey water, typically from baths and basins, and reuse it for WC flushing. Manufacturers typically claim water savings of around 30-40%, i.e. it is assumed that all WC water is supplied by grey water. The theory and practice of greywater systems is covered in chapter 3 and we will return to look at them in comparison with efficiency measures later.

BATNEEC

Greywater reuse does not currently make sense at the single household level with current technologies because of poor cost effectiveness, poor reliability, high life cycle impact and unreliable water savings. Blackwater recycling already occurs indirectly via sewage treatment and discharge to rivers or groundwater but more directly in a number of trial schemes such as the now abandoned Beazer Homes trial with Anglian Water.

Future

Technological breakthroughs are required and the author is sceptical about the future of domestic greywater systems in the UK. For water-stressed areas, blackwater recycling may have more potential on a larger scale. New technologies include MBR and SBR sewage treatment plants, which are capable of producing high quality effluent suitable for reuse after disinfection. Life cycle impacts must be considered, particularly energy use and toxic by-products of disinfection. Simple, direct greywater reuse in gardens has great potential in drier climates where summer water use for irrigation is very significant.

Where new developments beyond the reach of mains sewerage require on-site treatment, there may be a scale at which reuse becomes practical.

4.3.4 Rainwater (other than garden butts)

Discussion

Properly collected and stored rainwater is generally accepted as suitable for use in WCs, washing machines and for garden use, see chapter 2. Typically these account for around 50% of domestic use but as best practice WC and washing machine volumes have dropped by over 50% whilst bathing has increased this figure would need revising when considering new-build and major refurbishment works. Rainwater and greywater will be compared with efficiency measures later.

BATNEEC

A wide range of commercial systems is available.

Future

Current systems are not usually cost effective, especially at the domestic scale, and energy use for pumping equivalent volumes is generally higher than for mains water. Future technical developments could include low cost variable speed pumps. Tanks are a major system cost but are probably a mature technology unless innovative solutions for new-build can be found with, for example, shared function.

4.3.5 Taps

Discussion

Appropriate technologies differ for kitchens, bathrooms or commercial washrooms. The sensible approach is *sufficiency* with optimised flow rates and good ergonomics.

Some models incorporate features such as the Hansa Ecotop cartridge. For mains pressure supplies, aerators or laminar flow devices will eliminate splashing whilst regulating flow rates and providing the illusion of more water flowing. Savings will be higher for lower levels of user awareness. For commercial washrooms spray taps offer around 80% reduction in flow rate but correct specification and adjustment of flow is crucial for user satisfaction. Generally flow rates are set too high leading to splashing. Timed turn-off and electronic taps offer savings in commercial applications as well as hygiene benefit.

BATNEEC

A wide range of fittings are available, see EA Fact Cards. Regulated sprays and aerators allow easy specification of flow rates. Hot and cold must be clearly and indelibly marked and operation should be obvious to avoid wastage as users try to find which position provides hot water.

Future

The widespread use of standard threaded outlets on tapware would allow the use of sprays, aerators and innovations. 'Waterbrake' cartridges and integrated adjustment of flow rate and hot water flow could become standard features at little extra cost. Ideas such as flow sensitive spray fittings (Tapmagic) have great potential. Electronic taps may have application in health and commercial settings but their appropriateness for domestic use is unlikely to be justified on sustainability grounds.

4.3.6 Supply restrictor valves

Discussion

Flow restriction and pressure and flow regulation are mature technologies. Savings are variable and are dependent on user awareness. For new installations flow regulation is justified in terms of improved performance alone (balanced dynamic pressure, reduced splashing). Regulation by showerheads and aerators should also be considered where appropriate and can have efficiency advantages.

BATNEEC

Products are low cost and readily available.

Future

Wider use of existing technology, specification as part of building and Water Regulations. Regulators for pressures less than 1 bar are becoming available.

4.3.7 Urinals

Discussion

The Water Regulations require that flushing is limited in frequency and volume and should only occur when a building is in use. Many installations, even new ones, do not meet this base specification. Also many technically compliant installations are wrongly adjusted or not maintained and so default to continuous operation using arbitrary volumes of water.

Manufactures offering water saving solutions quote impressive savings but these are usually based on arbitrary volumes rather than correctly set flushing rates that meet the old Byelaws or New Regulations.

Purpose designed waterless urinals have been available for over 100 years and there are many models on the market. Most use disposables or require a maintenance contract. This makes commercial sense for manufacturers and suppliers of such systems and this has driven the technology. After much research with an available system, BRE patented a chemical and consumable free solution many years ago. They say that since it was so simple and lasted indefinitely it had little commercial potential. Our own research has backed up the BRE findings and led to a number of designs

BATNEEC

Waterless urinal operation can be achieved without the use of special chemicals or consumables.

Future

Waterless operation offers a number of advantages provided a system can be marketed that is cost effective and easy to maintain. From an environmental point of view water savings must not be offset by increased chemical consumption or other impacts and this has recently been achieved by commercial designs.

4.3.8 Waterless and vacuum toilets

Discussion

Current waterless toilets are not a simple direct replacement for the WC. For rural and suburban eco-houses and remote public toilets they can represent a best available technology but their widespread use is not considered likely in the UK at present. The EA fact cards provide some more detail and the book "Lifting The Lid" (Harper and Halestrap, 1999) is definitive for the UK.

Vacuum toilets are not normally recommended for simple water saving except in extreme situations such as aircraft and trains.

BATNEEC

All available dry toilet systems are zero water use but some require electricity. Vacuum toilets use about 1.2 litres per flush.

Future

Dry toilet designs are evolving but are mostly intended for rural sanitation. Dry toilets could have a significant benefit when viewed on a global-scale. Vacuum technology may have wider application but would require some technical problems to be solved if it is to be used on the domestic scale whether in individual dwellings or blocks of flats. Again cost and life cycle issues must be considered.

4.3.9 Water efficient WCs and displacement techniques/ retrofits

Discussion

WCs have traditionally represented the largest single use of water in dwellings and offices. As WC water use decreases and water use for bathing increases this balance will shift. Current regulations require WCs to flush with a maximum volume of 6 litres for a full flush and up to two thirds of this if a reduced flush is available. Actual flush volumes are almost always higher than labelled as the nominal volume is measured with the water supply turned off. In reality water enters the cistern during the flush. Similarly when considering dual flush the average flush cannot be calculated simply using normal assumptions of ratios between solids and liquids uses. Trials with 6/3 and 6/4 litre dual flush WCs have delivered average volumes of between 4 and a little over 6 litres per flush – more research is needed. Another factor that is hard to quantify is that of flush valve leakage. The problem is made worse in the UK by the low penetration of water metering which means that a customer faced with a leaking valve will, as with failing inlet valves, usually allow it to leak rather than pay a plumber to try and fix it.

All these factors must be considered when evaluating real world performance as two 6-litre WCs may have very different water use over their life. The matrix below compares the pros and cons of single and dual flush as well as valves and siphons. Real data is required to lift the debate beyond personal opinion and anecdotes.

Objective discussion of such issues is clouded by the politics and vested interests that surround this surprisingly emotive subject.

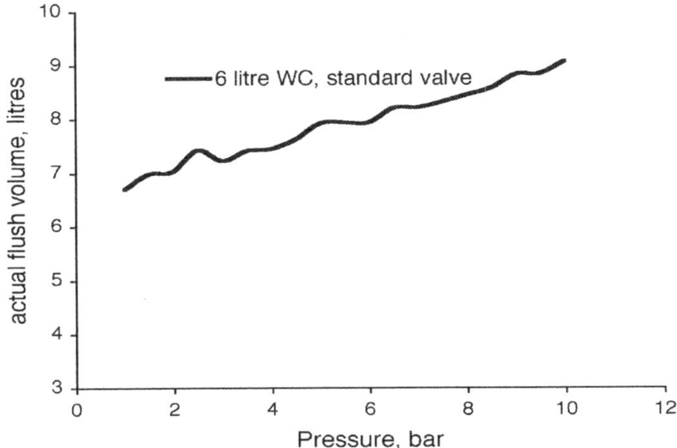

Figure 4.5 Flush volume against supply pressure for a nominally 6 litre WC. A delayed action inlet valve will maintain the nominal volume at all pressures. Data from WRc for 7.5 litre WC, scaled for 6 litre.

Nominal flush volume and performance

Issues of actual flush volumes aside it is now illegal to install a new WC with a nominal flush volume greater than 6 litres. Poor everyday experience with older WCs using 9 and 13 litres and more recent low flush 7.5 litre models might not inspire confidence in 6 litre models. Obviously the more water that is put down a given pan, the better it will be cleared and further the contents will be carried along the drains. Experience however suggests that there is no clear correlation between the <u>design</u> flush volumes of a range of WCs and their performance. Some designs will release a deluge of 13 litres but still fail to clear the pan.

Figure 4.6 shows the number of standard test balls (as a percentage) that were flushed out of the pan for a range of WCs tested at Brunel and Heriot Watt Universities between 1980 and 2002. As with the washing machine data, caution is needed but similar conclusions can be drawn. The better performing WCs are clearly designed to be efficient in their function whereas some of the less efficient models rely on larger volumes of water to do the job by attrition.

Clearly this is only one test and tests only measure what can be standardised rather than simulating real world requirements. We can expect performance for a given pan to improve as flush volume increases. The important observation is that flush volume alone is not a reliable indicator of flush performance.

Table 4.3 Matrix showing pros and cons of WC technologies.

Valve	Siphon
• Fast flush. • Easy operation. • Dual flush easy to distinguish. • Will eventually leak – hard to spot. • Unfamiliar to UK plumbers. • Mechanisms can stick open.	• Leak-free. • Robust. • Familiar to UK plumbers. • Parts widely available. • Dual flush less elegant. • Flow rate tends to be less.
Dual-Flush	Single-flush
• Potential for water saving. • Double flushing possible. • Half flush may be insufficient • for women's public toilets. • Users may try both buttons. • WC more likely to be used as bin?	• No user education or understanding required. • Simpler mechanism.

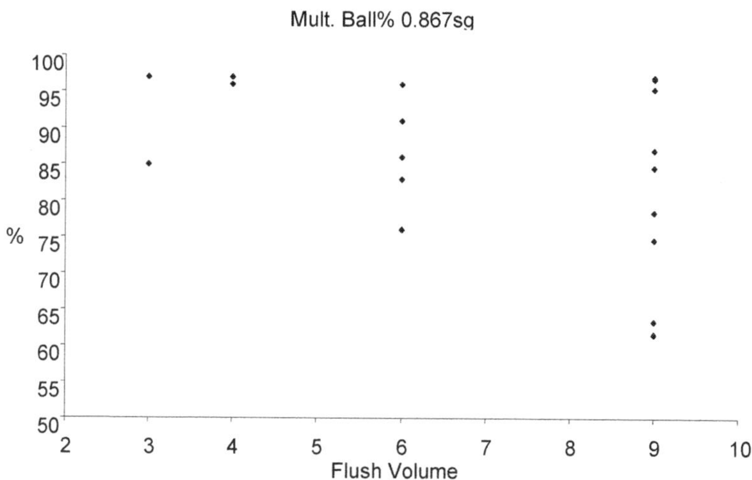

Figure 4.6 Number of plastic test balls of specific gravity 0.867 (as a percentage) cleared from the pan with a full flush for a number of WCs at their design flush volume. (Data from student experiments at Brunel and Heriot Watt, 1980–2002, courtesy of Prof. John Swaffield).

Displacement devices and retrofits

Where an older WC uses more water than it needs a displacement device can be fitted to reduce the water use (EA fact cards).

Currently illegal dual flush retrofits are available for siphon flush WCs and their legality is being reconsidered at the time of writing. Trials have shown average savings of 27% (Southern Water, 2000) but all the issues and limitations of dual flush apply.

BATNEEC

At the time of writing very few WCs have been listed with WRAS under the new Regulations. Independently tested and models are available from 6 to 4.5 litre single flush and 6/3 litre dual flush and others are going through the approvals process. Delayed action inlet valves are available which solve the issue of actual flush volumes being higher than when tested. Leak free siphon WCs are available with 6 to 4.5 litre single flush.

Future

Emphasis must be placed on actual flush volumes over the WCs life and there is room for regulatory control without which changes are unlikely to happen in the UK.

4 litres (full flush) is generally thought to represent a lower limit for use with existing gravity drainage but flush boosters are possible. These collect a number of flushes, possibly with greywater as well, and discharge them as a single larger flush ensuring good drain carry. This would reduce requirements to what is needed for pan clearance and scouring, perhaps 2 – 3 litres with suitable design.

Leak free and leak detecting inlet and flush mechanisms are possible but the industry is very price sensitive and their development is unlikely without regulatory pressure.

4.3.10 Showers and baths

Discussion

Water efficient showers are a complex subject. In the UK we have electric showers, gravity fed and mains pressure hot water systems and a wide range of user expectations. The rest of the world tend to have mains pressure hot water systems so it is a simple matter to specify a shower head of a given flow rate and for users to find something that suits them.

Most 'water saver' showers introduce air or atomise the water drops to improve wetting for a given flow rate. The result feels like a 'power shower' but with perhaps 4–9 litres of water per minute rather than 12–20 that might be delivered by 'power showers'. This is still more than many electric showers and some gravity fed showers will deliver. For safety reasons, flow regulators and water saver showerheads should not be fitted to electric showers without consulting the manufacturer. As the smaller droplets cool quickly users may experience cool feet due to the temperature drop between showerhead and tray (Fiskum 1993).

Whilst undertaking reproducible performance tests for WCs is a major technical and legal challenge it is perhaps trivial compared with developing such objective comparisons for shower performance which is judged in terms of comfort and rinse performance. Other important but difficult to quantify factors which will effect water consumption include ease of temperature and flow control.

Baths are rather simpler but the water use is difficult to assess as they are used in different ways. Capacities are usually given to the overflow. Most UK catalogues specify the water volume without a person whilst some European manufacturers consider Archimedes and allow for an 'average person' so quoted figures may not be directly comparable.

The thermal mass of most baths is negligible compared with the water content but heat-loss will influence the amount of top-up for long soaks.

Bath shape will influence water depth for a given volume.

BATNEEC

More research is needed. Small baths are available but may not always be acceptable.

Future

Efficient but comfortable shapes, better insulation, low thermal mass.

4.3.11 Plumbing systems

Discussion

Factors other than the appliances and terminal fittings also influence water efficiency in a building. For example water metering, supply pressure, hot water pipe sizing, length and choice of hot water system will all have an effect. Current recommendations suggest sizing water pipes and then going up a size to provide a margin of safety in terms of flow rate. With mains pressure water systems small pipes can be used and successful systems have been installed with 8 and 10mm microbore for kitchen sink and showers. Baths require a higher flow rate but the dead-leg is not an issue. Available design graphs and tables do not cover high flows in small pipes.

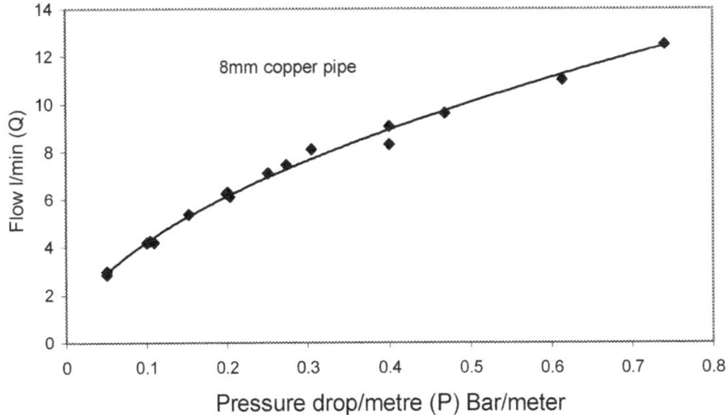

Figure 4.7 Results of tests to develop sizing charts for small bore pipe (Elemental Solutions).

BATNEEC

Water metering, optimised hot water dead-legs, optimised plumbing layout and thermal store combi boilers (if a combi is used). Also flow and pressure regulation (see section 6). Leak detection may be a cost effective measure for larger buildings where leaks may go unnoticed and may be installed to prevent water damage with a small potential water saving as a bonus.

Future

Leak detection integrated with remote meter reading? Research and guidance for use of microbore pipe for hot water delivery.

4.4 EFFICIENCY VERSUS REUSE AND HARVESTING

An important point to consider is the relative importance and benefits of efficiency measures, rainwater harvesting and greywater reuse. Some questions that can be answered for general and specific situations are:
- Can efficiency measures provide a similar magnitude of water savings to reuse?
- How can most water be saved for a given capital cost?
- How does the life cycle impact of efficiency and reuse options compare?
- What other factors might influence the choice of approach?

Economics

A simple model is proposed for the purpose of comparing a number of options using the same base assumptions. Whilst simplistic it does go beyond the limits of simple payback whereby a cistern displacement device (e.g a simple brick) will always 'beat' an efficient WC because the cost is so low (see chapter 9 for more sophisticated economic analysis). If we consider the net worth of a measure after n years we can plot the straight line graph:

$$W_n = (S - r)n - C \qquad\qquad (4.1)$$

Where:

W_n	- net worth of a measure over n years
S	- annual saving
r	- any increased annual running cost
C	- capital cost in the same units as W_n, r and S

Interest on borrowed capital, discount rates and inflation are ignored in this model but a comparison of the financial return can be plotted to show the income that could be had from investing the cost of the capital expenditure (see Figure 4.8).

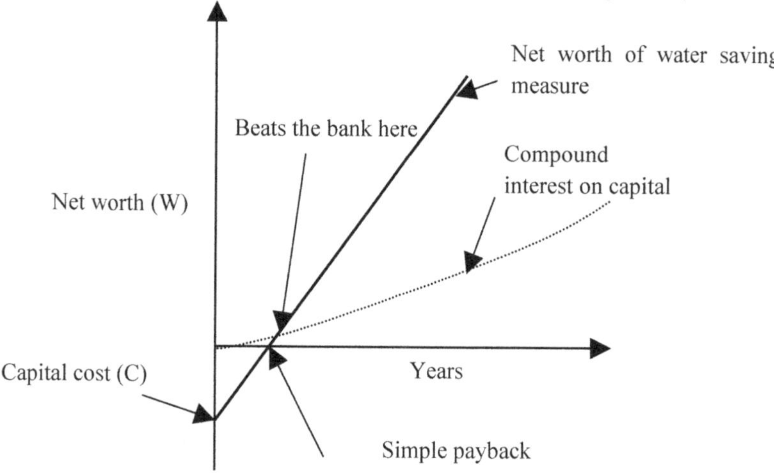

Figure 4.8 Net worth with time for a water efficiency measure. The compound interest curve compares the growth of interest on the capital cost if invested rather than in water efficiency. The gradient shows the annual saving.

Using this format we can compare a number of measures as shown in Table 4.4 and Figure 4.9.

Table 4.4 Data and assumptions for Figure 4.9

Measure	% of Water Saving		Water saved (m³/year)	Annual cost (£)	Capital cost (£)	Annual savings (£)
4 litre WC	55%	35%	34.6	0	300	52
6 litre WC	33%	35%	20.7	0	150	31
Sava flush	11%	35%	6.9	0	2	10
Rainwater system, big roof	70%	53%	66.6	35	1500	65
Greywater system	70%	35%	44.0	40	1500	26
Water butt	50%	6%	5.4	0	20	8
Other						0
4.0% compound interest	4%		*Investment*		300	

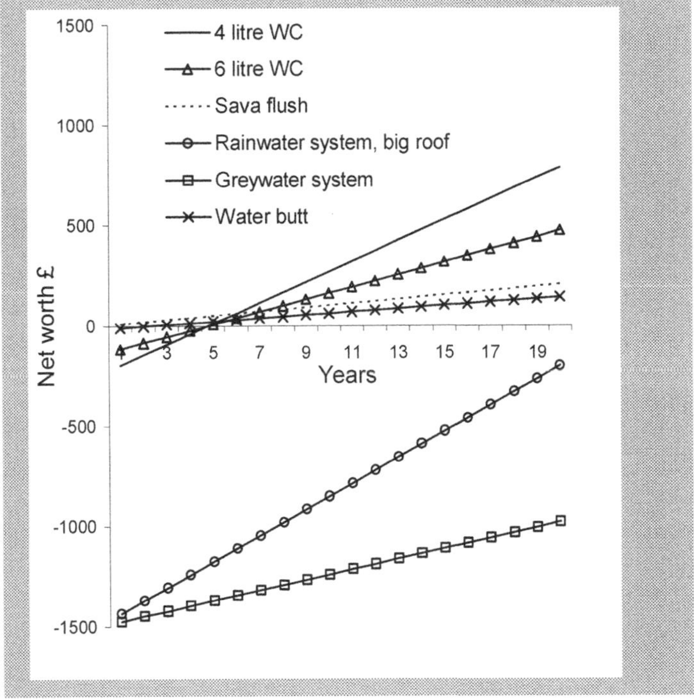

Figure 4.9 Comparing the net worth of a number of measures using the assumptions shown in Table 4.4. [1]

---

[1] Running costs for the rainwater system are assumed to be £5/year for electricity and £300 every 10 years for pump or other component replacement (conservative). A large (c.a. 100m²) roof and reasonable rainfall (about 800mm) are assumed. Water is used for WC, washing machine and garden use. Water and sewage costs are

When running costs, albeit optimistic ones, are included in the payback, rain and greywater systems are far less cost effective than efficiency measures.

For many situations the rain yield is less than required for WCs and washing machines. Similarly with greywater systems, greywater production does not always match demand for WC flushing. This has been shown in a number of studies (EA, 2000; Essex and Suffolk Water, 2001). The graph assumes that rainwater discharged to the sewer does not incur a charge, a current loophole that some water companies would like to close.

Another model that allows cumulative measures to be plotted shows that for domestic situations, if the budget available for a reuse system is spent on efficient appliances then greater savings can usually be obtained for less cost although not all the measures will be economic (Grant 2001). Efficiency measures also tend to reduce energy consumption and other costs.

4.4.1 Environmental impact

Grey and rainwater systems almost always use energy for pumping and even UV disinfection. Commercially available greywater systems also use disinfectants such as chlorine or bromine. Other considerations include the cradle to grave impact of tanks, pumps, electronics, pipes etc. All these impacts might apply to mains water and a common assumption is that decentralised or autonomous systems are environmentally preferable since water does not have to be treated to drinkable standards and pumped over large distances. In practice small-scale systems have a significantly greater impact than the mains. Where water is scarce and measures such as desalination are being considered, then reuse and recycling may be viable. Similarly where there is a large demand for low-grade water for irrigation matched with a need for a high quality effluent then a virtuous circle may be created.

Another consideration is the effect on aquatic pollution if halogen treated greywater is discharged to sewers and rainwater is diverted from soakaways to sewer via the WC.

Crettaz *et al.* (1999) and Dixon and McManus (Chapter 6) compare rainwater harvesting and water efficient WC scenarios using life cycle assessment tools with perhaps surprising results. Their analysis shows that the domestic rainwater

---

£1.50/m³. Running cost for the greywater system is assumed to be £20/year for chemicals and electricity and £200 every 10 years for pump or other component replacement (conservative). It is assumed that 70% of WC water use is met by greywater (optimistic). Capital costs of £1,500 for installed rain and grey systems are very optimistic especially for retrofit situations. Compound interest is on £300 (tax free), i.e. the assumed cost of a 4 litre WC.

reuse scenarios had a significantly higher environmental impact than the efficiency scenarios.

Clearly LCA interpretation is partly art as well as science but initial results challenge the common assumption that, whilst not economic, rainwater reuse is at least an environmentally sound option.

4.4.2 Potential to offset water supply infrastructure

One of the drivers for water demand management is the financial, environmental and social impact of creating new resources such as reservoirs. It is unlikely that even widespread uptake of rainwater systems would have significant effect on peak summer demand when supplies are most stretched. Greywater and blackwater recycling could theoretically reduce this demand but is not currently a least-cost solution.

4.4.3 Effluent reduction (and stormwater attenuation)

Grey and blackwater reuse can reduce effluent volumes, which may be an important consideration on some sites. As mentioned above, consideration must be given to the chemical quality of this effluent before considering such reductions to be environmentally beneficial.

Rainwater reuse is often considered to reduce stormwater discharge and planning has been obtained on poorly drained sites with the condition that rainwater reuse is implemented. In practice, whilst total volumes of stormwater will be reduced, soakaways, balancing tanks and SuDS systems (see Chapter 5) cannot normally be downsized when rainwater reuse is implemented. This is partly to cover failure of pumps and filters (perhaps due to extreme weather conditions) and partly due to the need to cope with prolonged rain when soils are saturated and rainwater tanks are full. Flash summer storms may well be retained but these are not the critical events for river flooding or soakaways in heavy soil.

4.5 CONCLUSIONS

Distinguishing between 'water conservation' and 'water efficiency' makes discussions about barriers, drivers and other issues more meaningful.

Water and energy efficient products such as low flush WCs and 'A' rated white goods can perform as well or better than those using more water. Typically efficient appliances have tended to cost more and so the concept of payback is often applied. However the higher cost of some efficient appliances does not always need to be offset by water savings as the higher cost may be due to other features or qualities. At the same time there is growing evidence that efficient goods are beginning to be supplied for the same price as inefficient ones.

Water efficiency is still not a major positive consideration for consumers. People are wary of low water use items, which are assumed to be worthy, but of low performance. Performance measures as used in 'Which?' reports and energy labels could provide a step towards challenging such simplistic assumptions.

In the UK, regulatory pressure is likely to be required to enforce technical changes.

Some technologies can be easily retrofitted whilst others are likely to be limited to new installations and refurbishment.

Efficiency measures appear to be more cost effective and more environmentally benign than recycling or rainwater harvesting. Improved technologies and specific applications may change this balance.

Water savings and performance can easily be countered by bad design, poor installation and lack of maintenance.

4.6 REFERENCES

Crettaz, P., Jolliet, O., Cuanillon, J. M. and Orlando, S. (1999) Life cycle assessment of drinking water and rain water for toilet flushing. Aqua 48 (3), 73-83.

EA (2001) A scenario approach to water demand forecasting. Environment Agency, Worthing.

EA (2001) Conserving Water in Buildings, 'fact cards'. Environment Agency, Worthing.

EA (2003) The Economics of Water Efficient Products in The Household. Environment Agency, Worthing.

Fiskum, L. E. (1993) Shower Test. BYGGFORSK, The Norwegian Building Research Institute, Oslo.

Grant, N. J. (2001) Presentation at the National Water Conservation Group, DEFRA Dec 7th 2001.

Harper, P., Halestrap, L. (1999) Lifting the Lid. Centre for Alternative Technology, Machynlleth.

Keating, T., Lawson, R. (2000) The Water Efficiency of Dual Flush Toilets. Southern Water.

5

Water conservation and sewerage systems

John Blanksby

5.1 INTRODUCTION

5.1.1 Aims

This book is about the efficacious use of a valuable resource, water, in order to ensure that supply and distribution systems are sustained into a future which will be driven by issues such as climate change and growing demand. The effective and efficient use of water will contribute to all three pillars of sustainability; economic, environmental and social and there is no need here to reiterate the benefits of minimising the use of potable water in our homes and workplaces. However, unlike other utilities, the water supplied to our buildings does not disappear. It becomes contaminated and requires treatment. Where on site treatment is physically possible and economically viable, there remains a general need to reduce the amount of water in the treatment process. However the majority of the wastewater in the UK and the developed world is collected and treated off site, in many cases at some considerable distance, and in this case the wastewater is also the medium for transporting waste solids which enter the collection system. These collection systems, or sewerage systems as they are traditionally known in the UK are as important as our water supply systems in ensuring the public health of the nation. Their construction resulted in a step change in life expectancy and their effective operation is a major contributor to the welfare of our communities. The aim of this chapter is to remove a potential barrier to the introduction of water conservation methods within our urban areas by:

- providing a brief insight into how sewerage systems are designed and operated to those involved in water conservation and building drainage, but not versed in sewerage.
- enabling those involved in water conservation and building drainage to take into account the downstream impacts of their actions.
- facilitating the introduction of water conservation techniques into our urban areas by helping to develop realistic expectations based on the requirements for the operation of our sewerage systems.

As in all cases where change is being introduced, it is the responsibility of those initiating the change to demonstrate that they understand the potential impacts of those changes on others. Real understanding will reduce the number of problems caused by the change. Therefore, the demonstration of real understanding will smooth the pathway for the change.

5.1.2 Scope

Sewage is a complex mixture of water, dissolved substances and suspended solids, both inorganic and organic. The water within the system is the medium which transports the dissolved substances and suspended solids and the quantity of water available for transport may vary between dry and wet weather and diurnally in dry weather. The introduction of water conservation methods may affect the amount of water available in both wet and dry weather. However, the total amount of dissolved substances and suspended solids are largely unaffected by many of these methods. Consequently there can be a significant impact on the performance on the sewerage system, in terms of carrying capacity of the water and the concentration of dissolved substances and solids.

This chapter focuses on the movement of sediments and gross solids through different types of sewerage systems. Space dictates that it cannot provide a comprehensive review, but it identifies the key aspects of these processes that might be affected by water conservation measures and thus provides a starting point for those working in the area of water conservation who wish to investigate the issues further.

5.1.3 Development of combined and separate drainage systems

Although ancient civilisations constructed and maintained drainage systems, it was not until the mid 19th century that urban populations grew so large that it was necessary to provide formal drainage systems on any scale. Prior to the industrial revolution, the majority of the population lived in small towns and villages and London was the only major conurbation in the UK. Some culverting of watercourses had taken place so as to make movement in the cities more easy, but those piped drains that had been constructed were for surface water only, as solid waste was disposed of by different routes.

Improved understanding of the causes and prevention of disease, and more enlightened attitudes resulted in the steady improvement of our burgeoning urban areas. Public water supply systems were constructed, sewerage systems installed and highway systems developed. The sewerage systems were based around the surface water drainage systems that had previously been constructed. Initially communal water closets and latterly individual households were connected to the surface water sewers converting them into combined sewers. These networks were improved and enhanced as the urban areas grew and roof and highway drainage was also directed into the new sewers.

The result of this was that rivers, which had previously been relatively clean, rapidly became highly polluted and a source of disease themselves. As a response to the situation, sewage treatment processes were developed and new sewers were constructed, intercepting the discharges into the rivers and carrying the dry weather flows to the treatment plant. However, the treatment plant could not handle storm flows and so these flows were diverted to the rivers at the points of interception. The devices that control this process are called Combined Sewer Overflows (CSO). CSOs were also built at points where the sewerage systems were so overloaded that flooding occurred. CSOs can be a major source of pollution to the aquatic environment.

In the early years of the twentieth century separate drainage systems were installed in more rural areas and were also used in green field developments where no existing drainage system existed. Separate drainage systems have two pipes, one for carrying sanitary waste, known as the foul sewer and one for carrying surface water, known as the surface water sewer. These systems can be effective in controlling the flows to treatment, but they can be subject to wrong and cross connections and the surface water draining from streets tends to be highly polluted. Therefore it is possible for these systems to be more damaging to the aquatic environment than combined systems.

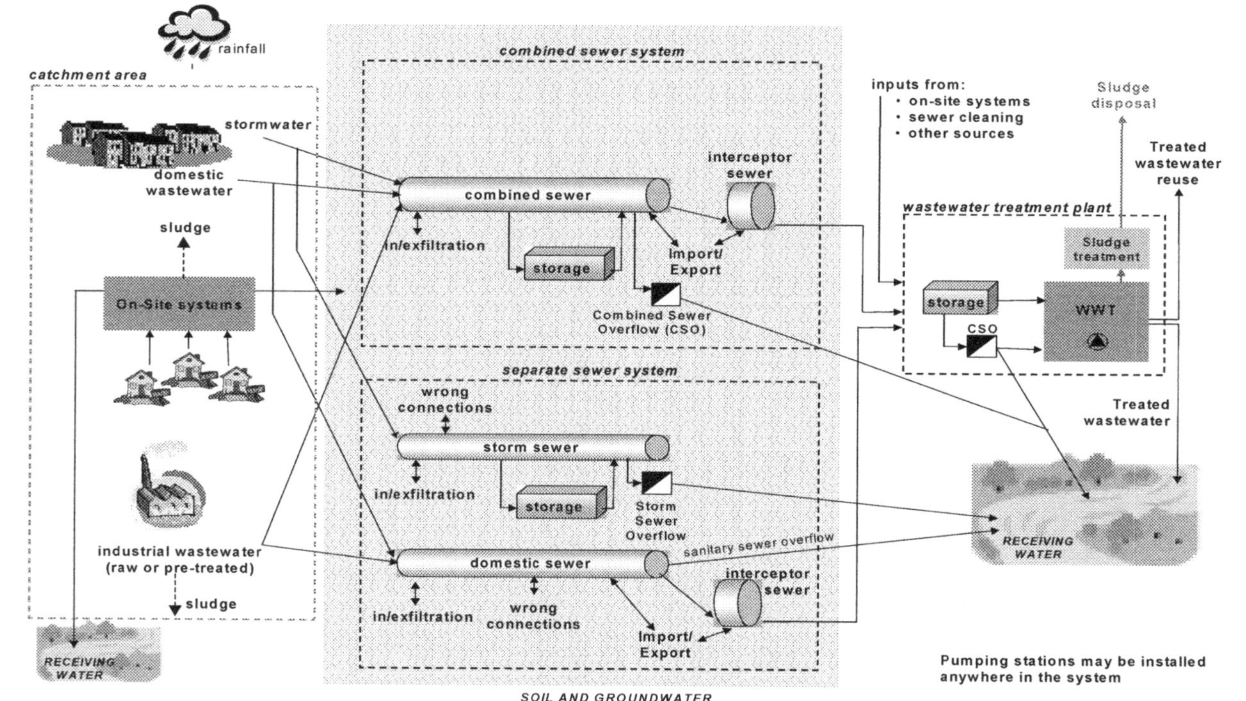

.Figure 5.1 Diagrammatic representation of sewerage system (Matos *et al.*, 2003).

Ongoing development in urban areas has caused overloading of systems which were constructed over 100 years ago. The old approach of relieving the system by creating new overflows is now being superseded by other approaches including flow reduction and flow attenuation, the former being removal of flows from the system altogether and the latter being the use of storage to limit flows passed downstream.

The result of this activity is that we have complex and diverse sewerage systems with many interactions that can be represented by Figure 5.1

5.1.4 Design of sewerage systems

The vast majority of sewage is water, and traditional design processes reflect this in that pipe diameters and gradients are determined using classical fluid dynamic approaches. However, engineering design does recognise the solids component and simple empirical rules are applied to ensure that pipes are "self cleansing". In the recent past, two documents set out the design rules for new sewers and drain. These are the Building Regulations (HMSO, 1991) which deal with private sewers and drains and Sewers for Adoption (WRc, 2001) which deals with public sewers. In 2002, the government harmonized these documents to ensure that private sewers would be constructed to the same standard as public sewers. This measure ensures that private sewers are constructed to adoptable standards and paves the way for future adoption. The design standards that are set out in the Building Regulations and Sewers for adoption have developed incrementally over the years and represent current best practice and are common with those set out in the European Standard EN752-5: Drain and Sewer Systems Outside Buildings. The nature of these documents is that they are applicable to peripheral sewers. However, in core areas, practical necessity means that they are not always applied. This is particularly the case in long, flat trunk sewers and in sewers affected by flow control devices, which increase depths in sewers at low rates of flow. In these cases, more sophisticated methods of representing the transport, deposition and erosion of sediments and gross solids are available. The reader is referred to Butler & Davies (2004) for further detail on sewer design.

5.2 FLOWS AND LOADS IN SEWERS

5.2.1 Flow and load in foul sewers and combined sewers and drains in dry weather

Foul sewers are designed to carry a small multiple of dry weather flow so as to provide adequate capacity for the peak diurnal rate of foul sewage, peaks in trade effluent discharges and a degree of infiltration. It would be typical for foul sewers to have the capacity to carry four times the average dry weather flow and this means that the pipe will never run full, providing that there are no wrong connections from surface water drainage systems and that the pipe is constructed to a reasonable standard of workmanship so as to exclude too much infiltration.

At the periphery of the sewerage system, the flow is discontinuous. This is because of the use patterns within individual households. Moving down the system, there is a point where a combination of properties connected and peak usage in mornings and evenings results in continuous flow at these times. Even further down the system, there is continuous flow throughout the day.

Variation in water usage throughout the day results in diurnal variation of flow. The peaks are generated by the aforementioned patterns of water usage and the trough occurs in the early hours when the majority of people are asleep. The peaks are relatively larger at the periphery of the sewerage system than at its outlet. There are two reasons for this: First the effect of attenuation of the individual waves discharged from appliances. Second the effect of differing time of travel from the many different properties served by the system. These two effects are reflected in the design flows for domestic properties in the Building Regulations and can be seen in Figure 5.2.

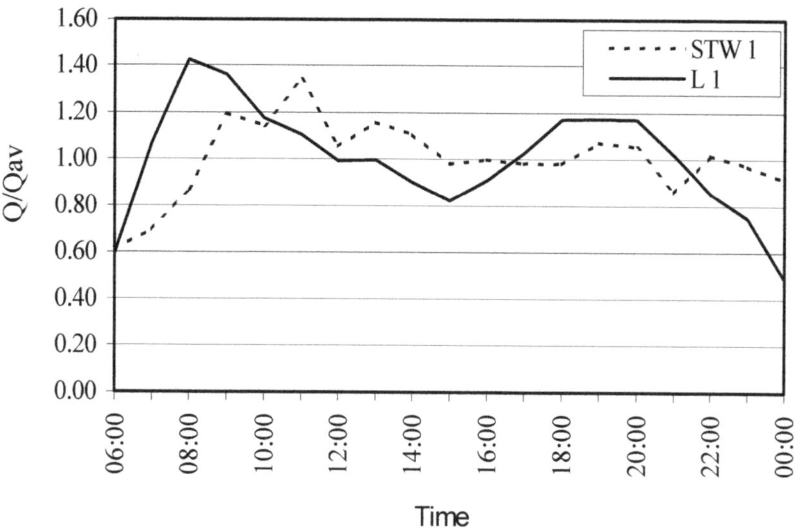

Figure 5.2 Dry weather diurnal flow (Q), expressed as a proportion of average discharge (Q$_{av}$), at different locations in sewerage system (Data courtesy of Yorkshire Water Services Ltd).

In Figure 5.2, discharges are expressed as a proportion of the average discharge. The peak discharge at S1, the inlet of a large city sewage treatment works occurs three hours later than that at L1, situated some 600 metres from the periphery of a local catchment. It can also be seen that the peak discharge is less at S! and the range of flows is less. The peak and range of flows at a point upstream of L1 would be greater than those at L1.

The load in foul sewers also varies diurnally. The load includes dissolved and solid organic and inorganic matter from domestic, industrial and commercial premises. The dissolved matter and fine suspended particles are readily conveyed through the sewerage system by the water. However, the conveyance of the gross solids is more complex and depends on the relationship between the specific gravity of the particulate, the size and quantity of the particulate and the quantity and velocity of the water available for transport. In addition, solids may catch on features within the sewer. Figure 5.3 shows the diurnal variation in flow and load at three locations in different drainage systems. Each location is approximately 600 metres from the periphery of the system, but the drainage systems have different gradients and serve different socio economic groups. However, there is considerable consistency in both the flow and loads. The loads have been determined by capturing solids in mesh sacks within the sewerage system and comprise faeces, toilet paper, sanitary products and other material.

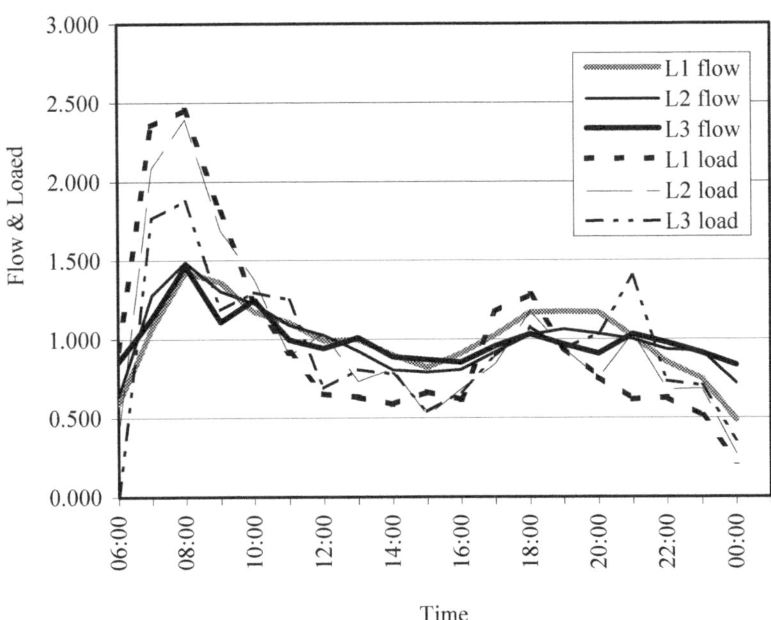

Figure 5.3 Dry weather diurnal flow and loads, expressed as the proportion of average discharge, in different sewerage systems (Data courtesy of Yorkshire Water Services Ltd).

The peaks and troughs of the load curves are more accentuated than those of flow curves. This is because of toilet usage in the early morning and during the evening. The proportions of different materials within the gross solids matrix alters as it passes down the system. At the periphery of the system, faeces and tissue form the majority of the material, but these tend to break down into finer particles as they pass down the system and through pumping stations, so that in the core regions, sanitary products form the major proportion of the material. The breakdown of faecal material and tissue is shown in Table 5.1, which comprises data collected at the same location L1 as in Figures 5.2 and 5.3 and at the downstream sewage treatment works, some 6km away.

Table 5.1 Changes in proportions of sewer solids through the sewerage system (Data courtesy of Yorkshire Water Services Ltd.)

Material	Faeces	Fine material	Sanpro	Other	Total
Proportion at L1	0.3	0.6	0.1		1
Proportion at STW	0.15	0.45	0.25	0.15	1
Mass at L1	75	150	25		250
Mass at STW	15	45	25	15	100

The table shows the proportion of different materials collected at L1 and the sewage treatment works. At L1 faeces and fine material including tissue form 90% of the material collected. At the STW, the proportion of faeces and fine material has reduced and other material from the larger catchment including industrial and commercial premises has been introduced. Assuming that the mass of sanpro products does not alter and the unit mass of sampro products is 25, it can be seen that there is a significant loss of total mass of gross solids as the sewage passes down the system and that this is due to an 80% reduction in faecal material and a 67% reduction in fine material which include tissue. The data presented in the table cannot be viewed as definitive, because it was collected for different studies with different objectives, but it indicative of the breakdown of faecal material and tissue.

The data used to inform the above was collected in dry weather. This means that it confined to gross solids from domestic sources. In addition, sewerage systems contain contaminated residual solids resulting from the washoff from urban surfaces in wet weather. These residual solids generally comprise heavier organic and inorganic sediments which cannot be transported in dry weather when velocities are low compared with those in wet weather. These solids may collect in significant quantities in large diameter sewers with slack gradients and low velocities. In this case they form sinks of polluted material which may be remobilised during the rising limb of ensuing storms.

The discontinuous nature of flows at the periphery of the system affects the transport of different materials in different ways. Littlewood and Butler (2002) describe research into the transport of three types of solids (sanitary towels, "Westminster solids" representing faecal stools, and toilet tissue) and a combination of Westminster solids and toilet tissue in different diameter pipes with different size flushes. Two transport mechanisms were noted:

- 'floating' when the solid is small relative to the pipe diameter and the flush wave. In this case, the solid is transported at the same velocity as the flush wave without effect to the wave, and
- "sliding dam" when the solid is large compared with the flush wave and pipe diameter. In this case, the flush wave builds up behind the solid, which acts as a dam in the base of the pipe. When hydrostatic head and the flow's momentum overcomes the friction between solid and pipe wall, the solid begins to move along the bottom of the pipe. The effectiveness of the dam in building up the hydrostatic head is dependent on the shape of the solid.

Littlewood and Butler concluded that "the solid transport experiments showed that for toilet tissue alone, Westminster solids alone, and a combination solid consisting of a Westminster solid with 5 or 10 sheets of toilet paper placed upstream, the solid transport capabilities of a particular wave were increased by reducing pipe diameter, and that the increase continued into the 50mm diameter pipe". For the sanitary towel, however, the optimum diameter was the 100mm diameter pipe. The sanitary towel did not move at all in the 50mm diameter pipe, and moved poorly in the 75mm diameter pipe. The Limiting Solid Transport Distance (LSTD) for different sized pipes is illustrated in Figure 5.4 which shows the movement of a medium sized Westminster Solid with a specific gravity of 0.95 resulting from a series of three litre flushes.

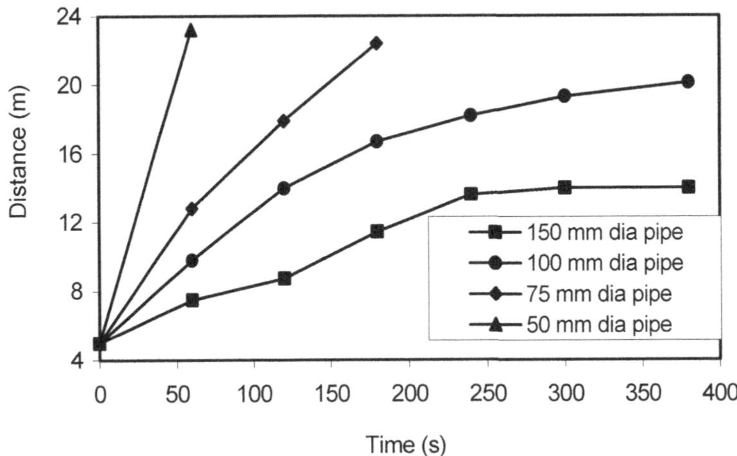

Figure 5.4 Limiting solid transport distance with reducing pipe diameter: 3l(a) flush, MB solid (Littlewood and Butler, 2002).

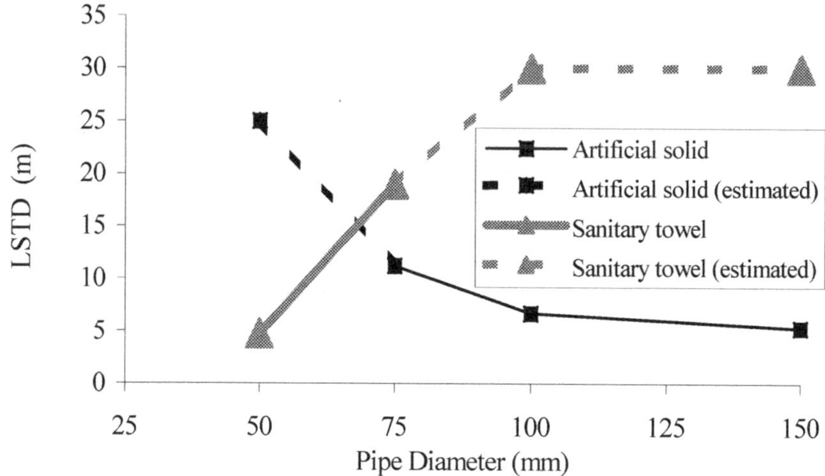

Figure 5.5 Limiting solid transport distance vs. pipe diameter (Littlewood and Butler 2002).

"The introduction of toilet tissue to the Westminster solid increased the carrying capacity of a wave, for a particular pipe diameter. This is because the movement mechanism that occurs is the sliding dam type, where the movement of the solid depends on the size of the pool that builds up behind the solid. The addition of toilet tissue increases the size of the dam that forms, and so increases the movement that occurs". The benefit of reducing the pipe diameter is clear to see. However this has to be considered in conjunction with the propensity for sanitary solids such as sanitary towels to ground in small diameter pipes. The combined LSTD for Westminster solids and sanitary towels is illustrated in Figure 5.5.

"There is scope to decrease WC waste pipe diameter to cope with low volume flushes, but this have to be carried out in conjunction with measures to prevent disposal of large synthetic solids". Littlewood and Butler also noted that flush size could also have an effect on transport. However, tests were only carried out at a fixed gradient, so further work is required to complete understanding. Littlewood and Butler's findings demonstrate that is possible to implement water conservation measures such as reducing the volume of toilet flushes, providing that large fibrous sanitary products are prevented from being flushed. This has benefits in that it will reduce the need for screening and disposing of saturated material as part of the treatment process. However, the minimum diameter for a drain carrying WC effluent is 100mm. This means that introduction of low flush toilets into existing properties may cause blockages, especially in larger properties where there is a greater distance between connections.

In addition to these mechanisms, solids can also be caught on displaced joints and other pipe defects. Blanksby *et al.* (2002) identified that blockages within the public sewer system were highly skewed towards the small diameter pipes at the periphery of the sewer system. They also identified that the majority of blockages, once cleared, did not block again and that just because a defect existed, it was far from certain that a blockage would occur. However it is clear that blockages are more likely to occur in smaller diameter pipes, and that the smaller the quantity of water to transport the solids, the more likely there is to be solids deposited in the pipes and against defects.

By reducing flush size and the discharge from appliances such as automatic washing machines, there may be a propensity to cause more frequent blockages. This may be particularly the case where there is a low density of housing which have longer pipe runs between connections. In this case, it is possible that there is insufficient energy to carry the solids downstream to be low the next connection, where a new wave takes over. This will inevitably result in a build up of solids and a blockage forming.

Pumped sewers and vacuum sewerage systems

There are clear energy efficiency benefits in reducing the volume of sewage transported. However, the storage time in collection chambers in existing systems will increase if water conservation measures are introduced. Although this should not result in problems to the sewage treatment plant providing it is sufficiently far downstream to allow re-aeration to take place, the increased residence time in collection chambers and rising mains may lead to septicity and associated problems of odour and production of hydrogen sulphide gas. These have operational and safety implications and should be considered and acted on.

5.2.2 Flow in combined sewers in wet weather

During wet weather, combined sewers may carry many times the flow rate occurring in dry weather. The shape of rising and falling limbs of flow hydrographs and peak flows depend on the characteristics of the upstream drainage area served by the sewer and the rainfall which varies in intensity and duration. Characteristics of the drainage area include drained impermeable area, the degree of separation of the drainage system the configuration of the upstream drainage system including ancillaries such as CSOs, tanks, pumping stations (PS) and flow controls. This makes responses to rainfall complex and catchment specific. Therefore it is only possible to make generalisations about wet weather flows and loads and their impacts.

Rainfall – The flow entering the drainage system is directly proportional to the rainfall falling. Thus when the rainfall is slight such as a drizzle, then the response will be small and when it is large the response will be large.

Drained impermeable area – The flow entering the drainage system is also directly proportional to the drained impermeable area. Some runoff may also emanate from permeable surfaces, but this is generally small unless the ground is saturated and the rainfall is intense.

The degree of separation of the drainage system – If there is a high degree of separation of the drainage system, then much of the runoff will be directed to surface water sewers. This reduces the response of the combined system to rainfall. Where there is little separation, there will be large response.

Sewer ancillaries - CSOs generally have a setting of approximately 6 times average dry weather flow. The setting is the flow to treatment when the overflow begins to operate, so unless the rainfall generates flows greater than 5 times average dry weather flow, it is unlikely that overflows will begin to operate. However once it does operate all the flows in excess of 6 times dry weather flow will spill out of the system into the receiving waters. This restricts the flow in the downstream sewer.

Pumping stations generally operate with duty/standby pumps. The duty pump switches on when the level in the wet well reaches a predetermined level. It then pumps at a more or less constant rate until the wet well level is reduced to a second predetermined level. In wet weather, as the flow rate increases, the duty pump will fail to handle the incoming flows and the water level in the set well continues to rise. The standby pump will cut in and pumping will then continue at a higher rate of flow until its switch off level is reached

Tanks store part of the flow. Depending on the nature of the tank, on or off line and its control regime the tank will store part of the rising limb of the event or will peak lop the event.

Flow controls will reduce the discharge downstream of the control and may cause reduce velocities upstream of the control as depths and hence cross sectional areas of flow are artificially increased.

5.2.3 Load in combined sewers in wet weather

As flows increase during the rising limb, sewer solids deposited during dry weather will be taken up into the flow. The timing at which the different materials are taken up depends on the specific gravity of the particles, the velocity needed to overcome the friction forces between particulates and the sewer walls and the depth of flow. Generally lighter gross solids will be the first to be taken up into the flow and the heavier inorganic sediments will be the last. Conversely as flows decrease at the end of an event the heavier particles will be the first to be deposited.

As solids are taken into the flow they produce a flush. The size of flush can depend on a number of factors relating to the upstream network, and the rainfall. At the periphery of the system these include the gradient of the pipes, the diameter and the distance between connections which provide the dry weather flushes which move the solids downstream, and the volume and frequency of the flushes in dry weather. The duration of the dry periods between rainfall affects the build up of pollutants on urban surfaces and of sediments in sewers. However the relationship between the inter event dry weather period and the build up of gross solids is not so well established.

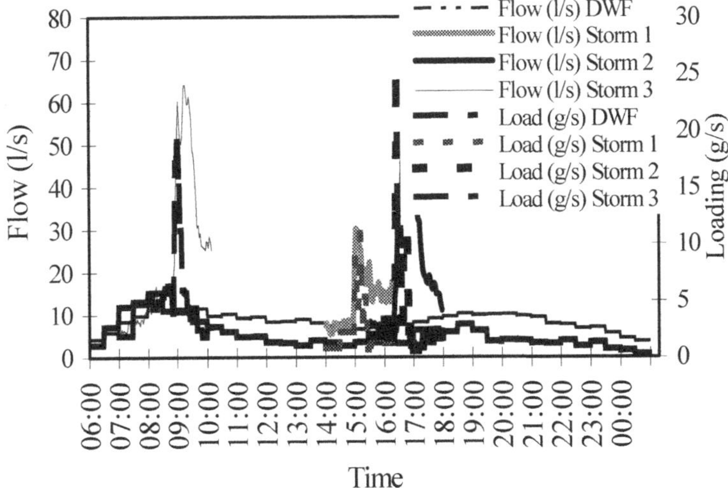

Figure 5.6 Wet weather flow and loading at L1 (Data courtesy of Yorkshire Water Services Ltd).

Figure 5.6 shows the response to three storm events at Location L1 The figure shows that the solids loading responds rapidly at the onset of the rising limb of the flow and that at this location the storm loading may be up to ten times the dry weather flow loading. It should also be noted that these are relatively minor events and the storm response is only ten times the dry weather flow. At the end of each event the loading rate dips down below the dry weather flow loading rate indicating an initial period of deposition at the end of the storm. However the mass of solids deposited during this period is only a small proportion of the load flushed during the event, suggesting that there is a build up of material over time.

5.2.4 Flows and load in surface water sewers in wet and dry weather

By their nature, surface water sewers only operate during periods of rainfall and the flows that they carry is a function of rainfall intensity at any particular time and the drained impermeable are connected to the sewer. Surface water sewers tend to operate by gravity, but on occasions flow may be pumped. The only organic material entering the sewer comes off urban surfaces, but this may carry a considerable pollutant load from road surfaces and even roofs. The majority of the larger material comprises street litter and road sweepings.

5.2.5 SUDS

Until recently, lack of clarity about the performance of SUDS has resulted in developers in England and Wales believing that they are a panacea for any site with potential drainage problems. In Scotland, where up until know, there has been a more public acceptance of the need for SUDS, it has been recognized that SUDS will produce environmental benefits, but they have only limited benefit in terms of flood control. The publication of the Framework for Sustainable Drainage Systems (SUDS) in England and Wales by the National SUDS Working Group in May 2003 has gone a long way to rectifying this situation by recommending design criteria for SUDS (EA, 2003). The framework complements the SUDS Design and Best Practice Manuals produced by CIRIA (2000,2001). Section 6 of the document suggests that the capacity of SUDS structures should be sufficient to contain the flows resulting from a 30 years return period storm. In addition it suggests that surface pathways within the development should be sufficient to provide flood protection within buildings against 100 – 200 year return period events. These criteria are concomitant with the design rainfall criteria of Sewers for Adoption presented in Table 1b above. However, the framework goes further and provides a basis for future drainage design. It suggests that from the perspective of river protection a SUDS drainage system should maintain greenfield runoff rate for rainfall event with 50% annual probability (2 year return period) and that for downstream river flood protection the drainage system should maintain greenfield runoff rate and volumes for a 1% annual probability event (100 year return period). The framework also suggests that 90% of total annual rainfall should receive some form of treatment for pollution protection.

At present the framework has been published for consultation and these design criteria will be subject to critical analysis, together with proposals for institutional arrangements they provide a basis for the implementation of SUDS techniques.

5.3 PRESSURES ON URBAN DRAINAGE SYSTEMS

Sewers are public assets with a long life. The June 2003 returns (OFWAT, 2003) reported the total length of critical sewer in England and Wales as 79,848km out of a total length of adopted sewers of 308,159km. The average annual length of critical sewers renovated or replaced was 225km. This means that the average life expectancy of critical sewers is in the order of 330 years. The average life expectancy of smaller diameter, peripheral, non critical sewers is considerably. It may be that the true asset life may be less than this as current deterioration rate may accelerate, but even so, the long asset life means that the serviceability of sewers is vulnerable to changes in circumstances that were not considered at the time when they were designed and constructed. Fortunately for ourselves the designers of our major city combined sewer systems made provision for a reasonable degree of growth, otherwise we would be experiencing more flooding and pollution of rivers from CSO discharges than we currently do. Nevertheless, the current level of expenditure in improving CSOs and reducing discharges to receiving waters and the renewed interest in urban flood prevention is indicative of the problems of exceeding system capacity.

The latest pressures on urban drainage systems are those of societal and climate change (OST, 2004). The degree of climate change that the UK will experience over the next hundred years or so will be dependent on global rather than UK behavior. However it is likely that to some degree or another we shall experience wetter winters and dryer summers, but with short periods of intense rainfall. Sewers for Adoption and the SUDS Framework coupled with the refusal of insurers to provide cover for properties with inadequate flood protection will ensure that new systems are fit for purpose. Providing the design standards are adequate for climate change impacts and that adequate provision for householders increasing the impermeability of their property is made, there is no reason that new surface water drainage systems should not perform satisfactorily. Also, providing adequate provision is made for the conveyance and management of sewer solids, foul sewers should experience no more pressures than at present.

It is our existing urban drainage systems that are most at threat from increased rainfall and more intense development. During the past forty years there has been a tendency for sewer flows to reduce in inner city areas as slum clearance and industrial decline has taken place. However, this situation is being reversed by commercial development and a perceived need to construct high density, low rise, and housing. These trends may increase the drained impermeable areas and the loading from domestic use. With increased rainfall intensities and reduced return periods, and longer inter event dry periods in summer there is the potential for an increase in frequency and location of flooding and for more frequent and more heavily polluted CSO discharges.

The options for managing this situation lie on a continuum. At one end is the wholesale replacement of otherwise structurally sound sewers, which will be both disruptive and expensive. At the other is the adoption of methods of controlling surface water flows at source, both by providing storage and surface pathways which do not cause flooding and are not otherwise hazardous. It is in this context that the impact of *water conservation* methods should be considered. In some cases a particular method may be beneficial or benign, in others it may cause problems in the downstream sewerage system or a perceived benefit may not in reality occur. The following section considers the use of different types of water conservation methods and how they are likely to impact on downstream sewerage systems, particularly those in new developments which will be constructed as separate systems and those in our existing urban centres which are predominantly combined.

5.4 POTENTIAL IMPACT OF WATER CONSERVATION METHODS ON PERIPHERAL AND CORE DRAINAGE SYSTEMS AND ON TREATMENT PLANT

Water conservation is a term used to cover a broad range of techniques which are driven by a desire to reduce domestic, commercial and industrial use, thus minimizing overall demand on the environment and maximizing use of existing water supply and distribution assets to cover growth in future demand. In chapter 4 of this book, Nick Grant identifies five general categories of techniques

- Water *conservation*; doing less with less.
- Water *efficiency*; doing more with less.
- Water *sufficiency*; enough is enough.
- Water *substitution*; replace water with something else
- Water *reuse, recycling and harvesting*; a potentially virtuous circle.

Using these categories and examples of their application used by Grant, Table 5.2 describes the potential benefits and problems in the performance of urban drainage systems. The table is divided into five parts describing the effects of the five categories on foul, combined and surface water systems. The descriptions are qualitative because size of the impacts is dependent on the degree to which the techniques are used. The adoption of a single technique in one household will be small whereas wholesale adoptions of techniques across whole estates and communities will have significant impacts.

The impacts of water conservation techniques are largely beneficial. There may be a propensity to increase the deposition of sediments in flat sewers at the core of the sewerage system in dry weather and this may affect the performance of sewer ancillaries and pumping stations. However the cost of this may be balanced by reduced flows to treatment

Water efficiency techniques have the potential to improve the operational effectiveness of peripheral drains and sewers by improving flushing. However, there is potential for sediment deposition in core sewers. These may be offset by savings by the reduction of flows to be treated

Table 5.2a Impacts of water conservation techniques

Technique	Foul sewers	Combined sewers	Surface water sewers
Do not water lawns.	This should not have any impact other than reducing ingress of solids where runoff currently enters the sewerage system. In the case of foul drainage systems this could occur at wrong connections		
Do not wash the car (as often).		There will be a reduction in the volume of water to be pumped and treated and a reduction of pollutants entering the drainage system. However this may be countered by increased wash off from dirtier cars during wet weather.	Do not wash the car at all where the wash waters enter the surface water drainage system. This causes dry weather pollution of receiving waters
Take shallow baths and short showers.	This reduces the total volume of water in the sewerage system with little impact on solids or pollutant loading. Discharges from baths and showers are not generally at a sufficient rate as to have a beneficial effect on the movement of solids in peripheral sewers and so the reduction in discharge is unlikely to have any serious detrimental effect. At the core of the system, there may be problems relating to sediments in flat pipes and these may be marginally worsened. However there will be an overall reduction in any pumping costs and in the hydraulic capacity required at treatment plant.		

Table 5.2b Impacts of water efficiency techniques

Technique	Foul sewers	Combined sewers	Surface water sewers
Hydraulically efficient WCs	These may speed the emptying of WCs increasing the velocity of the flood wave in peripheral sewers and increasing the effectiveness at transporting solids, reducing the propensity for blockage. At the core of the system there will be less benefit as the flood wave subsides. There may be some additional deposition of sediment in flat sewers, but reduction in flow volume will be beneficial in reducing pumping costs and treatment plant capacity.		
Hydraulically efficient wash basins, baths and showers	Increasing the speed of emptying of wash basins and baths may make a minor contribution to the flushing of peripheral sewers. However this may be at the expense of allowing additional solids into the sewerage system. There will be little benefit in core area sewers, but reduction in flow volume will be beneficial in reducing pumping costs and treatment plant capacity.		
Hydraulically efficient white goods	Benefits may be obtained by flow reduction at pumping stations and treatment plant, but flow rates should be maintained so as to maintain adequate flushing of peripheral sewers		
Garden design, drought tolerant plants and grasses, mulch, absorbent surfaces	This should not have any impact other than reducing ingress of solids where runoff currently enters the sewerage system . In the case of foul drainage systems this could occur at wrong connections		
Fix leaks.	This will result in savings in pumping and treatment costs, although there is a potential for sediment deposition in flat sewers to increase		

Table 5.2c Impacts of water sufficiency techniques

Technique	Foul sewers	Combined sewers	Surface water sewers
Dry toilets	These techniques will be of major benefit as they will reduce the amount of solids in the sewer system. The solids will be dry, therefore alternative treatment methods will be required, but the disposal of screenings will cost significantly less that current practice. Incineration will be made easier and transport and tipping costs will be reduced as the water content of the screenings reduces from typical current values of 70% towards 30% or less.		
Waterless urinals			
Vacuum drainage systems			
Air for industrial cleaning processes	Contaminated industrial wastewaters will be reduced and solid wastes will be disposed of through the solid waste disposal system		
Broom rather than (or before) wash down of floors	Contaminated wastewaters will be reduced and solid wastes will be disposed of through the solid waste disposal system		

Table 5.2d Impacts of water substitution techniques

Technique	Foul sewers	Combined sewers	Surface water sewers
Reduced flush WCs	This may have a significant deleterious effect on the transport of solids in peripheral sewers, resulting in more frequent blockage. To some extent this may be overcome by increasing the rate of flow of the flush. There are potential cost savings through the reduction of pumping and hydraulic capacity of treatment plant		
Ergonomics and flow regulation	The rate of flow from wash basins, baths & showers is generally low, so there will be little impact on the flushing of peripheral sewers. There will be little benefit in core area sewers, but reduction in flow volume will be beneficial in reducing pumping costs & treatment plant capacity.		
Bath size			
Control			
Reduce garden watering	This should not have any impact other than reducing ingress of solids where runoff currently enters the sewerage system . In the case of foul drainage systems this could occur at wrong connections		

Table 5.2e Impacts of water reuse, recycling and harvesting techniques

Technique		Foul sewers	Combined sewers	Surface water sewers
Water reuse	Greywater irregation		All these techniques reduce the volume of water carried through the sewerage system to treatment without any impact on the peak flushes which transports the solids in the peripheral sewers	Dry weather pollution of receiving waters where there is potential for runoff to enter surface water gullies
	Process water reuse			
	Shared bathwater			
Water recycling	Greywater recycling		Reduction in flow volume to treatment with associated cost savings. However interception of washing machine wastes may result in reduced flushing of sewers in peripheral sewers. Sediment problems in flat sewers may be exacerbated.	
	Blackwater recycling		Providing that solids are also removed, treated and disposed of via the solids waste system this will be beneficial. However, if only the water is intercepted there will be a potential for frequent blockage of peripheral sewers	
Water harvesting	Water butts		The impact depends on the design objectives of the harvesting system. Harvesting systems are designed to maximize the collected water and hence there is a tendency to collect water as soon as possible during an event and then to overflow any excess. This means that any surface pollutants washed off on the rising limb of storms will be contained in the storage vessel (water butt or tank) The excess flows will be relatively clean and will be beneficial to the surface water discharge. However, if the storage vessel is full, it will have little impact on the peak runoff and therefore no benefit to the capacity of the drainage system. Storage designed to reduce peak flows in downstream sewers only start to fill once a predetermined flow rate is exceeded. This means that it will not fill during lower intensity rainfall events. Therefore less water is harvested. This means that larger storage vessels are required if the full benefits of rainwater harvesting are to be gained, but this may be physically or economically impracticable. This is not to say that rainwater harvesting is detrimental to surface water drainage system. However, those systems which collect water at the onset of rainfall may require some form of treatment to remove pollutants washed off surfaces and they are of little benefit to the carrying capacity of the drainage system	
	Rainwater harvesting			

Water sufficiency techniques have the greatest potential to cause problems in peripheral sewers. This is because the capacity to flush solids through piped systems may be reduced. This may be overcome in new developments by reducing the diameter of drains to 75mm or 50mm. This should be done in conjunction with the disposal of sanitary protection products via the solid waste system, but currently this reduction is contrary to the provisions of the building regulations. In the case of existing properties, the drainage system should be surveyed and an assessment of the impact of these techniques on its operation should be made.

Water substitution techniques generally have no adverse effects on drainage systems as they either involve local treatment or in the case of housekeeping techniques, the volume of solids entering the drainage system is significantly reduced. Care should be taken where vacuum systems are to be discharged into existing sewers.

Water reuse, water butts and rainwater harvesting have no significant adverse affect on the drainage system. Greywater recycling may have adverse impacts in peripheral drains and sewers by reducing the water available for flushing. Blackwater recycling should only be considered if the solids are to be removed from the drainage system and treated locally, otherwise there will be a significant potential for blockage.

The message is clear, in new developments where the drainage system can be designed to cater for the flow regime, there are clear benefits of introducing water conservation technologies. However, the impact of water conservation technologies on existing drainage systems has the potential for both benefits and problems. Whether the techniques are being introduced to a single property, or an infill/brownfield development is being carried out, it is essential to assess their potential impacts on the performance of drains and sewers and ensure that potential problems are dealt with.

In the case of new developments, it is worth considering the impact of failure of the local or decentalised treatment technologies employed as part of the water conservation process, or occupants rejecting conservation techniques after a period of time and reverting to traditional technology. Although communities can be happy with the introduction of these technologies (Hedberg, 1999), failures have been identified in many cases (West, 2003) and these may be hazardous to health where the treatment processes are part of domestic reuse applications (Schaart, 2003). The ultimate fallback position for failed local or decentralised system is to take the development into the conventional sewerage system, in a similar way to first time sewerage schemes are currently constructed. However, this may present significant technical and economic consequences. These consequences will be site specific and it will be worthwhile for developers and their advisors to quantify them and to ensure that appropriate management arrangements are made so that risk of failure is reduced to an acceptable level.

5.5 CONCLUSIONS

Water conservation measures may have a significant impact on downstream drainage systems if they are adopted on an estate or community wide basis.

The main potential adverse effects will be in increasing the propensity for deposition of gross solids in peripheral foul and combined sewers in dry weather and also in increasing sediment deposition in flat core area sewers. Although the design of outlets from WCs, wash hand basins, baths and white goods may mitigate the impact on peripheral sewers. The gross solids and sediments may cause blockages and limit the carrying capacity of combined sewers in wet weather. Flushes of gross solids and sediments in combined sewer systems in wet weather may affect the performance of ancillaries such as CSOs and pumping stations, also screens and screenings handling plant at inlet works.

Beneficial effects will be the reduction of pumped flows and hydraulic capacity at treatment plant. Also dry treatment systems will be of wider environmental benefit providing that there is a suitable disposal route for the final product. Perceived benefits of flow reduction resulting from water harvesting will not occur unless storage vessels are properly designed.

Designers of water conservation schemes should be aware of the potential problems that may occur as should regulators and water service providers. It is unlikely that problems will occur at household level, but even so the potential impacts should be reviewed, particularly when there are long peripheral lengths of private drains or sewers where flow is discontinuous and there are very few connections.

At a larger scale there is a requirement for further understanding of the impact of water conservation measures on the transport of gross solids and sediments. This need is complementary to that for similar understanding of social and climate change impacts on the transport of gross solids and sediment.

5.6. REFERENCES

Blanksby, J., Khan, A. and Jack, A., (2002) Assessment of cause of blockage of small diameter sewers. *Proceedings of international conference Sewers Operation and Maintenance*, Bradford, 26th – 28th November 2002. ISBN: 1 85143 213 2.

Butler, D. and Davies, J.W. (2004). *Urban Drainage*. 2nd Edn., Spon Press, London.

CIRIA (2000) *Sustainable urban drainage systems - design manual for England and Wales - Report No. C522.* Construction Industry Research and Information Association, London.

CIRIA (2001) *Sustainable urban drainage systems - best practice manual for England, Scotland, Wales and Northern Ireland - Report No. C523.* Construction Industry Research and Information Association, London.

EA (2003) *Framework for sustainable drainage systems (SUDS) in England and Wales*, National SUDS Working Group, Environment Agency, May 2003, http://www.environment-agency.gov.uk/commondata/105385/suds_book.pdf.

Hedberg, T. (1999) Attitudes to traditional and alternative sustainable sanitary systems. *Water Science and Technology* **39** (5), 9-16.

HMSO (1991). *Building Regulations 1991: Part H-Drainage and Waste Disposal.* The Stationary Office, http://www.safety.odpm.gov.uk/bregs/building.htm.

Littlewood, K. and Butler, D. (2002). Influence of diameter on the movement of gross solids in small pipes, *Proceedings of international conference on Sewers Operation and Maintenance*, Bradford, 26th – 28th November.

Matos, R., Ashley, R M., Cardoso, A., Molinari, A., Schulz, A., Duarte, P. (2003). *Performance Indicators for Wastewater Services.* IWA Publishing.

ODPM (2002) Building Regulations, Part H, Drainage and Waste Disposal, 2002 edition, The Office of the Deputy Prime Minister, http://www.odpm.gov.uk/stellent/groups/odpm_buildreg/documents/page/odpm_breg_600283.hcsp.

OFWAT (2003) *The June return 2003.* Office of Water services, http://www.ofwat.gov.uk/aptrix/ofwat/publish.nsf/Content/junereturn.

OST (2004) *Foresight Future Flooding.* Office of Science and Technology, http://www.foresight.gov.uk/fcd.html.

Schaart, N. (2003) Sustainable urban drainage and public hygiene; a risk analysis. (In Dutch. Org. title: Duurzame waterketen en volksgezondheid; een risicoanalyse). M.Sc. thesis, Delft University of Technology, section sanitary engineering, Delft January 2003.

West, S. (2003) *Innovative On-site and decentralised sewage treatment reuase and management systems in Northern Europe and the USA.* Report of a study tour. 16th March, Sydney Water Australia.

WRc (2001) *Sewers for Adoption.* 5th Edition, WRc July.

6

An introduction to life cycle and rebound effects in water systems

Andrew M. Dixon and Marcelle McManus

6.1 INTRODUCTION

The environmental benefits gained from introducing measures to reduce water demand may be diminished, or eliminated due to life cycle environmental impacts and rebound effects. Life cycle environmental impacts are those that occur anywhere in the life cycle of a product or service system. The life cycle includes extraction of raw materials, their processing, manufacture of goods, sale, use, and ultimately end of life – the stage at which the product no longer fulfils its function. Rebound effects are those that occur as a response to an action or initiative. In an environmental context, this can mean that a reduction in environmental impact in one part of the system can lead to unplanned (and often unexpected) increases in environmental impact in another part of the system.

This chapter comprises an introduction to research on and around the subject of life cycle impacts of water systems including a look at the so-called rebound effect. Its purpose is to raise awareness of life cycle and rebound effects and promote discussion on the topic in general within the context of water demand management.

The first section introduces the concept of life cycle thinking and its application in Life Cycle Assessment (LCA). The results of a preliminary review into research into life cycle issues and the water system are described. The second section focuses on the rebound effect and how this phenomenon relates to the water system.

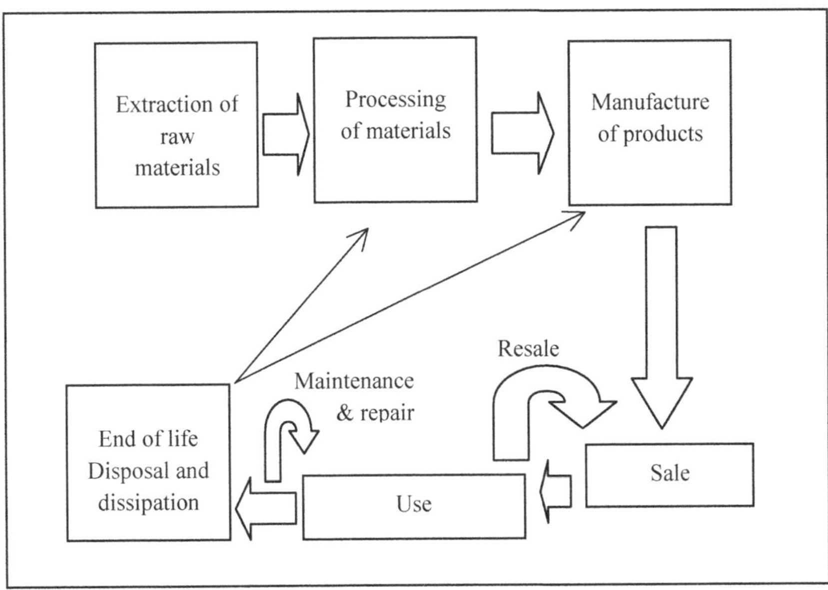

Figure 6.1 Typical product life cycle.

6.2 LIFE CYCLE THINKING

Life cycle thinking requires one to consider a product, service or system at all stages throughout its 'life', from 'cradle to grave'. Figure 6.1 illustrates a typical product life cycle, including arrows indicating maintenance resale and re-use of materials and components. Perhaps most importantly, life cycle thinking promotes a view of the world as an interconnected system rather than a reduced collection of objects. There are environmental impacts at each stage in the life cycle and the distribution of impacts varies from product to product. For example the greatest environmental impact (around 90% of the total impact) of a washing machine occurs in its use phase (Electrolux, 2004). There are very many products related to water demand management and each has its own life cycle which intersects with the many other product lifecycles in the wider economy. There is a number of tools and methodologies that promote life cycle thinking, of which life cycle assessment (LCA) is one. LCA takes account of technical and scientific aspects of the environmental impacts of product system life cycles and is described in the ISO14040 series of standards. Life cycle thinking is a mainstay of design methodologies such as eco-design, whose strategies include re-addressing human needs, dematerialisation and working in partnership with suppliers and customers.

6.2.1 Life Cycle Assessment

LCA methodology is described in the ISO 14040 series. It is a technique for assessing the environmental aspects and potential impacts associated with a product, by:
- defining goal and scope of the assessment i.e. Why is the study being carried out? What are the system boundaries? What is the functional unit by which different options are compared?
- compiling an inventory of relevant inputs and outputs of a product system, e.g. energy and material flows, emissions to air and water (see Figure 6.2)
- evaluating the potential environmental impacts associated with those inputs and outputs, e.g. effects to air, land and water systems – soil contamination, global warming, eutrophication,
- interpreting the results of the inventory analysis and impact assessment phases in relation to the objectives of the study, e.g. normalising impacts by comparison of the pollution caused by an average European. Impacts can be expressed in terms of the different impacts examined (e.g. ozone depletion, global warming etc) or aggregated as a single unit. The latter option is not recommended in the ISO Standards but is often included in studies for easy comparison.

Figure 6.2 shows the inputs and outputs of materials and energy at various stages of the life cycle incorporated in the inventory stage of the LCA. Life cycle thinking and LCA have become more widely used in recent years due to the more holistic approach to environmental management use. With other environmental management techniques it is possible to make improvements in one part of the life cycle whilst unwittingly increasing the impacts in another.

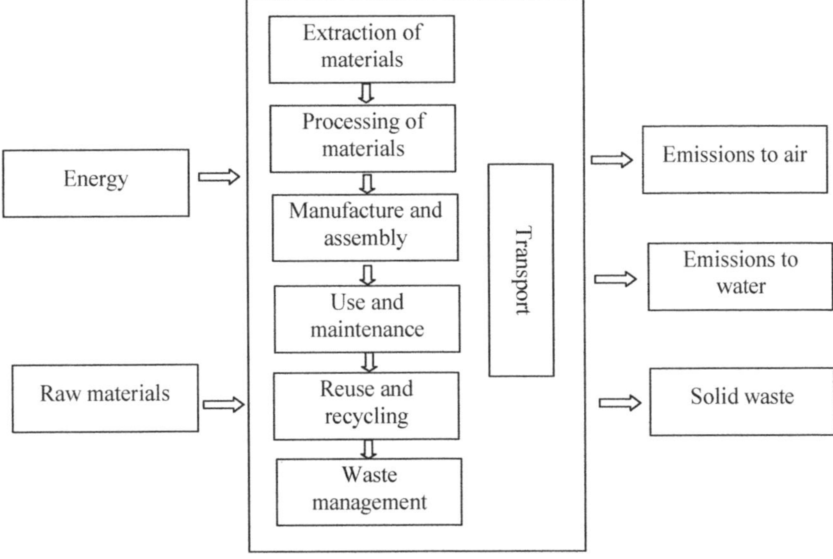

Figure 6.2 Flows of energy and materials and emissions accounted for in an LCA.

6.2.2 Recognised limitations of LCA

Of course, there are still some limitations of the life cycle approach. These limitations, as mentioned in ISO 14040, are as below:

- The nature of choices and assumptions made in LCA may be subjective,
- Models used in LCA are limited by their assumptions and may not cover all impacts and applications,
- Results of LCA's focussing on global issues may not be applicable to local conditions,
- Accuracy of the LCA is dependent on the availability and quality of data,
- Lack of spatial and temporal dimensions in the inventory data introduces uncertainties in the impact assessment results.

LCA is part of the environmental management toolbox and generally, information developed in an LCA study should be used as part as a much more comprehensive decision making process. Comparing results of different LCA studies is difficult and only possible if all assumptions made are the same for the LCA's considered and the same boundaries have been drawn. For further information on LCA, reference can be made to the ISO 14040 series or journals and websites included at the end of this chapter.

6.3 LIFE CYCLE ASSESSMENT AND WATER SYSTEMS

There has been relatively little published work on life cycle impacts of water systems. The majority of this research focuses on wastewater systems rather than demand management, and a brief overview of research in these areas is provided in the following paragraphs.

6.3.1 LCA and demand management

The following text describes some examples of LCA carried out on demand management systems.

Rainwater re-use vs. low flush toilets

The aim of Crettaz *et al.* (1999) was to quantify the life cycle environmental impacts of rainwater re-use systems and other systems that reduced WC flush water use. Their paper is structured around the four steps of LCA, goal definition, inventory, impact assessment and interpretation analysis. The researchers considered around 160 different processes and sub-processes representing the drinking water distribution and treatment systems as well as the installations required for rainwater re-use and wastewater treatment. The report considered emissions to air, water and soil as well as energy consumption. The authors concluded that low flush WC's were favourable for all environmental problems compared with conventional WC's.

Perhaps surprisingly, in this case rainwater re-use scenarios were found to be unfavourable in terms of environmental and economic impact. Rainwater re-use only becomes favourable if the electricity requirement for water treatment is very high. As a result they recommend that the installation of low flush toilets should be promoted, and the pollutant transfer to the water table should be minimised using appropriate infiltration techniques e.g. sand filters. If rainwater re-use is to be advocated it should use a low pressure, energy efficient pump, pumping to the garden typically requires a higher pressure pump which increases environmental impacts. Crettaz *et al.* recommend that future research should look at rainwater re-use for laundry, optimisation of storage tank size, an estimation of uncertainty in LCA results and a further evaluation of the transport of heavy metals through the environment.

Strategies for water supply at the housing estate scale

Van der Hoek *et al.* (1999) describe how LCA was used to help identify water system strategies for a new sustainable housing estate in Amsterdam, The Netherlands. The researchers considered several scenarios, which included rainwater re-use, greywater re-use and dual supply systems using surface water. The rainwater and greywater systems had the highest life cycle impacts largely due to energy consumption. In the scenario with lowest impact residents used water from a local lake in its dual supply system. Lower quality water from the lake would be treated and used for toilet flushing and laundry whilst the mains distribution system would be used for drinking water and be available for the fire service in the case of an emergency. The results may have been different for other locations or if other treatment measures had been considered and if renewable energy sources were to be utilised, this emphasises the critical nature of defining scope and boundary conditions in an LCA study.

Residential storage tanks

Research undertaken in Australia (James, 2003), has shown the importance of using LCA in water management. The study examined the use of residential rainwater storage tanks to provide water for different domestic applications. Two sizes were compared with not having a tank at all. The larger tank was used for toilet flushing and watering the garden and the smaller was used just for watering the garden. The research showed a 30% reduction in mains water usage for the large tank and 8% for the smaller tank. In addition, the storage tank systems diverted nitrogen from the stormwater drainage system into the local garden soil system thus reducing downstream eutriphication. The LCA also showed that the energy used in the system (energy for materials used to produce and run the tanks over their life cycle), is greater than that used by the traditional system. However, it was considered that the scale of the energy and material impacts for the tanks were relatively low compared with overall Australian energy use, while the water savings were significant for the area, this is in contrast to the first example which based its recommendations on the energy and material related impacts, thus highlighting how similar studies can end up with different conclusions based on the weighting criteria that they employ.

6.3.2 LCA and wastewater systems

LCA has been used more extensively to explore the sustainability of wastewater systems (Emmerson *et al.*, 1995; Ashley *et al.*, 1997; Balkema *et al.*, 1997; Dennison *et al.*, 1998; Mels *et al.*, 1999; Brix, 2000; Lundin and Morrison, 2002). Emmerson *et al.*, (1995) recommend that LCA should be more widely used in the water industry as a decision aid in environmental policy and in environmental improvement activities. This is echoed in the work of other authors listed and throughout this chapter.

Key points extracted from the papers on LCA and water systems include:
- Results of LCA's are sometimes unexpected, or difficult to anticipate
- LCA's are data intensive requiring extensive, high quality information
- The boundary conditions of an LCA must be defined with care since they may have a significant effect on the results. Key issues are spatial boundaries, timescale over which life cycle comparison is made, scale at which comparison is made and the study detail level
- LCA's are necessarily specific and limited. They relate to physical issues and do not account fully for other spheres of value such as bio-diversity, public amenity, habitat, and aesthetics.

6.4 REBOUND EFFECTS

Simply put, rebound effects are those that occur as a response to an action or initiative. In an environmental context, this can mean that a reduction in environmental impact in one part of the system can lead to unplanned (and often unexpected) increases in environmental impact in another part of the system. Consider the following statement:

> *'Sustainability concepts that rest on the idea of resource or energy improvements due to technological progress tend to overestimate the actual saving effects because they ignore the behavioural responses evoked by technological improvements that lead to rebound effects'* (Binswanger, 2001)

Rebound effects relate to life cycle effects in that both effects are at the system scale. Rebound effects particularly relate to the use stage of the product life cycle and often link together the use stages of the life cycle for many different products. The rebound effect has often been investigated in the context of energy efficiency, although it can be applied broadly to any resource, even time. A thorough review and analysis of the rebound effect including a review of various empirical studies has been carried out by Binswanger (2001) and is recommended reading as an introduction to the subject.

It is accepted that the rebound effect exists, yet its significance is in dispute. Several authors (Binswanger, 2001; Herring, 2000; Sanne, 2000; Moezzi, 1998) have developed the concept of rebound effect so that its significance can be better understood. The principal types of rebound effects identified by Greene *et al.* (in the context of vehicle travel) (in Herring, 2000) are listed below and illustrated with examples from the water system.

- Direct rebound effect – e.g. installing a water efficient shower and consequently spending longer showering 'because it uses less water'
- Indirect rebound effect – money saved on the water bill is spent on further consumption (e.g. a more distant holiday)
- General equilibrium effects – involving producers and consumers and representing the result of myriad adjustments of supply and demand in all sectors.

It has been argued that the direct rebound effects tend to be relatively low, depending on the system, whilst the indirect effects can be significant and even negate any intended environmental gains. The significance of this effect in the water sector may depend on whether customers are metered and how the pricing structure is set up. The third effect, so-called general equilibrium effects are very difficult to predict, although results of input-output analysis may prove useful in improving understanding of these effects.

6.4.1 Rebound effects on time savings

In addition to economic and resource based rebound effects, the rebound effect can be considered in the context of time saving (in Binswanger, 2001). A device may save time but if one uses this 'saved time' in a more energy intensive manner then the resultant environmental impact is increased. This concept sheds new light on certain activities previously thought of as being environmentally friendly. For example, is taking a long hot bath better than a short sharp shower if the time saved is spent driving a car? Is watering a garden by hand with a hose more environmentally friendly than using a trickle irrigation system, since using a hose prevents you from simultaneously consuming more resources, whilst the trickle irrigation lets you simultaneously go shopping and purchase resource intensive knick-knacks? These sorts of effects are very difficult to predict and to manage; the examples may seem trivial yet rebound effects are serious and real considerations. Nevertheless, it is clear that we must think beyond the boundaries of conventional sector classifications and adopt a more holistic view of our economy and its environmental impacts. This requires research to take an interdisciplinary approach; only in cross sector and cross-disciplinary partnerships can these problems be addressed.

6.5 CONCLUDING REMARKS

Water management must be considered and practised in the context of the wider economy if the full life cycle environmental impacts are to be accounted for.

Design and implementation of demand management measures will evoke a behavioural response that may lead to a net increase in environmental impact. These impacts may be outside of the water sectors traditional boundaries.

A preliminary review revealed that little research has been carried out into life cycle thinking and assessment in the field of water conservation and demand management. Research involving LCA of wastewater management and treatment is more extensive. No evidence was found of research into the environmental impacts of rebound effects in the water system. Research into environmental impacts of rebound effects is most often focussed on energy use.

This chapter serves as an introduction to the concept of LCA and rebound effects. A debate on what this might mean for the water industry professional is required. Questions about the relevance of life cycles and rebound effects should be raised and answered. Implications for the water industry should be considered; what can and should we do about life cycle and rebound effects?

6.6 REFERENCES

Ashley, R. M., Souter, N., Butler, D., Davies, J., Dunkerley, J. and Hendry, S. (1999) Assessment of the sustainability of alternatives for the disposal of domestic sanitary waste. *Water Science and Technology* **39** (5), 251-258.

Balkema, A J., Preisig, H.A., Otterpohl, R., Lambert, A.J. and Weijers, S.R. (2001) Developing a model based decision support tool for the identification of sustainable treatment options for domestic wastewater. *Water Science and Technology* **43**(7), 265-270.

Balkema, A.J., Weijers, S.R. and Lambert, F.J.D. (1997) On methodologies for comparison of wastewater treatment systems with respect to sustainability. *Options for Closed Water Systems-IAWQ international conference*, Wageningen.

Binswanger, M. (2001) Technological progress and sustainable development: what about the rebound effect? *Ecological Economics* **36**, 119-132.

Brix, H., (2000) How green are aquaculture, constructed wetlands and conventional wastewater treatment systems? *Water Science and Technology* **40** (3), 45-50.

Crettaz, P., Jolliet, O., Cuanillon, J. M. & Orlande, S. (1999) Life cycle assessment of drinking water and rainwater for toilet flushing. *Aqua* **48** (3), 73-83.

Dillenc (1999) Selection and evaluation of a new concept of water supply for "IJburg" Amsterdam. *Water Science and Technology* **39** (5), 33-40.

Electrolux (2004). Life cycle environmental impact of a washing machine and dishwasher. http://www.electrolux.com/node463.asp.

Emmerson, R. H. C., Morse, G.K., Lester & Edge ,D. R. (1995) Life-cycle analysis of small-scale sewage treatment processes. *Journal of the Chartered Institution of Water and Environmental Management* **9** (3), 317-325.

Dennison, F.J., Azapagic, A., Clift, R. and Colbourne, J.S. (1998) Assessing management options for wastewater treatment works in the context of life cycle assessment. *Water Science and Technology* **38** (11), 23-30.

Dennison, F.J., Azapagic, A., Clift, R. and Colbourne, J.S. (1999) Life cycle assessment: comparing strategic options for the mains infrastructure - Part I. *Water Science and Technology* **39** (10-11), 315-319.

Herring, H. (2000). Is energy efficiency environmentally friendly? *Energy and Environment* **11**(3), 313-326.

ISO 14040:1997, *Environmental management, Life cycle assessment, Principles and framework.* International Organisation for Standardization.

James, K. (2003) Water Issues in Australia - An LCA perspective. *International Journal of Life Cycle Assessment* **8** (4), 242.

Jelsma, J. and Knot, M. (2001) Designing environmentally efficient services; a script approach. In *Proc. of Towards Sustainable Product Design 6*, Amsterdam, 164-169.

van der Hoek, J.P., Dijkman, B.J., Terpstra, G.J., M. J. Uitzinger and M. R. B. van Dillen, (1999) Selection and evaluation of a new concept of water supply for Iljburg, Amsterdam. *Water Science and Technology* **39** (5), 33-40.

Lundin, M., S. Molander & G. M. Morrison (1999). A set of indicators for the assessment of temporal variations in the sustainability of sanitary systems. *Water Science and Technology* **39** (5), 235-242.

Lundin, M., Bengtsson, and Molander, S. (2000) Life cycle assessment of wastewater systems: Influence of system boundaries and scale on calculated environmental loads. *Environmental Science and Technology* **34** (1), 180-186.

Mels, A.R., van Nieuwenhuijzen, A.F., van der Graaf, J.H.J.M., Klapwijk, B., de Koning, J. and Rulkens, W.H. (1999). Sustainability criteria as a tool in the development of new sewage treatment methods. *Water Science and Technology* **39** (5), 243-250.

Moezzi, Mithra, (1998). The Predicament of Efficiency. *Proc. ACEEE Summer Study on Energy Efficiency in Buildings*, Energy Enduse Forecasting and Market Assessment Group at Lawrence Berkeley National Laboratory, August, 1998, http://enduse.lbl.gov/info/ACEEE-Pred.pdf.

Roeleveld, P.J., Klapwijk, A., Eggels, P.G., Rulkens, W.H. and van Starkenburg, W. (1997) Sustainability of municipal wastewater treatment, *Water Science and Technology* **35** (10), 221-228.

Sanne, C. (2000) Dealing with environmental savings in a dynamic economy, how to stop chasing your tail in the pursuit of sustainability. *Energy Policy* **28**, 487-495.

Simon, M. and Dixon, A. (2001) A comparison of ecodesign potential in UK manufacturing industry sectors. *Proc. of International conference on engineering design*, ICED 01 Glasgow, August 21-23.

Sombekke, H.D.M., Voorhoeve, D.K. and Hiemstra, P. (1997). Environmental impact assessment of groundwater treatment with nanofiltration. *Desalination* **113**, 293-296.

Tillman, A.-M., Lundström, H. And Svingby, M. (1998) Life cycle assessment of municipal waste water systems. *International Journal of LCA* **3** (3), 145-157.

Vaze, P. and Balchin, S., (1996) The Pilot United Kingdom Environmental Accounts, Office for National Statistics. *Economic Trends* issue 514, August.

6.6.1 Further information on life cycle assessment

Relevant websites (correct at time of publication):

- http://www.epa.gov/ORD/NRMRL/lcaccess
- www.pre.nl (Pre consultants and vendors of SimaPro LCA software.)
- http://www.eiolca.net is an interesting site which uses input-output modelling to predict how the US economy will react as a whole, sector by sector according to change in business in one sector.

Relevant journals include:

- International Journal of Life Cycle Assessment
- Industrial Ecology
- Journal of Cleaner Production

7

Developing a strategy for managing losses in water distribution networks

Stuart Trow and Malcolm Farley

7.1 INTRODUCTION

7.1.1 Understanding water losses

Globally, water demand is rising and resources are diminishing. Water loss from the pipe network, always the *bete noire* of the operations engineer, has long been a feature of operations management, even in countries with a well-developed infrastructure and good operating practices. However, it takes on a new dimension in developing countries, where a combination of poor infrastructure, poor sanitation, and intermittent supplies often poses a serious health risk. In water-scarce countries there is also a temptation to augment treated water with untreated in an attempt to satisfy demand.

Yet not all losses are the result of poor infrastructure and leaking pipes. *Apparent* losses from the network, and excessive use or misuse of water, are often the result of local customs, combined with low tariff structures or inadequate metering policies. These losses, and the overuse of water, can often be reduced by the introduction of demand management and water conservation programmes alongside initiatives to tackle leakage and improve the pipe network. Together, these programmes form a strategy for restoring a potentially huge lost resource.

The key to developing a water loss strategy is to gain a better understanding of the reasons for losses and the factors which influence them. Then techniques and procedures can be developed, and tailored to the specific characteristics of the network and local influencing factors, to tackle each of the causes in order of priority. Whilst there is no quick and easy answer to reducing losses, much can be learned from the recent experiences of the UK water industry. Urged on by the regulators to develop 'more robust' procedures for analysing and controlling losses, industry practitioners now have in place a number of techniques for understanding, measuring, and monitoring losses within the distribution network. Most water utilities have also introduced programmes to encourage customers to use less water in parallel with programmes to address real and apparent losses.

7.1.2 The IWA Water Loss Task Force

The International Water Association (IWA) Water Loss Task Forces have developed concepts and methodologies for quantifying, and making international comparisons of, the components of water loss. One such methodology is the *Water Balance* and the computer software which has been developed to support the calculation (see section 7.2.2). The Task Forces have also developed a performance measure, the *infrastructure leakage index* (ILI) (Hirner and Lambert, 2000; Lambert and McKenzie, 2002) for making international comparisons of water losses in a fair and consistent manner. The ILI takes into account a utility's operational practices as well as its performance in terms of the volume of losses. These parameters include pressure management, infrastructure condition, leakage activity and leak repair programme. The ILI, which is the ratio of the annual volume losses to the level of unavoidable losses present in all systems, provides a much more accurate measure of utility performance than the traditional method of expressing water losses and non-revenue water as a percentage of water supplied.

The IWA concepts for water loss management are fully explained in *Losses in Water Distribution Networks - A Practitioner's Guide to Assessment, Monitoring and Control* (Farley and Trow, 2003). This chapter sets out some of the IWA concepts and approaches to developing a strategy for leakage management.

7.2 UNDERSTANDING THE NETWORK

A diagnostic approach, followed by the practical implementation of solutions which are practicable and achievable, can be applied to any water company, anywhere in the world, to develop a water loss management strategy. However, consultants working in developing countries will invariably face a slower pace, greater financial constraints, less developed infrastructure, lower levels of skills and technology, and political, cultural and social influences. These all have an influence on the scope for managing losses and demand, and affect the pace at which changes can be brought about. But the aim should always be to make some improvement to the current operational practice, by transferring information, new skills, and motivation – to always leave something behind besides a report. There are many international case studies (Farley and Trow, 2003) which exemplify the pitfalls, but also the successes, of introducing a strategy in developing countries as well as in those which have all the benefits of a well-developed infrastructure and sound operating policies. The UK National Leakage Initiative (1991-1994) (WSA/WCA, 1994) provided a model for the concept of understanding losses and developing solutions and strategies for managing water loss, and the techniques for supporting it.

7.2.1 Prioritising the tasks

The first step in developing a water loss strategy for a water utility, as in any manufacturing process, is to pose a few simple questions about the system characteristics, the production process, and the operating practices - then use the available tools and mechanisms to suggest appropriate solutions. Typical questions are:

- How much water is being lost?
- Where is it being lost from?
- Why is it being lost?
- What strategies can be introduced to reduce losses and improve performance?
- How can we maintain the strategy and sustain the achievements gained?

Table 7.1 summarises the tasks required to address each question.

Table 7.1 Tasks and tools for developing a strategy

QUESTION/SOLUTION	TASK
How much water is being lost? - Measure components	WATER BALANCE - Improved estimation/measurement - techniques - Meter calibration policy - Meter checks - Identify improvements to recording - procedures
Where is it being lost from? - Quantify leakage - Quantify apparent losses	NETWORK AUDIT - Leakage studies (reservoirs, transmission mains, distribution network) - Operational/customer investigations
Why is it being lost? - Conduct network and operational audit	REVIEW NETWORK PRACTICES - Investigate: - historical reasons - poor practice - quality management - procedure - poor materials/infrastructure - local/political influences - cultural/social/financial
How to improve performance? - Upgrade the network - Design a strategy and action plans	STRATEGY DEVELOPMENT - Update records systems - Introduce zoning - Introduce leakage monitoring - Address causes of apparent losses - Initiate leak detection/repair policy - Design short/medium/long term action plans
How to sustain performance? - Ensure sustainability with appropriate staffing and organisational structures	TRAINING AND O&M - improve awareness - increase motivation - transfer skills - introduce best practice/technology - community involvement - Water conservation/demand management programmes - monitor action plan recommendations - introduce O&M procedures

Table 7.2 The water balance and its components

<table>
<tr>
<td rowspan="11" style="writing-mode: vertical-lr; text-orientation: sideways;">System input volume (corrected for known errors)</td>
<td rowspan="4">Authorised consumption</td>
<td rowspan="2">Billed authorised consumption</td>
<td>Billed metered consumption (including water exported)</td>
<td rowspan="2">Revenue water</td>
</tr>
<tr>
<td>Billed un-metered consumption</td>
</tr>
<tr>
<td rowspan="2">Unbilled authorised consumption</td>
<td>Unbilled metered consumption</td>
<td rowspan="9">Non-Revenue Water (NRW)</td>
</tr>
<tr>
<td>Unbilled un-metered consumption</td>
</tr>
<tr>
<td rowspan="7">Water losses</td>
<td rowspan="2">Apparent losses</td>
<td>Unauthorised consumption</td>
</tr>
<tr>
<td>Customer metering inaccuracies</td>
</tr>
<tr>
<td rowspan="5">Real losses</td>
<td>Leakage on transmission and/or distribution mains</td>
</tr>
<tr>
<td>Leakage and overflows at utility's storage tanks</td>
</tr>
<tr>
<td>Leakage on service connections up to point of customer metering</td>
</tr>
</table>

7.2.2 The water balance

The first two questions, 'how much?' and 'where from?' can be answered by conducting a water balance. Table 7.2 shows the components of water lost from the network, which is the difference between system input volume and authorised consumption. Water losses consist of real losses and apparent losses. Real losses are the annual volumes lost through all types of leaks, bursts and overflows on mains, service reservoirs and service connections, up to the point of customer metering. They can be severe, and may go undetected for months or even years, the volume lost depending largely on the characteristics of the pipe network and the leak detection and repair policy practised by the company:

Apparent losses consist of unauthorised consumption and all types of metering inaccuracies.

Abbreviated definitions of principal components of the IWA water balance are as follows:

- System Input Volume is the annual volume input to that part of the water supply system
- Authorised Consumption is the annual volume of metered and/or non-metered water taken by registered customers, the water supplier and others who are implicitly or explicitly authorised to do so.
- Non-Revenue Water (NRW) is the difference between System Input Volume and Billed Authorised Consumption.

- Water Losses is the difference between System Input Volume and Authorised Consumption, and consists of Apparent Losses and Real Losses
- Apparent Losses consists of Unauthorised Consumption and all types of metering inaccuracies
- Real Losses are the annual volumes lost through all types of leaks, bursts and overflows on mains, service reservoirs and service connections, up to the point of customer metering.

The components of the water balance should always be calculated as volumes before any attempt is made to calculate performance indicators. The separation of Non-Revenue Water into components – Unbilled Authorised Consumption, Apparent Losses and Real Losses – should always be attempted. Software is available, based on component analysis, to help the practitioner measure and/or estimate each component of water balance (Liemberger and McKenzie, 2003).

Because of the wide diversity of formats and definitions used for water balance calculations internationally (often within the same country), there has been an urgent need for a common international terminology. Drawing on the best practice from many countries, IWA Task Forces on Water Losses and Performance Indicators recently produced an international 'best practice' standard approach (Farley and Trow, 2003) for water balance calculations with definitions of all terms involved.

7.2.3 Network review

The third question - 'why is it being lost?' - can be addressed by carrying out a review of the network management and operating practices. Again, like an audit of any manufacturing process, this should identify policies and procedures which need changing or improving, as well as those which are being done well.

Having identified the 'how? where? and why?' of water losses the utility can then begin to address the last two questions - 'what strategies and policies can be introduced to reduce losses and improve performance' and 'how can the strategies be sustained'? This entails not only introducing equipment for measurement and monitoring flows, leakage control equipment and leak repair policies, but also education and awareness programmes and a fully operational O&M policy.

The water balance and network review should help in tackling priority issues. First of all, how significant are the apparent losses? Is meter accuracy (supply and customer) an issue - is there significant meter error to justify a meter calibration or checking programme? How significant is unauthorised consumption or theft? Or is the big issue leakage? If so, dividing the supply zone into smaller manageable *district metered areas* (DMAs) is the likely solution. These areas can be used to monitor the increase in bursts and leaks, directing the operator to work in the most cost effective part of the network (see Section 7.4).

7.3 DEALING WITH REAL LOSSES (LEAKAGE)

Water loss occurs in all distribution systems - only the volume of loss varies. This depends on the characteristics of the pipe network and other local factors, the water company's operational practice, and the level of technology and expertise applied to controlling it. The volume lost varies widely from country to country, and between regions of each country. The components of water loss, and their relative significance, also vary between countries. One of the cornerstones of a water loss strategy is therefore to understand the relative significance of each of the components, ensuring that each is measured or estimated as accurately as possible, so that priorities can be set via a series of action plans.

Real losses comprise leakage from pipes, joints and fittings, from leakage through service reservoir floors and walls, and from reservoir overflows. Real losses can be severe, and may go undetected for months or even years. The volume lost will depend largely on the characteristics of the pipe network and the leak detection and repair policy practised by the company:

- the pressure in the network
- the frequency and typical flow rates of new leaks and bursts
- the proportions of new leaks which are 'reported'
- the "awareness" time (how quickly the loss is noticed)
- the "location" time (how quickly each new leak is located)
- the repair time (how quickly it is repaired or shut off)
- the level of "background" leakage (undetectable small leaks)

7.3.1 Developing a leakage strategy

The methodology which should be followed for setting leakage targets and for deriving an appropriate strategy for managing leakage is dependent on the operating practices and local characteristics of individual networks. Figure 7.1 illustrates a generalised leakage management strategy, and the following sections highlight some of the steps to follow. These apply to any water service provider.

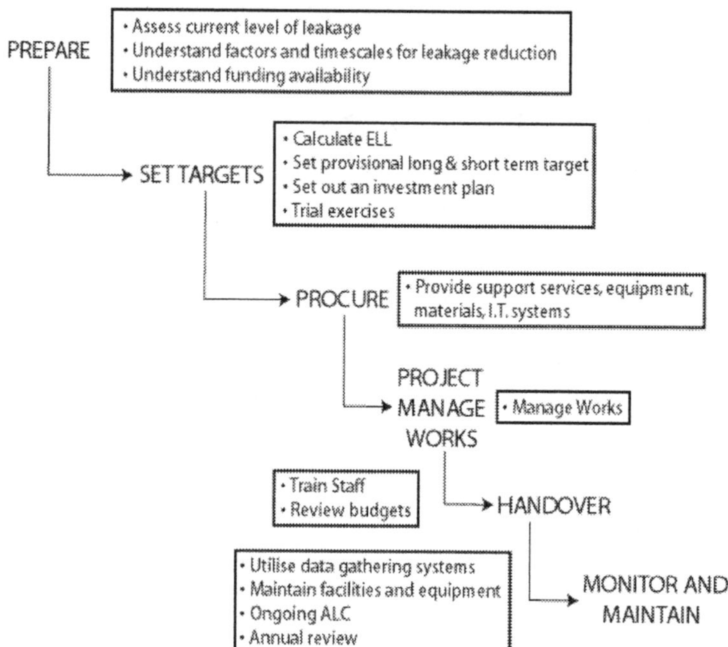

Figure 7.1 – Steps in developing a leakage management strategy.

7.3.1.1 Establishing funding requirements

An appropriate leakage reduction plan can only be set with due regard to the funding requirements. Even if leakage targets can be shown to be economic, the work has to be funded up front in order to achieve a pay back over a longer term – in some cases over 20 years. Funding could come from raising charges to customers, from government sources, from international grants and loans, or by accepting a lower level of profit during the leakage reduction work. It is important for the sponsors of the leakage programme to involve the water suppliers' financial managers at a very early stage, as resolving the funding issues could take just as much time and effort as the technical aspects of the strategy

7.3.1.2 Reviewing the organisation structure

Where a major leakage reduction programme is planned it will be not be efficient or fully effective unless the organisation structure is reviewed and changed to take account of the new demands placed upon it. Whenever the workload warrants the appointment of a leakage manager, this is the preferred course of action. The leakage manager should then have responsibility for delivering the leakage reduction programme along the strategy agreed by the directors. It is essential that the leakage manager be given the support of the directors, and that if possible the same person remains in post during the leakage reduction stage. That person, who should be committed to the project, will become a focal point for leakage management activity and will be able to coordinate the various aspects of the programme.

There will inevitably be times when the programme is not going according to plan. During such periods it is important to differentiate between two stages of leakage management:

- Leakage reduction down towards a set target. This stage should be regarded as a project involving capital works, and other transitional costs, which can be managed as a project in a similar way to building a new treatment works. The project should be managed by someone with good project management skills, who does not necessarily have to be a leakage expert. However, the team should incorporate experts or external advisors, and also involve operational staff who will be responsible for maintaining leakage once it is reduced.
- Maintaining leakage at the target level. This stage should be incorporated into the ongoing management of the water supply organization, and managed in a similar way to the operation of the treatment works. During this stage there is less need for a dedicated leakage manager, but if the organisation is sufficiently large there is a case for a leakage coordinator to ensure consistency of approach across the region.

These two stages may run concurrently in the same organisation. A supply zone will be handed over from operational staff to the project management staff, and once the leakage reduction work is complete it will be handed back. Different zones will be at different stages during the overall project until work has been completed across the region.

7.3.1.3 Setting targets

The most important aspect of any leakage strategy is the leakage target. What level of leakage should the water supplier aim for, and what level should be maintained in the longer term? Leakage is synonymous with waste, therefore a company which wastes the product it is meant to supply must come under scrutiny to investigate what else it wastes, for example customer income. This in turn can lead to setting the price of the product in excess of what is reasonable.

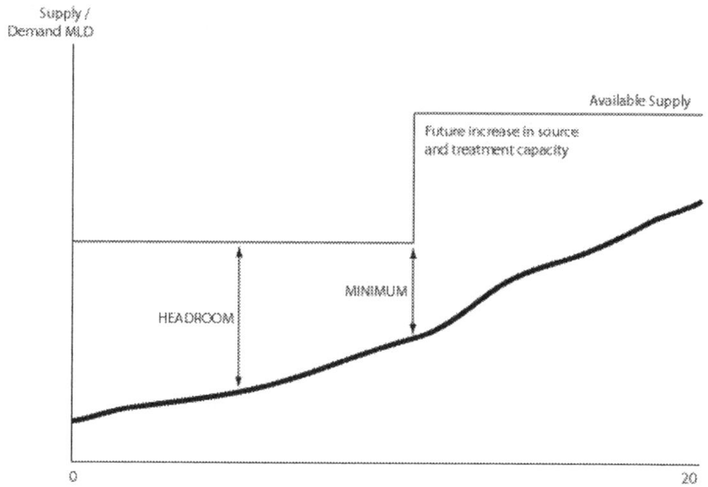

Figure 7.2 The effect of increasing demand and supply on available 'headroom'.

Every water supplier would like to eliminate leakage from water distribution systems. It serves no purpose, is wasteful, and has a deleterious effect on several aspects of the operation of a well-run water supply network. Leakage will add to the cost of producing and distributing water. It will add to the capacity requirement for storage systems, treatment works, and mains sizing. However, for the vast majority of water distribution systems, leakage is something which cannot be eliminated completely. *There will always be a level of leakage which has to be tolerated, and which has to be managed.*

When considering alternative ways of bridging the gap between the future need for water into an area, and the current availability of water, there are usually two principal methods:

- Supply augmentation. This may mean adding reservoir or pumping capacity, increasing treatment capacity, bringing in water from an adjacent area
- Reducing the future need for water by leakage reduction and demand management.

Each of these methods will increase the forecast 'headroom' i.e. the difference between the available supply and the projected demand, and therefore reduce the risk that supplies will become insufficient to meet customer demand. This is shown in Figure 7.2. However, when comparing the economics of the two methods there is a general principle which should be appreciated.

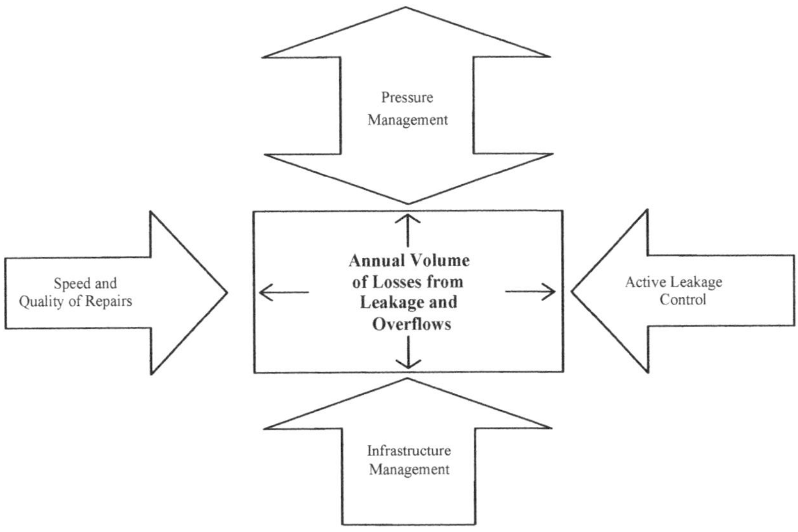

Figure 7.3 The Four Pillars of Leakage Management.

Reducing leakage costs money. However, unlike supply augmentation, there is less scope for economies of scale. In fact the reverse tends to be true. Everything to do with leakage reduction follows a law of diminishing returns; the more effort that is put in, the less will be the impact in terms of water saved. Figure 7.3 shows the prime techniques for leakage management, which have been referred to as the four 'pillars'. Each of these follows a similar law of diminishing returns (Figure 7.4).

Active Leakage Control

When first undertaking leakage detection and repair work, leaks will be relatively easy to find. A backlog may have built up due to under investment in previous years resulting in fewer leaks being found and fixed than occur in any given year. However, once the more obvious mains and service bursts have been found, then a higher level of effort has to be put in to reduce leakage by a similar volume.

Pressure Management

The most cost effective schemes are those which cover a large area, and which make a significant impact on average pressures. An example would be the installation of a pressure reducing valve on the branch from a trunk main to cover a whole town. Once such schemes have been completed, the next stage may be to install PRVs in conjunction with district metering. In the extreme, some water suppliers have installed PRVs on supplies to districts of less than 200 properties, or even on individual properties. The cost of the scheme reduces much less than the benefit obtained due to the reduction in the area covered, and so the schemes become less cost effective.

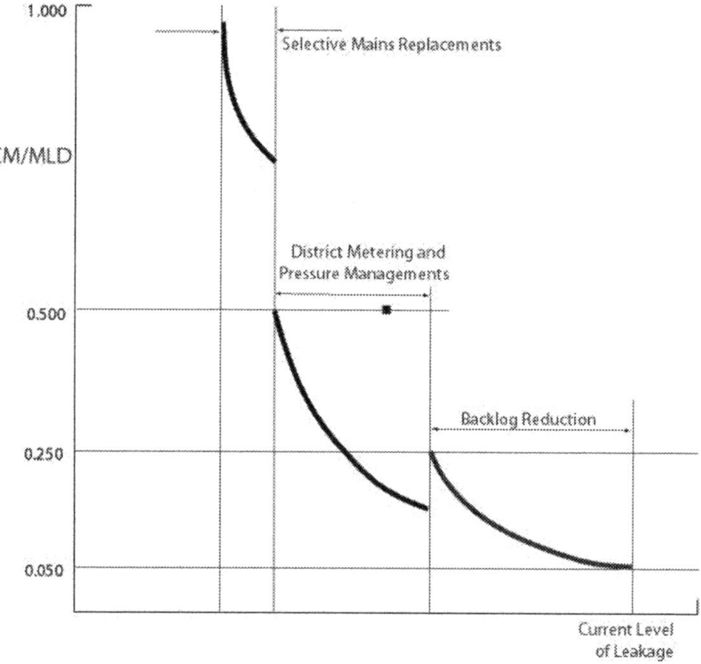

Figure 7.4 Diminishing returns from leakage management measures.

District metering

When installing zonal and district metering there is a tendency to favour areas which can be metered without the need for additional facilities to be installed in the network. Areas will be created using natural breaks in the network, along the lines of main roads, rivers and canals, and over undeveloped land.

The aim is to provide single feed district meter areas (DMAs) which are supplied through only one meter installation. This will tend to minimise the number of valves which have to be shut in order to create a discrete area. The number of areas which can be created in this way will depend on the layout of the distribution network. The cost and the benefit will be similar. As the number of properties contained within DMAs increases, there will come a point at which adding further properties incurs a higher unit cost.

Mains and service renewal

Replacing an old water main with a new installation will reduce leakage on the main. If water mains are being replaced for reasons other than leakage control, for example water quality problems, then the benefit to the leakage engineer should also be taken into account. When mains replacement is being used as a primary measure for leakage reduction, targeting studies should be carried out to determine which areas, and which mains within those areas, have the highest burst frequency (number per kilometre per year) and which have the highest levels of background leakage. If targeting is carried out effectively, it is inevitable that the initial schemes will be more cost effective than the later ones. So, mains replacement will also follow a law of diminishing returns.

Speed of repairs

Reducing the time it takes to repair a leak will reduce the volume of leakage. However, once the repair time is reduced below a certain threshold, the unit repair cost will tend to rise due to standby, call out and overtime payments to staff, or supplementary payments to contractors to make additional repair teams available.

7.3.1.4 Economic level of leakage

For any water distribution system there is a level of leakage below which it is not cost effective to make further investment, or use additional resources, to drive leakage down further. In other words, the value of the water saved is less than the cost of making the further reduction. This point is known as the *economic level of leakage* (ELL). Leakage targets based on ELL must therefore be specific and dynamic.

Figure 7.5 shows the generalised relationship between expenditure on leakage management operations, and the unit production costs of water as a function of the level of losses. The key to a successful strategy is to collect sufficient factual data to allow this relationship to be understood for each supply zone.

In order to understand the estimation of ELL, it is necessary to appreciate how water is valued. This will vary from one region to another, and also within areas of the same region.

Short term ELL

In the short term there are a number of key parameters which govern the actual level of leakage. These are effectively fixed, and include:
- The average pressure in the system
- The condition of the mains and service pipes
- The facilities available for collecting data (i.e. district metering and telemetry).

So, at any particular time, the only parameter which can be changed quickly to have an impact on the level of leakage, is the number of personnel out looking for leaks and then repairing them. Leak location and repair is sometimes called *active leakage control* (ALC). There is a steady state situation in which the marginal cost of the ALC effort is equal to the marginal cost of the water saved by adopting that ALC policy.

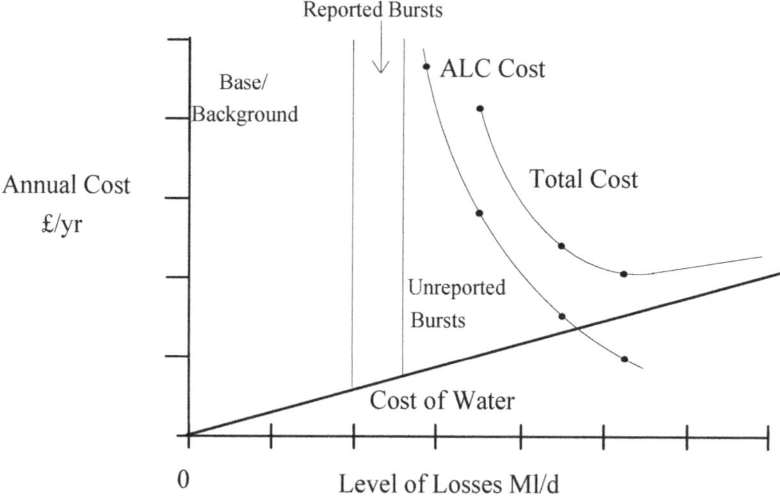

Figure 7.5 General relationship between operating costs and the level of losses.

Long Term ELL

In the longer term, investment in facilities such as district metering, telemetry, pressure management and mains renewal will have an impact on the short term ELL. The reduction in the short term ELL and consequently the savings and costs associated with the change can be compared to the investment cost of making the change. Investment costs are sometimes called transitional costs, i.e. they represent the cost of making the transition from one steady state to another.

Short term ELL is based on an economic analysis which estimates the optimum level of ALC effort taking account of the costs of ALC and the short term value of the water in the supply zone. Long term ELL is based on a form of investment analysis taking account of the following questions:
- What is the current level of leakage?
- What is the short term ELL?

- How will the short term ELL change with the investment under consideration?
- What is the saving in water losses and the change in ALC resources from the proposed investment compared with the current policy?
- What is the cost of the proposed investment?
- What is the return on the investment?

The answers to these questions will allow the water supplier to decide an investment policy using normal investment decision criteria.

7.3.1.5 Setting provisional leakage targets

Setting leakage targets goes beyond the stage of calculating ELL. Targets must be specific to a particular supply zone, and the overall target for the water supplier should then be the aggregation of each zonal target. Different procedures based on the analysis of ELL will apply in each zone.

Figure 7.6 illustrates the influencing factors. External influences can also significantly affect the target for leakage compared to the ELL. Such factors include:

- Comparisons with similar water suppliers. It is inevitable that water suppliers will compare their current leakage level and their calculated ELL with that of other water suppliers working in the same country or within the same geographical region of that country
- International comparisons
- Political influences. The true ELL for a company with plentiful supplies and a poor infrastructure may equate for instance to 35% of total demand. However, the company may come under pressure from its customers or from government or peer pressure to reduce leakage because it is perceived to be too high.

The most significant external factor affecting leakage targets is the impact of regulation on the water supplier. The nature of the regulation will depend on the general control and ownership of the water supply organization in a particular country. Regulation creates expectations from several parties which have to be managed:

- Customers expect the water supplier to control the prices they pay for their supply
- Government economic regulators also expect operating and investment costs to be justified
- Shareholders expect the water supplier to be managed efficiently and to produce a return on investment

- Environmental regulators and pressure groups seek to avoid further abstractions of raw water which deplete stream lakes and rivers, and the protest against the building of dams
- Drinking water inspectors protect supplies, and seek to control any works which may have an adverse effect on water quality.
- National governments aim to manage water supplies generally to ensure there is a general sufficiency for public health and for economic development.

Although targets may not be mandatory, there may be a. requirement for leakage data to be submitted annually or at other regular intervals. If targets are set internally, without any external influence, there is a need to understand why targets have been set, i.e

- is there a current water shortage generally, or in a particular region, or is a shortage forecast in the near future?
- have targets been set to bring the supplier into line with other similar orgainisations, or has the country as a whole become embarrassed about high leakage levels making it a political issue?

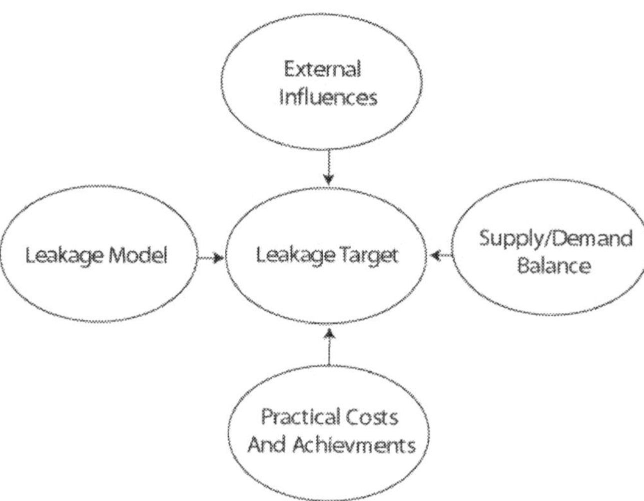

Figure 7.6 Factors influencing the target for leakage and losses.

Whether targets have been set on external or internal influences, they will have a time-scale attached to them. A general target may be to reduce leakage by 20% within 5 years, or it may be to achieve a value for leakage in litres per property per day by a given year, that year having some other significance. Time-scales must be realistic.

There is no easy solution to controlling leakage, even if substantial finance is made available for a one-off exercise within a constrained timescale. Leakage management is a painstaking process, and projects which aim to make a substantial reduction in leakage in a short space of time, are unlikely to succeed in the long term. Attempting to achieve too much too soon will lead to inefficiencies. There is a natural pace to a leakage reduction plan which varies from one region to another depending on a number of different parameters.

Where no particular internal or external influence exists and there is simply a general desire to reduce leakage down to an economic level, then it is recommended that a provisional long term target should be established as a goal to work towards. This goal should be ambitious, but it should not be impossible to achieve. The short term target should be set with reference to the long term target. A reasonable approach is to aim to achieve between 50% and 80% of the long term reduction within say 5 years. This time-scale is chosen as a reasonable one in which to set up the necessary facilities, to award contracts, and to carry out the initial works. A time-scale of between 4 and 7 years is reasonable; any less is too ambitious, and any more will not be as economic. The costs of leakage reduction will still be the same, but there is the cost of supplying the excess leakage after year 7.

Leakage will tend to be reduced over time following the "S" curve shape. The early leakage management expenditure may show little return on the investment, and there is a risk that staff will lose faith. However, if the effort is maintained (usually after 1 to 2 years), then leakage will begin to reduce quickly giving significant reductions in years 3 and 4. In the final stages, the rate of reduction will slow down due to the laws of diminishing returns explained earlier.

7.3.1.6 Setting up procedures to collect data

Before any major expenditure is made on leakage reduction, it is very important that procedures should be established to collect all relevant data. Data collection should be established in a hierarchical way, so that by aggregating the data for DMAs within a supply zone, it is possible to provide average data for that zone. Similarly, supply zone data should aggregate to the company average values. In some cases it is necessary to create un-metered DMAs - these are areas which are not actually covered by a district meter, but for which the water supplier does have relevant data (e.g. number of properties and length of mains in the area) to make an estimate of leakage by other means. They usually have a defined boundary.

Leakage management is data hungry, and so investment in software systems and staff resources to manage the data can be significant. However, without such systems, there is a risk that investments will be targeted ineffectively and leak detection and repair operations will be carried out inefficiently. Therefore, although the cost of establishing good data management systems represents a high initial outlay, there will be a return on that investment in the longer run.

7.3.1.7 Establishing trial exercises

These demonstrate benefits in a small area before introducing them company-wide. Whenever possible, demonstration exercises should be undertaken to show the techniques and also to deliver some early benefits in the leakage reduction programme. The exercises should not be regarded as separate from the mainstream programme. They should be planned as an integral part of it and carried out by the same team which will deliver the main body of work.

Pilot exercises may focus on a particular aspect of the leakage management programme, e.g. pressure control, or they may be all embracing to test the integrated strategy in a limited geographic area, e.g. a particular supply zone.

7.3.1.8 Importance of company specific data

Whichever method is used to model leakage, or to derive a strategy, there is an inevitable dilemma. The strategy will be most important when the water supplier first embarks on a major leakage reduction exercise. To be accurate and reliable, the leakage made and the investment plan should be based on good quality data specific to that organization, and a well thought out investment programme in which there is a degree of confidence that the expenditure planned will give the required benefits. However, if little or no work has been undertaken, then there will be no specific data. No data means that use has to be made of default values and assumptions and so there is less confidence in the modelled results.

Leakage modelling has been carried out over a number of years now, and the BABE concepts have been used in several countries around the world, with very different operating environments, and different policies, levels of service and configurations of their water distribution networks. As such, the original UK data have been enhanced, and the effect of these different global situations has resulted in a wealth of new data.

The solution is to take gradual steps (Figure 7.7), using data from each step to make changes to the model and to the strategy based on the lessons learnt. It is almost impossible to set out an effective leakage reduction plan and to stay with it from start to finish.

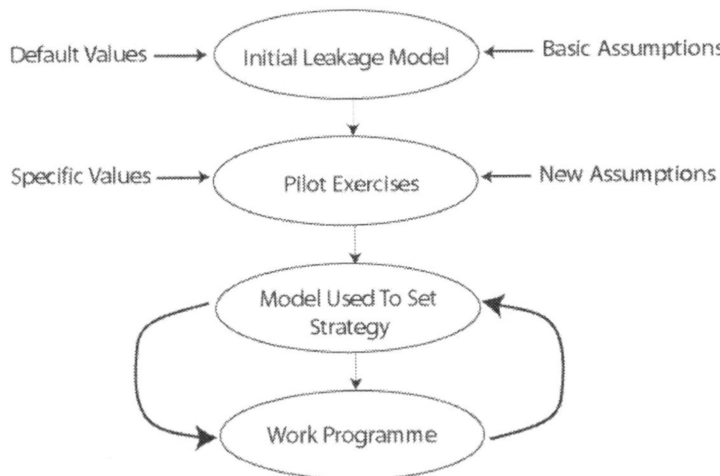

Figure 7.7 Using specific data to replace default values and assumptions.

7.3.1.9 Introducing the strategy

Perhaps the most important aspect of the leakage management strategy is that it is understood and followed by all parts of the water supply organization. There is a need to gain the commitment and cooperation of staff in a number of different departments, and in regional offices and depots of a large diverse organization. Effective leakage management requires an input from a number of different personnel, and unless they are all committed, the implementation of the programme will not be efficient, and it may then be difficult to maintain the infrastructure which has lead to the lower leakage levels.

As an integral part of the strategy, careful consideration should be given to:
- A launch event such as a major seminar.
- Education and training of all staff, not just those directly involved with delivering the leakage management programme.
- A 'carrot and stick' approach has been found to be successful. Staff responsible for the ongoing operation of the water distribution system, are rewarded for proper maintenance of the leakage management infrastructure, whereas disciplinary action is taken against those who wilfully neglect to take account of the need to keep these systems in place.
- Public relations.

7.4 LEAKAGE MANAGEMENT

The leakage reduction works will comprise the four pillars of leakage management shown in Figure 7.3.

7.4.1 Active leakage control (ALC)

Leakage management can be classified into two groups - passive control and active control.

Passive (reactive) leakage control

Passive control is reacting to reported bursts or a drop in pressure, usually reported by customers or noted by the company's own staff. The method can be justified in areas with plentiful or low cost supplies. Often practised in less developed supply systems where the occurrence of underground leakage is less well understood, it is the first step to improvement (i.e. to make sure all visible leaks are repaired)

Active leakage control (ALC)

The main methods of ALC are:
- regular survey
- leakage monitoring

7.4.1.1 Regular survey

This is a method of starting at one end of the distribution system and proceeding to the other using one of the following techniques;
- listening for leaks on pipework and fittings
- reading metered flows into temporarily-zoned areas to identify high-volume night flows
- using clusters of noise loggers

7.4.1.2 Leakage monitoring

This technique requires flows into zones to be monitored to measure leakage and to prioritise leak detection activities. Leakage monitoring requires the installation of flow meters at strategic points throughout the distribution system, each meter recording flows into a discrete district which has a defined and permanent boundary. Such a district is called a District Meter Area (DMA). This technique has now become one of the most cost-effective activities (and the one most widely practised) for leakage management. Figure 7.8 shows a typical DMA.

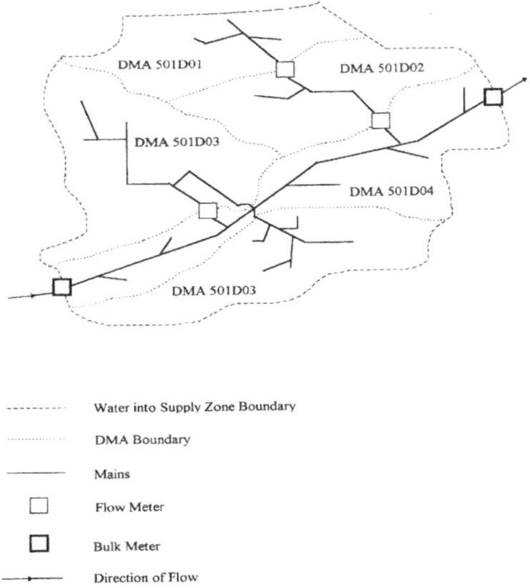

Figure 7.8 A typical DMA.

The duration of water being lost can be kept to a minimum by analysing the DMA flow data so that the operator is aware as early as possible that bursts or an accumulation of leaks have occurred. The total volume of water lost is:

rate of flow x the length of time over which the break/burst has occurred

So the quicker the operator can analyse DMA flow data the quicker he/she can act to identify the area containing the burst or leak and locate it. This, together with speedy repair, limits the total volume of water lost.

7.4.1.3 Analysis of DMA flows to estimate leakage

Best practice analysis of DMA flows requires the estimation of leakage when the flow into the DMA is at its minimum - this typically occurs at night when customer demand is also at its minimum and therefore the leakage component is at its largest percentage of the flow.

Techniques based on component analysis - e.g Bursts and Background Estimates (BABE) - are available to analyse the minimum night flow and estimate the level of leakage and the relative volumes of background and burst volumes (Figure 7.9).

The analysis of leakage is based on the minimum night flow, which can be recorded and analysed continuously night after night with the use of data loggers and appropriate software. This analysis enables the leakage practitioner to monitor the DMA or groups of DMAs for the occurrence of new bursts and to activate repair (Figure 7.10).

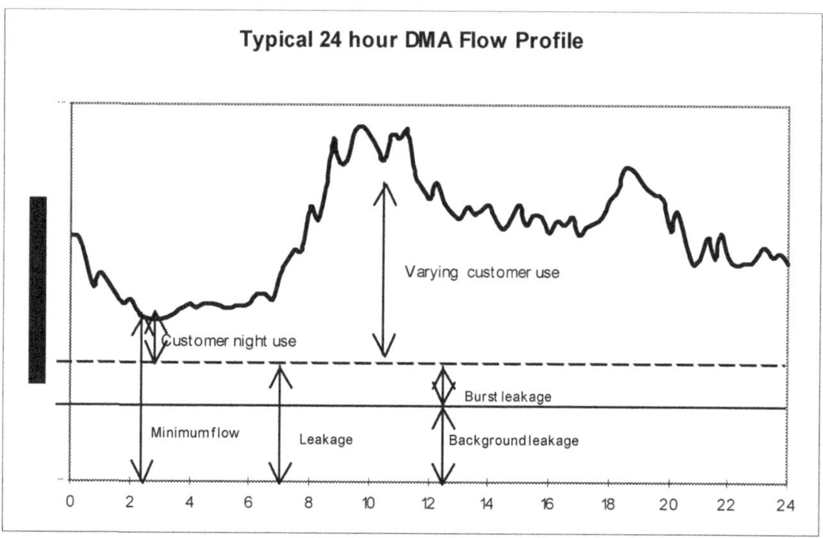

Figure 7.9 A typical 24 hour flow profile of the components of leakage and customer use.

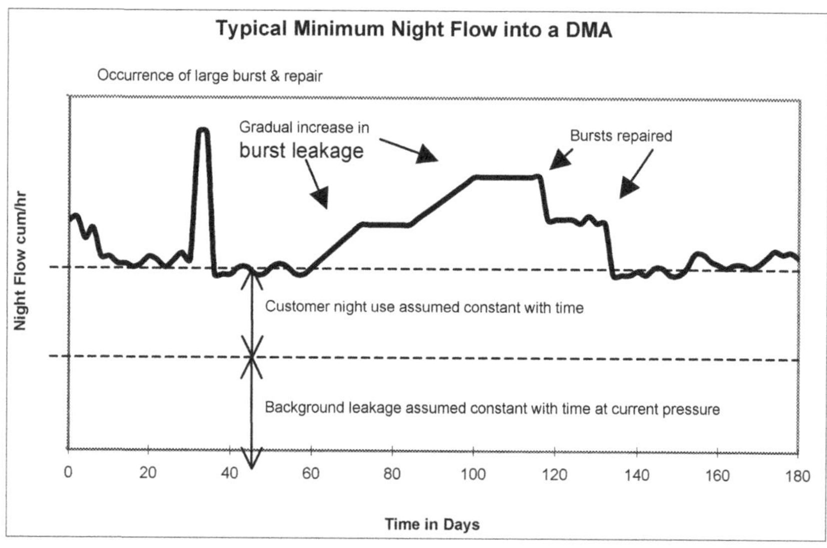

Figure 7.10 A typical DMA night flow profile.

This calculation, based on an aggregate of night flow data converted to annual leakage, is a 'bottom up' approach. The calculation can be confirmed by a 'top down' assessment of leakage. This analysis requires an assessment of customer use, which is subtracted from the total water put into supply, to derive leakage. In most instances this leakage volume, usually measured over a 12 month period, is compared with the aggregate of leakage from DMAs in the same area.

7.4.1.4 Choosing a policy

The most appropriate leakage control policy will be mainly dictated by the characteristics of the network and local conditions, which may include financial constraints on equipment and other resources. Staffing resources are relevant, as a labour intensive methodology may be suitable if manpower is plentiful and cheap. Where leaks fail to appear at the surface, however, a more intensive policy of leakage monitoring is required.

The main factor governing choice, however, is the value of the water, which determines whether a particular methodology is economic for the savings achieved. A low activity method, such as repair of visible leaks only, may be cost-effective in supply areas where water is plentiful and cheap to produce. On the other hand, countries which have a high cost of production and supply, like the Gulf States, can justify a much higher level of activity, like continual flow monitoring, or even telemetry systems, to warn of a burst or leakage occurring. In most developing countries the method of leakage control is usually passive, or low activity, mending only visible leaks or conducting regular surveys of the network with acoustic or electronic apparatus.

7.4.1.5 Managing DMAs

The volume lost from a leak is a combination of the awareness time and the time taken for location and repair:
- Awareness time – the average time from the start of a leak until the water company becomes aware of its existence
- Location time – the average time taken to locate the position of the leak
- Repair time - the average time for the company to shut off and repair the leak.

The main effect of an ALC policy is reducing the average *duration* of leaks, although repair times are unaffected by the choice of an active or passive policy. The awareness time is influenced by the data capture method:
- Telemetered flows – less than 1 day
- Monthly night flow measurements – 14 days
- Regular inspections – half the interval between inspections

The location time will be influenced by the nature and extent of monitoring systems, but mainly by the number of staff available and the equipment and technology at their disposal.

The technique of leakage monitoring is considered to be the major contributor to cost-effective and efficient leakage management. It is a methodology which can be applied to all networks. Even in systems with supply deficiencies leakage monitoring zones can be introduced gradually. One zone at a time is created and leaks detected and repaired, before moving on to create the next zone. This systematic approach gradually improves the hydraulic characteristics of the network and improves supply.

There is a clear distinction between leak detection and leak location. Detection is the 'narrowing down' of a leak or leaks to a section of the pipe network. Leak detection activities may be carried out routinely, i.e. as a 'blanket' survey of the network, or in precise areas of the network, guided by the analysis of DMA data. Leak location is the identification of the position of a leak prior to excavation and repair, although finding the exact location cannot be guaranteed. Location surveys can be carried out with or without prior leak detection activity.

There are a number of techniques to detect where leakage is taking place in the network. These include:

- sub-dividing DMAs into smaller areas by temporarily closing valves or by installing meters (see Figure 7.11)
- variations of the traditional step-test (valve isolation technique)
- the use of leak localisers
- sounding surveys

Technology and equipment for leak detection and location is discussed in more detail in Section 7.5.

There are two fundamental issues in managing DMAs:
- How often should areas be revisited?
- How much effort should be expended in each area before moving on to the next area?

The time period between inspection of a particular district is known as the intervention interval. The average intervention interval is a key element of the overall strategy. It should be set using economic considerations. It will have a major impact on the level of resources required

Many aspects of leakage management follow laws of diminishing returns. This is also the case here. Once the initial leaks and bursts have been found, from what is often known as the first pass survey, then the more time which is spent in a district, the lower will be the return in terms of the number and quality of bursts which are found.

If the area in question is not covered by district metering, it may still be possible to apply the same principles at a supply zone level. The larger area will mean it is difficult to identify individual leaks, and it will make leak location more difficult. However, a balance has to be struck between the investment cost of the district metering, and the ongoing cost of meter reading, and the savings which result from a more efficient operation. These areas are subject to what is known as a policy of "regular sounding", or regular survey. The frequency of the survey can be determined on economic principles and the strategy will include the number of staff required to maintain this inspection level. As it is difficult to estimate leakage in these areas they tend to be surveyed in rotation rather than on some form of prioritization. Economic frequencies of regular survey generally vary from a few weeks to about two years.

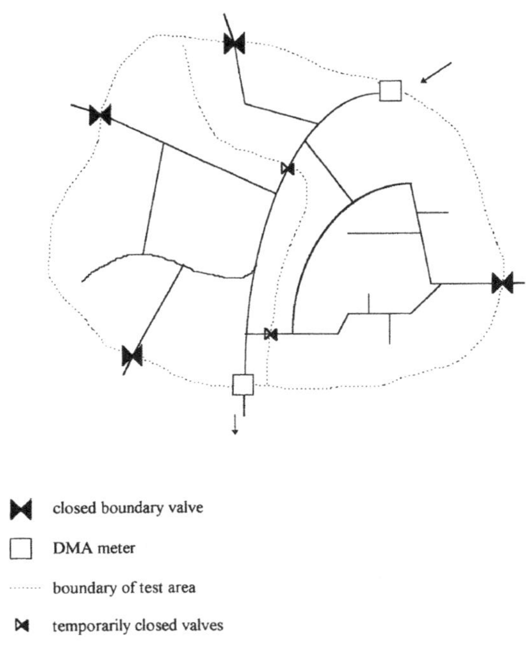

Figure 7.11 Dividing up a DMA.

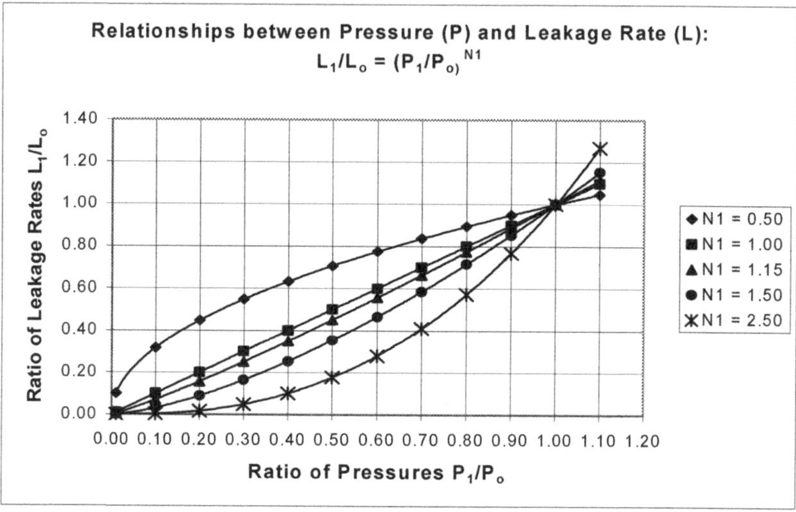

Figure 7.12 Relationship between pressure and leakage rate for different "N1" values. (N1 is determined from the proportion of fixed and expanding leakage paths in the distribution system).

7.4.2 Pressure management

The rate of leakage in water distribution systems is a function of the pressure applied by pumps or by gravity head. There is a physical relationship between leakage flow rate and pressure, which has been proven by laboratory tests and by tests on underground systems. Burst rates are also a function of pressure. The strength of the relationship, and the quantification of it, is not as well understood as the relationship between flow rate and pressure. However, there is still considerable evidence to show that burst frequency is very sensitive to changes in pressure.

Pressure management is one of the fundamental elements of a well-organised leakage management strategy. It should be an integral part of the strategy because it impacts on several other aspects:

- If pressure is reduced, the rate of increase in leakage will reduce. Therefore, there is an impact on the level of leak detection resources required.
- The flow rate from all leakage paths (bursts and background leaks) will reduce, as shown in Figure 7.12.
- The data used to calculate leakage targets and economic levels of leakage should be revised when pressure management is introduced
- Reducing pressure may make existing leaks more difficult to find, because they make less noise, or do not come up to the surface.

- Reducing pressure can reduce some types of consumption. Any consumption from devices connected direct to mains pressure will give a reduced flow rate at reduced pressure.

There are several benefits of pressure management, and if it is designed and maintained well, there are few, if any, disadvantages. The benefits include:

- Reduction of frequency of bursts

Data from one UK water supply company (Figure 7.13) shows the reduction in burst frequency before and after pressure management was installed in an area. The data set is limited in size, but it indicates that a unit reduction in pressure will give a 3 or 4 times reduction in burst frequency e.g. reducing pressure from 80m to 40m (a 2:1 reduction) will reduce the burst rate from 7 bursts per 100 properties per year to only 1. Of course there are many other factors which affect the burst frequency of mains. Therefore, it is difficult to obtain good quality data to prove the strength of the relationship.

Burst frequencies will be more reliable in larger areas, e.g. supply zone, but at that scale it is more difficult to make significant changes in pressure. Therefore most data is available at DMA level, where the burst rate is more erratic, and so it may take several years to determine the true benefits.

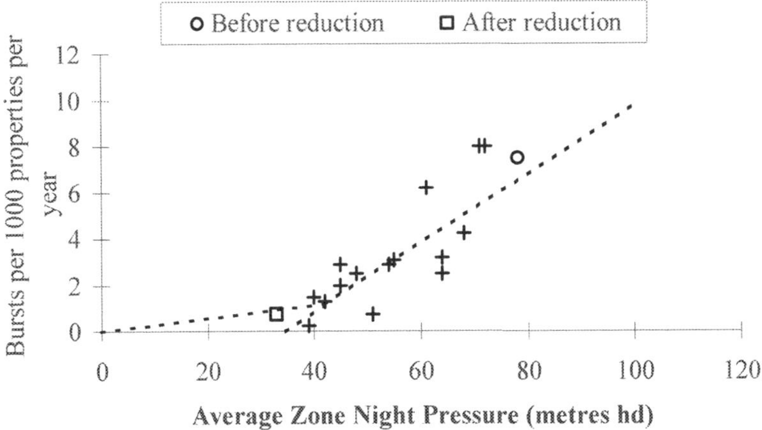

Figure 7.13 Relationship between Average Zone Night pressure (ANZP) and burst frequency for a sample of data from one UK water company.

- Provision of more constant supply to customers

Without pressure management, the pressure at customers' premises will be a function of the pressure of water where it enters the distribution system, less the head loss through the underground pipe network. Well organized pressure management regimes will result in a better understanding of the factors affecting pressure at customer premises, and will allow systems to be put in place to maintain pressures within specified bounds.

- Increased fire fighting capability

In a similar way, lack of pressure management may result in inadequate supplies from fire hydrants for fire fighting. Many water supply organizations avoid pressure management because they are concerned that it will reduce availability of fire fighting water, and result in disputes with the fire authorities. However, with modern technology and design techniques it is possible to reduce pressure (and therefore leakage) and also provide adequate supplies for fire fighting.

- Protection of the long-term life of the assets

Daily variations in pressure place stresses and strains on the pipe network. Damage can occur at joints, and fittings, and pressure splits can occur in some types of pipe. Damage may be due to a fatigue effect which takes place over a long period of time. The higher and more frequent the variation, the greater the chance of damage caused by pressure fluctuations. Pressure management will tend to smooth out these variations, resulting in less damage to the network, and a longer asset life expectancy.

7.4.3 Infrastructure management

The most significant factor affecting the level of leakage in a water network is the general condition of the mains and service pipes, and the service reservoirs. It is usually found that this is also the singularly most significant factor affecting the economic level of leakage for that network. The condition of the infrastructure is something which is inherited from previous regimes and generations, and it cannot be improved significantly without major capital investment in renewal and refurbishment works. It has been shown that expenditure on infrastructure improvement, even if it is targeted to the areas most prone to high leakage, is not a particularly cost effective method of managing leakage. Where improvements are being made for other purposes, such as the need to meet water quality parameters, or customer standards of service for interruptions to supply or minimum pressure standards, then the impact on leakage levels should be taken into account. However, where leakage is the primary problem then it is difficult to cost justify mains renewal works.

The condition of the infrastructure can be assessed in two ways:

- its propensity to burst. This will be governed by factors such as pressure and ground conditions and weather as well as the condition of the mains and service pipe fabric.
- Its propensity for background leakage. Again this is governed by pressure.

At first sight it would seem that these two parameters would be linked, and that mains in poor condition would be susceptible to both effects. However, it does not follow that a high burst rate means high background leakage and vice versa. Therefore a separate judgment has to be taken on the two parameters

Most water supply organisations regularly carry out work to renew or rehabilitate their water distribution networks. If they did not, the pipe network would continue to age and deteriorate, resulting in increasingly higher maintenance costs to carry out repairs, in order to maintain levels of service to customers. The primary justification for main renewal and rehabilitation is usually one of the following:

- The internal condition of the main is affecting the quality of the water delivered through it. This is often the case with corrosion of cast or ductile iron pipes, which have no internal protection.
- The internal bore of the main has reduced due to corrosion, or a build up of deposits, so that it is no longer capable of carrying sufficient flow
- The pipe wall has weakened and is no longer capable of withstanding the internal pressure of water, or it has insufficient beam strength to withstand traffic loading. This is often the case with asbestos cement pipes laid in aggressive ground.
- Some external factor has resulted in the main being unable to fulfil its current duty

It is unusual for mains to be replaced solely on the grounds of leakage reduction. Customer levels of service and operating costs are the primary drivers. In all cases, however, the impact on leakage should be assessed as part of the justification process for each section of main.

The impact on leakage will depend entirely on the extent to which the old main was contributing to the overall leakage level, and then on the technique chosen to rehabilitate the network. If mains are replaced with new ones, then leakage on the main will be reduced significantly, although not eliminated altogether. However, unless the service connections are also renewed, there could actually be an adverse effect due to increased pressure (due to the extra carrying capacity) causing leaks on services to flow at a higher rate. Mains relining can also result in higher leakage due to the scraping process causing damage to pipe joints, service connections, and the pipe wall. If the main is slip lined with a new plastic pipe, then this will not be a problem, but if the main is coated with cement mortar or epoxy resin, then the rehabilitated main may leak at a higher rate than before.

In overall terms the experience of the authors shows that unless main rehabilitation and renewal is targeted specifically to reduce leakage then it will have a neutral effect on leakage levels. The benefits gained in some projects, will be offset by the increased leakage caused by others, and in any event the small percentage of mains which are rehabilitated each year are statistically unlikely to be the source of the leakage problem.

It is often assumed that mains which are in poor condition, because they suffer from internal corrosion, or because they cause levels of service problems to customers, are also be the prime source of leakage. However, evidence from the UK suggests that this is not necessarily the case. Also, there is evidence to show that there is little or no correlation between burst frequency and background leakage. Areas with high burst frequency can have a low background leakage and vice versa. This may be due to the fact that high burst frequencies tend to occur in smaller diameter mains with low beam strength. Background leakage will tend to be more of a problem on larger diameter mains, and on service connections. Therefore, each section of main has to be assessed, and any policy based on generalisations is unlikely to be cost effective, and could actually produce little or no leakage reduction.

If main rehabilitation is to be part of the leakage management strategy then it should be targeted. The aim is to identify those mains which make the major contribution to leakage levels in a supply zone, and then to find the most appropriate technique to renew them. This investigation has a cost, and therefore a balance has to be found between the expenditure on analysis and design, and on the actual cost of the main replacement. If insufficient effort is made on the preparatory stage, then mains will be replaced with little benefit, whereas too much investigation will add to the overall cost unnecessarily.

The following steps should be followed.

1.   Identify those mains clearly in need of replacement
The first step is to examine records of mains failures and to consult with the local operations staff to identify those sections of main with a history of bursts and leaks, where repairs are carried out regularly. There will be a break even point depending on the value of the water, the frequency of bursts, the volume lost from each burst, and the cost of continuing-with repairs.

2. Identify areas of high leakage
The next step is to identify those areas which have a high level of leakage, after leakage detection and repair work has been carried out. This is best done for each DMA where they have been established. The areas should be prioritised according to their infrastructure condition factor, or simply in terms of the background leakage in litres/property/day or m^3/km/day. Each area should then be investigated to determine the primary sources of leakage. Any leaks which can be detected by the usual location methods will have already been identified, so some form of step testing or sub-metering should be used. The aim is to measure the leakage on sections of main. Ideally each street should be examined, but if this is not possible, then the leakage should be narrowed down to as small an area as possible.

3. Cost-benefit analysis
Within the DMA, each sub area can be assessed to determine whether it is cost effective to replace the mains to remove the background leakage. The leakage rate will vary considerably from one section of main to another. In carrying out the investigation, a burst which was not located by other means may actually be found and repaired, eliminating the need to replace the whole section of main.

4. Consider other benefits
When carrying out the cost benefit analysis it may be possible to bring other benefits into the equation and to give them a value. For example, if the replacement is aimed to reduce bursts there will be a benefit from not having to carry on with repair costs, and there is a benefit in terms of customer service.

5. Design the scheme
A package should be produced including a plan and all other relevant data.

6. Project management
It is vitally important to ensure that the scheme realises the projected benefits by good project management

7.4.4 Monitoring performance and maintaining progress

Once targets have been met, the key issue is then the ongoing maintenance and management of leakage at the reduced level. All aspects of leakage management requires constant effort if leakage is to be kept in a reduced state. Leakage never goes away, it is something which requires constant attention, otherwise leakage levels will rise and could return to the rate before the leakage reduction programme, wiping out much of the work and investment. If anything, leakage management becomes more difficult as this stage than during the reduction stage. The focus of attention on a major investment project has passed, further investment may be more difficult to obtain, and the ongoing work is seen as a cost burden which does not produce any benefit to the organisation.

An efficient and effective set of procedures must be put in place during the leakage reduction plan to ensure that once a set target has been reached, the leakage level is maintained at or near that target in future years. Leakage is like a spring, and unless downward force is maintained, it will bounce back. These procedures apply at three different levels:

- Strategic
- Tactical/Facilities
- Operational

Each of these is considered in turn.

7.4.4.1 Strategic monitoring

The overall indicator of performance on leakage management comes from the annual water balance calculation. However, the water supplier should not wait 12 months between carrying out these calculations. It is important that trends within the year are monitored, and that corrective action is taken if it appears that the annual target will not be met. The situation is similar to the financial management of a company, which has to file annual company accounts. Profit and loss figures will be produced more frequently to ensure the company is meeting monetary targets. It is recommended that water balance calculations be carried out every quarter, and possibly monthly where targets are critical or in areas where changes are being made to the regime which has applied in previous years.

7.4.4.2 Facilities monitoring and maintenance

The facilities which have been established during the reduction stage have to be maintained. This work tends to involve periodic inspections:

- District meters have to be calibrated or checked at regular intervals
- District boundaries have to be maintained
- Statistics such as property counts have to be kept up to date
- Pressure reducing valves will require monitoring and maintenance
- Equipment such as correlators will need regular calibration and periodic replacement

Records should be kept for each facility, and piece of equipment, similar to vehicle maintenance records, to show what maintenance work has been carried out and when. It is useful to maintain a file package for each DMA containing all relevant information which is updated with the results of the most recent survey work, changes to boundary valve positions, etc. Computer packages are available to store all this data electronically, possibly linked to the digitized mains records.

7.4.4.3 Operational Monitoring

Day to day management of leakage is a painstaking task, which requires ongoing monitoring of large volumes of data and information. Computer systems are available to store meter readings, pressure data, consumption data etc., which produce leakage values for DMAs based ion nightline information or regular (e.g. weekly) readings. These systems can also be developed to prioritise areas for leak detection surveys. While such systems help with the task of organizing leakage management operations, they also require maintenance

A key element of the ongoing monitoring is the assessment of productivity form the leak detection staff. During the reduction stage, it is possible to monitor expenditure in terms of the cost per Mld saved. However, when maintaining leakage at a set level then this is more difficult. Other measures have to be used, some of which are similar to those used in the reduction stage:

7.4.4.4 Use of new technology and operating practices

During the leakage reduction phase, the primary management incentive is achieving the leakage target. Once it has been achieved and maintained for a period (say 1 to 2 years) the next strategic aim should be to continue to achieve the same level of leakage at an every reducing operational annual cost. This will require investment in research and development and use of new technology, in order to gradually reduce staffing levels. Efficiency savings come from:

- Use of techniques which make ongoing ALC operations more efficient, such that the same result can be achieved for less effort
- Review of data and assumptions which go into the leakage calculation, whether these be based on night flow estimates or annual water balances. In many cases, some of the demand which is initially thought to be leakage turns out to be hidden customer use, operational use, or is due to meter inaccuracies
- Review of practices and staff levels to cover seasonal variations. The tendency for leakage to increase is often a seasonal phenomenon, but it is common for the same number of staff to be employed all year round. By investigating seasonal variations, it is possible to make savings by making changes to working practices, and duties at different times of

year, or by supplementing a permanent resource with contracted staff for a few weeks at a time when needed.

- Some companies monitor weather forecasts and use the data to predict the number of bursts which are likely to occur in the next few days. From this they decide how many leak location and repair staff to have on standby and for call out arrangements.

7.4.4.5 Undertaking an annual review to assess the effectiveness of the strategy

It is recommended that the leakage management strategy be kept under constant review, and that it is subject to some form of annual audit. This could mean a review by senior management, or it could involve external consultants. The review should include the following elements:

- Progress against target
- Changes to target due to lessons learnt
- Changes to assumptions and default data
- Investment made

7.5 TECHNOLOGY AND EQUIPMENT

The concept of monitoring flows into DMAs to monitor and interpret leakage is discussed in 7.4.1.This section examines the technology, both established and new, which is currently available for detecting and locating bursts and leaks.

7.5.1 Data capture and analysis

Modern metering technology and data capture techniques play a major part in the fast identification of a burst, and also in estimating the gradual accumulation of smaller leaks. Many water utilities have integrated DMA meter data, via telemetry, into their SCADA (supervisory control and data acquisition) systems. This approach has proved particularly effective when it is implemented together with a sophisticated analysis package that provides the leakage practitioner with guidance to those DMAs which require leak location work.

Flowmeters have also evolved - in the early 1980s when DMAs were first set up only mechanical meters were available. These meters were (and still are) robust, reliable and accurate. When a pulse generator is added they become electromechanical meters, capable of communicating flow information to a data logger. Electromagnetic (EM) meters, while having the advantage of being non-intrusive (no moving parts to be damaged by debris or stones), and being just as accurate, always had two major disadvantages - the flow range was not as good as mechanical meters, and most importantly, they were expensive. This has all changed. Modern EM meters now have a flow range which equals mechanical meters. Further, greater use by utilities and thereby greater production volumes, have brought down the costs. Coupled with their being non-intrusive, and having the capability to be buried (the meter body but not the electronics, which are linked to a surface box by cabling), these meters are now becoming increasingly used as DMA meters. One more major development has given them an 'edge' - battery powered units. With a battery life of 6-10 years, these units overcome the power supply problems faced in remote locations. Future developments will include linking them to global system for mobile communication (GSM) and general packet radio service (GPRS) technologies. There will also be a place in the future of DMA monitoring for ultrasonic meters. These are also non-intrusive, and well-proven in the field, but their high cost at present prohibits them from being used as anything other than supply meters or temporary check meters to verify the accuracy of supply and bulk meters.

Data loggers have also moved on apace. As well as taking advantage of GSM and GPRS facilities, to complement conventional telemetry systems, data logger manufacturers are producing optical and digital flow sensors to make loggers increasingly compatible with flowmeter signal output technology. Most loggers have an LCD display for instantaneous reading, or can be downloaded at the meter site using palmtop computers and 'blue tooth' communication technology

7.5.2 Finding the leaks

At the end of the day, useful as flow meter data are for indicating zones which contain unreported bursts or an accumulation of leaks, the leakage engineer still has to narrow down the area of leakage within the zone, and then, finally, has to pin-point the leak position. This has to be done with reasonable accuracy to avoid escalating excavation costs, and, the leakage engineer's worst nightmare, 'dry holes'!

The utility operator has a wealth of tools and equipment at his/her disposal with which to find bursts and leaks. The leak noise correlator, once considered the ultimate leak location tool, has now been around for in excess of 20 years, gradually changing in size from a unit that required a vehicle to carry it, to the modern range of portable hand-held units. Instead of depending on the noise level of the leak for its location, this instrument relies on the velocity of sound made by the leak as it travels along the pipe wall towards each of two microphones placed on conveniently spaced fittings. Hydrophones, placed in the water column, can also be used to enhance the leak sound in plastic pipes or large pipes. There is no doubt that the latest versions of the correlator, with the capability of frequency selection and filtering, can quickly and accurately locate a leak (to within 0.5 metres) in most sizes of pipe, provided there are sufficient contact points along the line of the main. As the instrument is portable it only requires a single operator.

Perhaps the most significant development in this particular manufacturing industry is the large number of makes of correlators now available internationally. Competition has resulted in lower-priced units from some manufacturers, some of whom have also developed a range of models of varying levels of sophistication. Some models are available at very low prices, but which do an adequate job for most situations.

In later years the correlator has been complemented by equipment for narrowing down areas of a DMA which contain a burst or number of leaks. As well as the traditional step testing (valve isolation) technique, which is still widely used, the technique of acoustic noise logging has become increasingly popular. A cluster of loggers, usually 6 or 12, is deployed in an area to be surveyed, each logger placed on a hydrant, meter or other surface fitting. Logger data can be analysed in situ, either by downloading on site, or by driving past each logger with a van-mounted receiver to pick up anomalous noise signals. Noises which are suspected of being caused by leaks can be confirmed, and the leak located using location equipment such as the leak noise correlator, ground microphone or acoustic sounding stick. Some noise logger systems also incorporate a multi-point correlation facility to provide 'instant' location of the leak position.

Despite the surge in leak location technology, the power and popularity of 'stick-sounding' is still as great as ever, and the technique is invariably used to confirm a leak site identified by a correlator. Sounding sticks are also invaluable in developing countries where these cheap and simple devices can be made from easily accessible materials such as wooden or metal rods. One of the latest leak location technologies, developed in 2003, takes the principle of sounding one step further. The instrument is a close-coupled surface array of eight linked sensors, embedded in an acoustic polymer mat, about 1.5 metres long. The mat is moved along the line of the main to confirm the position of a suspect leak.. Terming it 'advanced leak detection', the manufacturer claims a location accuracy of 20cm, reducing the chance of dry holes to 10%. At around 12000 US$, the equipment is comparable in price to those leak noise correlators at the lower end of the price range. The same manufacturer is also developing a UHF radio interferometer for leak location. By picking up changes in signal made by a leak the instrument will confirm a leak position where there is no clear acoustic signal.

There are a number of other location methods, both acoustic and non-acoustic, which can be brought into use when acoustic methods fail to find the leak. One of these is the tracer gas technique, using hydrogen, for those 'difficult leaks' often associated with non-metallic pipes, large diameter pipes, and some older service pipes where their position is not recorded, or which have not been laid in a straight line. This technique is not as dangerous as it sounds, as the gas used is industrial hydrogen (95% nitrogen). The gas, after being injected into a live main, comes out of solution and surfaces at the leak point, where it is 'sniffed' by a hand-held sensor.

Another technology which is fast becoming an alternative to the correlator for finding leaks in large transmission mains is in-pipe acoustic location. A microphone is inserted into a pressurised main through an air-valve. The microphone cable is calibrated to measure the distance from the entry point to the leak, which is identified and recorded as the microphone passes by. The only commercially available model of this equipment is the 'Sahara', developed by WRc, who provide the expertise and the technology as a contract service to utilities across the world. Because the sensor hears the leak noise from inside the pipe, the equipment is suitable for all types of material. At a cost of around 10 US$ per metre the equipment comes into its own for finding leaks where other techniques are not possible, such as strategically-laid pipes under main highways and rail tracks etc.

Finally, ground penetrating radar (GPR) is a technique that seems to be used with a great deal of success in some countries, but rarely used in others. The equipment is reasonably compact and portable, and picks up signals from 'anomalies' underground. These anomalies may be a void, a drain, a pipe etc, and of course, the disturbance to the ground made by a burst or a leak. The art is in the interpretation of the signal, and there are many experienced operators worldwide who are doing this on a daily basis. It is yet another weapon in the arsenal of leak detection equipment.

Leak detection and location technologies have been continuously evolving since the first data loggers and leak noise correlators were invented more than 20 years ago. What is significant is that the utility leakage engineer not only now has a wealth of equipment to choose from, and new technologies are continuously being developed and trialled, but also that competition is bringing down the costs of ownership. What is most pleasing is that practitioners in developing countries, once dependent on just the sounding stick for budgetary reasons, are now able to afford a 'package' of more advanced equipment at relatively low cost.

7.6 CONCLUSIONS

Water loss from distribution networks is a universal problem, requiring a management strategy which can be universally applied. Developing such a strategy requires a diagnostic approach – first identify the problem and its causes, and then use appropriate tools to address the problem, and to reduce or remove it. Management of leakage, or 'real losses', in the distribution network is an essential component of demand management and sits comfortably alongside other programmes to save water, such as addressing the causes of 'apparent losses', implementing a regulatory framework and introducing metering policies. The starting point for leakage management is to explore the cost effectiveness of each option for reducing leakage relating to the cost of water - the economics of leakage management

Much that has been written about leakage economics and strategies has been aimed at either the leakage specialist or the economic regulator. This chapter aims to make the practitioner aware of the key messages from the theory, and the issues which have to be taken into account, when developing appropriate policies and practices. The most important aspect of any leakage strategy is the leakage target. What level of leakage should the water supplier aim for, and what level should be maintained in the longer term? In an ideal world, every water supplier would like to eliminate leakage from water distribution systems. However, there will always be a level of leakage which has to be tolerated, and which has to be managed.

The chapter gives an overview of the prime techniques for leakage management, which have been referred to as the four 'pillars'. Each of these follows a similar law of diminishing returns on investment to make savings in leakage. The key issues for introducing these four elements have been discussed, together with the steps required to calculate the economic level of leakage.

A diagnostic approach, followed by the practical implementation of solutions which are practicable and achievable, can be applied to any water utility, anywhere in the world, to develop a leakage management strategy.

This chapter has focused on the economic issues of developing a leakage management strategy. However, other elements of the strategy require bringing the network up to a sufficient level for the techniques for assessment, monitoring and control to be put in place. Some of those techniques are programmes to address the identification of apparent losses - these may require social, cultural, political, and legal changes, and a longer term view. Other techniques are needed to address real losses - leakage detection and location programmes, purchase of equipment and network strengthening. These can be implemented in the short term, provided the required financial support is forthcoming.

What is common to all programmes is the need to maintain the motivation of staff, the understanding of what is being achieved, and the provision of skills and technology to sustain the programmes. This ensures that advances made in introducing a water loss strategy are sustained. Such issues include:

- ensuring appropriate staffing levels
- staff education and training
- operation and maintenance (O&M)
- assessing and monitoring performance

At some stage, in all organisations, it also becomes necessary to examine the policies for producing and delivering water. Some policies relate to managing elements of the infrastructure - pipework characteristics and condition, and the way in which it is operated and maintained - as well as upgrading and managing the infrastructure. Other policies are largely organisational - they relate to how the company views its relationship with its customers, and having the appropriate staffing and regulatory frameworks in place to deal with its main function - to produce and deliver water to its customers. Such policies are very subjective - they are influenced not only by the physical and local characteristics of the network, and the social and cultural attitudes of the customers, but by the structure of the company itself, whether public or privately owned, or public/private sector partnerships. In this case the organisation will have other drivers to consider, such as the interests of directors, shareholders, political and financial pressures, as well as customer and public perception. There are also increasing environmental risks of balancing new resources against the need to meet ever-increasing customer demand. Such policies include:

- demand management and water conservation
- regulatory and legal frameworks
- customer metering policy, tariff structures, and revenue collection.

7.7 REFERENCES

Farley, M. and Trow, S (2003) *Losses in Water Distribution Networks - A Practitioner's Guide to Assessment, Monitoring and Control,* IWA Publishing, London.

Hirner, W. and Lambert, A. (2000) *Losses from Water Supply Systems: Standard Terminology and Recommended Performance Measures.* IWA, www.iwahq.org.uk.

Lambert, A.O. and McKenzie, R.S. (2002) Practical experience in using the infrastructure leakage index. In *Proc. IWA Conf. Leakage Management – A Practical Approach,* Cyprus, November.

Liemberger, R and McKenzie, R. (2003) Aqualibre: A New Innovative Water Balance Software. In *Proc. IWA/AWWA Conference on Efficient Management of Urban Water Supply,* Tenerife, April.

WSA/WCA (1994) *Managing Leakage. UK Water Industry Managing Leakage, Reports A-J.* WSA/WCA Engineering & Operations Committee.

8

Demand management in developing countries

Kalanithy Vairavamoorthy and
M.A. Mohamed Mansoor

8.1 INTRODUCTION

8.1.1 Water crisis in developing countries

The available water sources throughout the world are becoming depleted and this problem is aggravated by the rate at which populations are increasing, especially in developing countries. In addition to increasing population, water resources are threatened by diminishing water quality caused by pollution, reduced quantity caused by overexploitation and denuding of water catchments areas, and the increase in water demand for agricultural use.

The threat to water resources has brought into focus the urgent need for planned action to manage water resources effectively as it is widely acknowledged that water is a major limiting factor in the socio-economic development of a world with a rapidly expanding population. The United Nations in their Millennium Declaration draws attention to the importance of water and water related activities in supporting development and eradicating poverty (UN, 2003).

Currently, some 30 countries are considered to be water stressed, of which 20 are absolutely water scarce. It is predicted that by 2020, the number of water scarce countries will likely approach 35 (Rosegrant *et al.*, 2002). More worrying is that it is the developing countries that face the greatest crisis and it has been estimated that by 2025, one-third of the population of the developing world will face severe water shortages (Seckler *et al.,* 1998). Figure 8.1 indicates the total non-irrigation water consumption (domestic, industrial, and livestock use) for the different regions of the world (Rosegrant *et al.*, 2002). This figure highlights that it is in the developing world where there will be a drastic increase in consumption. For example:

- On the continent of Africa by next year 12 African countries will be considered to be in a "Water Stress" situation. A further 10 African countries will be stressed by 2025 (a total of 20 out of the 29 countries). A total of 1.1 billion people or two thirds of Africa's population will be affected (Dzikus, 2001).

- At the current rate of population growth in India, combined with the growing strain on available water resources, India could well have the dubious distinction of having the largest number of water-deprived persons in the world in the next 25 years (Singh, 2000).

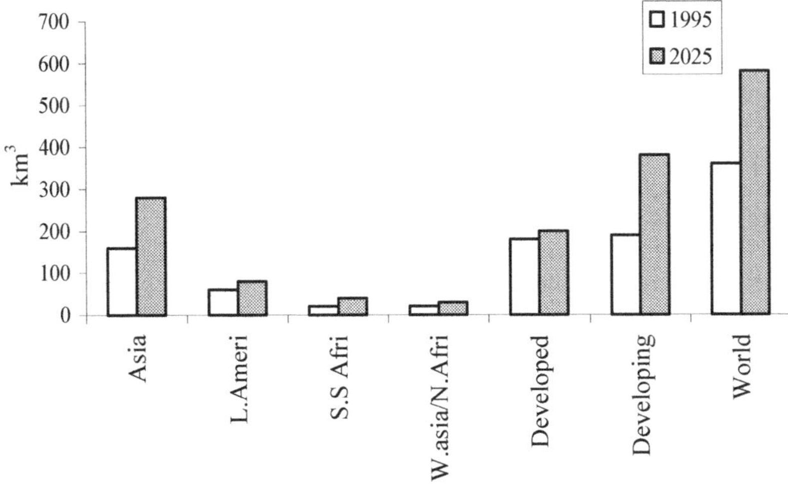

Figure 8.1 Total non irrigation water consumption by region (Rosegrant *et al.,* 2002).

8.1.2 Water scarcity in urban areas

The problem of water scarcity in urban areas is of particular concern. Migration of the rural population to urban centres has resulted in towns and cities expanding rapidly. For example between 1950 and 1990 the number of cities with populations of more than 1 million increased from 78 to 290 and this is expected to exceed 600 by 2025 (Serageldin, 1995).

The increasing concentration of populations in urban areas and the growth of large cities place a strain on existing public services and result in chaotic conditions in many towns and cities throughout the developing world. Water supply to cities is crucial not only for the survival of its inhabitants but also for the economic development of the country.

Awareness of what happens when there is a lack of water is all too apparent in many developing country cities, where poor living conditions and limited access to adequate water and sanitation significantly increase the health burden on the urban poor, who often constitute the very labour source that generates the cities wealth (UNESCO, 2003).

For example, eighty-five per cent of India's urban population has access to drinking water but only 20 per cent of the available drinking water meets health and safety standards. It is estimated that by the year 2050, half of India's population will be living in urban areas and will face acute water problems (Singh, 2000).

Providing adequate water supply to the rapidly growing urban populations is a serious problem for governments throughout the world. This was recognised in 'Cape Town Declaration', adopted by African Ministers in 1987:

" .... *there is serious concern at the inability of cities to provide safe drinking water to their populations which result in an increased burden of health care, reduced productivity and quality of life..........draw attention to the serious threat of depletion, pollution and degradation of Africa'' freshwater resources posed by the expanding urban areas in the continent.*" (UN-HABITAT, 1999).

The water stress condition that exists in many developing countries is not only due to source limitation. Other factors such as poor distribution through city networks and inequalities in service provision between the rich and the poor also contribute to this condition (UN-HABITAT, 1999). For example it has been reported that in India the water consumption ranges from 16 to 3 litres per day depending on the locality and the economic strata (Singh, 2000).

8.1.3 Supply driven approach

In general water supplies in cities are usually supply driven, meaning whenever there is a "shortage" the solution relies on the capital investment in new treatment and distribution networks. For this reason innovations in the water sector in terms of demand management are rather limited and water conservation measures are often perceived only as drought relief mechanisms that result in reduced service levels (UN-HABITAT, 1999).

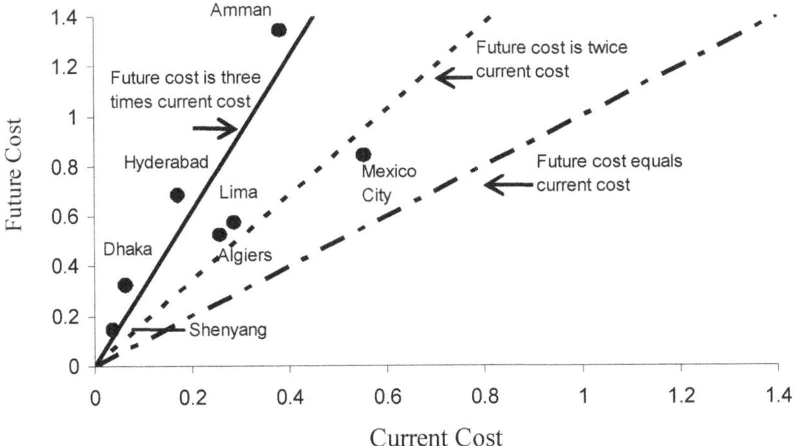

Figure 8.2 Current cost and projected costs of supplying water to urban areas (Serageldin, 1995).

But this supply driven approach adopted in most developing countries has been criticized as the cost of developing new sources or expanding existing sources is getting higher and higher as the most accessible water resources have already been tapped (UNESCO, 2003). For example, Beijing is having to abstract water from over 1000 kilometres away from the city and Mexico City may be forced to construct schemes where water is pumped to elevations of 2000 metres (Serageldin, 1995). The real cost of water per cubic metre in second and third generation projects in some cities have doubled between the first and second project and then doubled again between the second and third (Bhatia and Falkenmark, 1993). Figure 8.2 (Serageldin, 1995), shows the current cost and projected future costs of supplying water to urban areas. For example the figure shows that for a proposed groundwater based water supply scheme in Amman, Jordan, the average incremental cost was originally estimated at US$ 0.41 per cubic metre but then rose to US$ 1.33 per cubic metre as it became apparent that the groundwater resources were scarce and that surface water would be required to augment the scheme.

Hence as new source development is becoming limited the water sector in developing countries are now operating on a crisis management basis and adopting water conservation measures from dire need rather than good planning.

8.1.4 Growing interest in demand management

It is now being recognised by the water sector in developing countries that there is a need to manage the available water resources more judiciously and with care, there is a drive to become more proactive on water management practices.

In recognition of the growing interest in demand management in developing countries, this Chapter provides a brief description of some of the instruments available to support an urban demand management programme. In addition, the importance of institutional development and public awareness is emphasised. It should be noted that this paper recognises that interrupted (intermittent) water supply is the norm in most developing countries and hence where suitable appropriate references to this are made.

8.2 DEMAND MANAGEMENT (DM)

8.2.1 Definitions

Demand management focuses on measures that make better and more efficient use of limited supplies. It does not necessarily result in a reduction in the levels of service to the consumers. Water conservation can be defined as (DWAF, 1999a):

"The minimisation of loss or waste, the preservation, care and protection of water resources and the efficient and effective use of water."

And demand management as (DWAF, 1999a):

"The adaptation and implementation of a strategy (policies and initiatives) by a water institution to influence the water demand and usage of water in order to meet any of the following objectives: economic efficiency, social development, social equity, environmental protection, sustainability of water supply and services, and political acceptability."

As stated earlier, the 'supply-driven' approach adopted in most developing countries is not sustainable as the cost of developing new sources or expanding existing sources is getting higher. In such situations saving water is often the best 'next' source of water, both from an economic and from an environmental point of view. Hence it has been argued that for good water governance a mix of supply-side and demand-side approaches should be adopted (UNESCO, 2003).

A demand-side approach can be attractive to governments in that it can "buy time" by delaying the need for large capital investment in expansion of the water sector. In most cases, the savings achieved by delaying an investment can provide financial resources to more than cover the costs of implementing a comprehensive DM programme. In addition water conservation for industrial, agricultural and commercial users is always an attractive proposition as it almost always results in a reduction of operation costs.

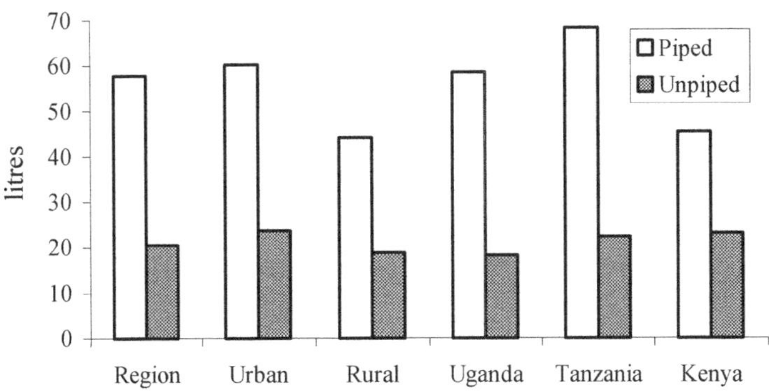

Figure 8.3 Mean daily per capita water use (Thompson *et al.*, 2001).

One of the major potential benefits of DM in developing countries is that it can provide a more equitable distribution of water by saving water in higher-income areas and providing greater quantities to low-income areas. Greater and more efficient use of water by the urban poor should result in improved health outcomes (UNESCO, 2003). Hence any DM approach in developing countries should focus on improving the conditions for the urban poor by securing better access to water and promoting hygiene. Figure 8.3 (Thompson *et al.*, 2001) shows the mean daily per capita water use for piped and unpiped consumers in East Africa. This figure clearly shows that water consumption is nearly a third less for unpiped users, who are in general from low-income communities. Figure 8.4 goes further and provides a breakdown of the amount of water used for different activities. What is interesting from this breakdown is that it is mainly the hygiene practices that get affected by the reduced quantities of water available to unpiped users. Therefore any DM programme must improve the access to water for low-income consumers so that improved hygiene and associated health benefits can be achieved.

The 'Cape Town Declaration', adopted by African Ministers in 1987 (UN-HABITAT, 1999) emphasised *"developing water management strategies at the regional, national and local levels which promote both equitable access and adequate supplies."*

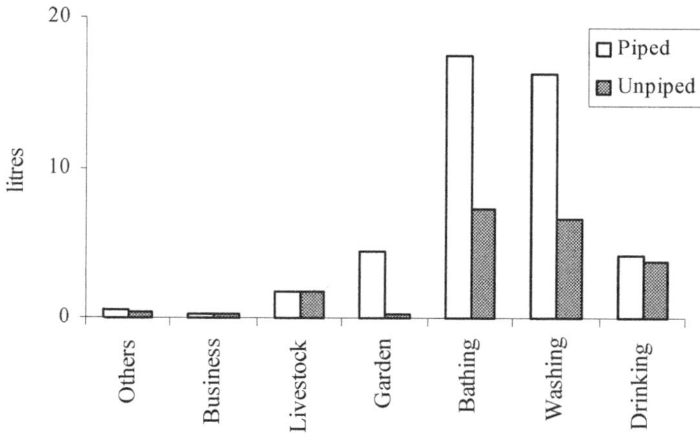

Figure 8.4 Mean daily per capita water use by type of use (Thompson *et al.*, 2001).

8.2.2 Instruments for demand management

To achieve the objectives of DM, a number of instruments have been developed. These instruments are interdependent and mutually reinforcing and the most optimal way they are applied will depend on the prevailing local conditions.

Instruments used in a DM programme include: intermittent water supply; water loss reduction (including leak detection and repair); comprehensive metering, changes in water pricing concepts, installation of water saving devices (retro-fitting), wastewater reuse, institutional development, and public awareness and educational campaigns.

There are other methods of increasing the availability of water such as rainwater harvesting but this is where lines between DM and supply management get blurred. Is water harvesting a supply or a demand technology? In this Chapter, rainwater harvesting is not considered, and readers are referred to Chapter 2 instead.

8.2.3 Demand management and consumer types

It should be noted that different instruments are applicable to different types of consumers (Wegelin-Schuringa, 1999).

High-income consumers

Applicable water DM options are in-house retrofitting and out-of-house water saving measures (garden, swimming pool). Water pricing measures may only be effective in combination with extensive awareness raising campaigns, as rich people tend not to save water because of cost.

Middle-income consumers

Very diverse with the conditions at the upper end of the category being very near to those of the high income category and the lower end near to conditions in low-income areas. The most effective water DM options for this group are water-pricing measures, especially increasing block tariff rates and an effective awareness raising strategy.

Low-income consumers

Rarely have individual connections, and generally use low volumes of water. Potentially it is these consumers who have most to gain from an effective water DM programme, as it should enhance their access to water supply services by providing increased availability of water and a more equitable distribution of this water. However, experience has shown that this can only be achieved with an active, community focused strategy.

8.2.4 Demand management programme - objective and goals

Before undertaking a DM programme it is important to establish objectives and goals of the programme. These should be tailored to meet local needs and hence should be developed together with a public participation process. Arlosoroff (1999) provides a brief description of typical goals that include: reduction of leakages and unaccounted for water; increase public awareness of methods to save water; reduce the water use of existing customers; reduce the water use of new consumers; initiate and complete the metering system etc.

It is important to note that the goals and objectives of DM do not refer only to the objective of efficient water resource management and ecological sustainability, but also to economic efficiency, social development and social equity. DWAF (1999a) provide comprehensive details of the objectives and goals of the "Water Conservation and Demand Management - National Strategy Framework", in South Africa. These include: "Objective: To promote social development and equity. Goals:

- Ensure social considerations in the planning process for water supply by all water institutions
- Ensure effective public and consumer representation in the water planning processes

- Identify and quantify direct and indirect social benefits of various demand management initiatives."

8.3 INTERMITTENT SUPPLY

8.3.1 Necessity rather than design

One of the most common methods of controlling water demand is the use of intermittent supplies. This is where the water is physically cut-off for most of the day and hence limiting the consumers' ability to collect water.

For instance, in South Asia it is estimated that at least 350 million people receive service as little as a few hours daily and nearly all Indian cities are reported to operate intermittent systems. Figure 8.5 (ADB, 1993) shows the average duration of water supplies for 8 major Asian cities. The situation is similar in other regions, and in Latin America alone more than 50 million inhabitants in ten of its major cities receive rationed supplies (Choe and Varley, 1997). In Zaria, Nigeria, in 1995, only 11% of the consumers with a piped supply received water, one day in two. Furthermore it is been reported that in Mombasa the average duration of the service is 2.9 hours per day (Hardoy *et al.*, 2001).

Intermittent supply is usually adopted through necessity rather than design and this results in serious problems with the supply including insufficient pressure (many areas have zero pressure); inequitable distribution of the available water; and a short duration of supply (see Figure 8.5), and water quality deterioration. These serious shortcomings associated with intermittent systems mean that wherever possible 24 hours continuous supply must be strived for. However, where 24 hours supply provision is not a realistic option, an attempt must be made to be proactive in the design of an intermittent system to ensure adequate service standards.

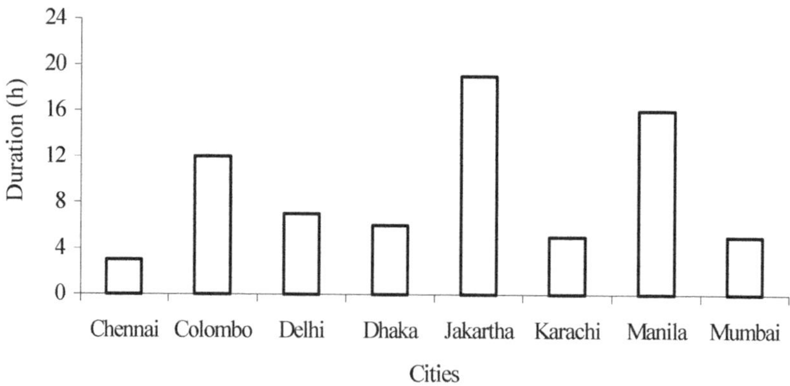

Figure 8.5 Water supply durations for selected Asian cities (ADB, 1993).

8.3.2 Problems with intermittent supply

The questions that arise are *why* do distribution networks perform poorly when operating under intermittent supply conditions?

Low Pressures

Intermittent systems are designed based on low per capita allocation, but with the assumption that demand will be spread over 24 hours, but in reality water is consumed in a short duration. Therefore the networks are undersized and the flows in pipes are much greater than anticipated. This results in severe pressure losses creating a generally low pressure in the network.

Inequitable Distribution of Water

Intermittent systems are generally water starved and consumers try to collect as much water as possible during supply hours (Vairavamoorthy *et al.*, 2001). The quantity they collect therefore is directly related to pressure at their outlets and since pressures vary greatly in the network the quantity they collect is inequitable (consumers in high pressure areas collect more water denying those in low pressure areas).

Water Contamination

This occurs in networks where there are prolonged periods of interruption of supply the pipes are empty for many hours of the day at which time pollutants can enter through leaks in the supply pipes (Vairavamoorthy *et al.*, 2001). The situation is particularly serious in cities with unsanitary excreta disposal where sewage flows in open ditches close to water distribution pipes. In Delhi, an intermittent supply and the proximity of water and sewage pipelines were the prime suspects of a paratyphoid fever outbreak in 1996 (Arti Karpi, 1996). In four Indian town districts between 27 percent and 76 percent of samples tested positive for faecal coliforms under intermittent supply (NEERI, 1994).

Consumers' Coping Costs

When the water supply is unsatisfactory, consumers incur a cost to cope with the situation (UNDP, 1999). Coping costs involved in an intermittent supply include above and below ground storage tanks, alternative water supplies, pumping, treatment facilities and also the inconveniences to the public tap users in terms of timings of supply (McIntosh, 2003). Low-income consumers cannot afford these facilities and hence pay through the time they spend in fetching water from public taps. The cost in time incurred by the poor consumer often exceeds the amount of what their wealthier neighbours pay for water from house connections. A study conducted in Dehra Dun (Choe *et al.*,

1996), a city of 300,000 inhabitants in the state of Uttar Pradesh in India, revealed that families pay significant costs to cope with the inconvenience of intermittent supplies regardless of their income levels. The estimated annual coping costs of households obtained from this study are summarised in Table 8.1. Similar findings were reported in Tegucigalpa, Honduras where coping cost for a poorest family with intermittent supply was around 180% of water tariff (Yepes *et al.*, 2001).

Table 8.1 Annual coping cost for Intermittent Supply in Dehra Dun (US$ at 1996/year)

Types of Expenses	HH Connection	Public Taps	Weighted Average
Water tariff	18	0	11.7
Total coping cost	13.3	69.3	22.1
Total annual cost of water	31.3	69.3	33.8

Source: Adapted from Yepes *et al.* (2001), summarised from Choe *et al.* (1996)

8.3.3 Improvements to intermittent supply

As stated above intermittent systems are often adopted in a reactive way and hence operate poorly. However, it must be recognised that where there is extreme water scarcity, adoption of intermittent supply is one of the most popular ways of conserving water. Therefore it is important to recognise this reality and highlight the problems associated with these types of systems, and provide recommendations and guidance to alleviate or minimise some of these problems.

Recently research work has been undertaken (Vairavamoorthy *et al.*, 2001 and 2004) to develop a series of design guidelines for intermittent water supply systems that:

- Improve equity in supply
- Improve water quality in intermittent systems.

It should be noted that these guidelines do not promote intermittent water supply. However, where 24 hours supply provision is not a realistic option, an attempt must be made to be proactive in the design of an intermittent system to ensure adequate service standards, in particular a more equitable distribution of the limited quantity of water and improved water quality.

8.3.4 Guidelines for improved equity in supply

The design guidelines (Vairavamoorthy *et al.*, 2001; Vairavamoorthy and Elango, 2002) are driven by a modified set of design objectives to be met at least cost. These objectives are:

Equity in Supply: Equitable distribution of the limited quantity of water is the keystone of the entire design process outlined in the guidelines and is a non-negotiable design objective. This objective is achieved mainly through effective pressure management.

People Driven Levels of Service – PDLS: The objectives are central to the design philosophy outlined in this guideline and can be regarded as the '*Golden Thread*' of the design with all components of the process feeding this thread. The PDLS are defined in terms of 4 parameters DTPO: Duration of the supply; Timings of the supply; Pressure at the outlet (or flow-rate at outlet); and Others such as the type of connection required and the locations of connections (in particular for standpipes).

It should be noted that all the above objectives are taken for granted when designing continuous systems, but are variables in the design of intermittent water systems. A major component of this new approach is modified mathematical modelling tools specifically developed for intermittent water distribution systems. These tools combined with optimal design algorithms with the objective of providing an equitable distribution of water at the least cost forms the basis of the new approach.

8.3.5 Guidelines for improved water quality

Currently a Department for International Development (DFID), funded research project is being undertaken at Loughborough University (WEDC), to improve the management and monitoring of water quality in urban intermittent water supplies by controlling risks of contamination (Vairavamoorthy *et al.*, 2004; Yan *et al.*, 2002). This project incorporates technical considerations in managing water quality in distribution networks and water user expectations of water quality in decision-making. There will be three distinct groups of outputs from this project. The first group will comprise of a guidance manual on how to undertake hazard assessments, identify critical control points and verification measures and incorporating socio-economic criteria and user expectation into a water supply assessment (Water Safety Plans). This manual will be designed to be accessible for all operational staff. The second group of outputs will be a contaminant transport and water quality modelling software to support detailed assessment and decision-making within a distribution system. This will be supported by a GIS based risk assessment tool that combines the water safety plan approach with socio-economic data to highlight priority areas for action.

8.4 WATER LOSSES

8.4.1 Water loss definitions

Water losses are inevitable in distribution systems and water institutions/utilities should strive to supply water efficiently and effectively by minimizing water losses. 'Water loss' and 'non-revenue water' have become well known terminologies in water DM and have replaced the terms such as 'unaccounted for water (UFW)' (Farley and Trow, 2003). The work performed by the International Water Association Task Forces on Water Losses is considered a major step forward in defining the "best practice" approach in assessing and presenting components of water loss (Farley and Trow, 2003). They define water loss as:

Water Loss = Real losses + Apparent losses

Where real losses cover leakage from pipes, joints and fittings and leakage from reservoirs etc. Whereas apparent losses consist of unauthorized connections (theft and illegal use) and metering errors. See chapter 7 for further details.

Water losses in the cities of developing countries are reaching extreme levels of up to 40-60% of the water supply (Arlosoroff, 1999). Figure 8.6 (WHO, 2000), shows the mean UFW for large cities in different regions of the world and Figure 8.7 (ADB, 1997), shows the mean UFW for 8 major Asian cities. Note that the 'unaccounted for water' reported in these figures are due to water losses (as it excludes unbilled and unauthorised consumption). In many cases the water loss indicators shown in these figures reflect the efficiency of the management of the water supply system. Any reduction in water losses requires coherent action to address not only technical and operational issues but also institutional, planning, financial and administrative issues (WHO, 2000).

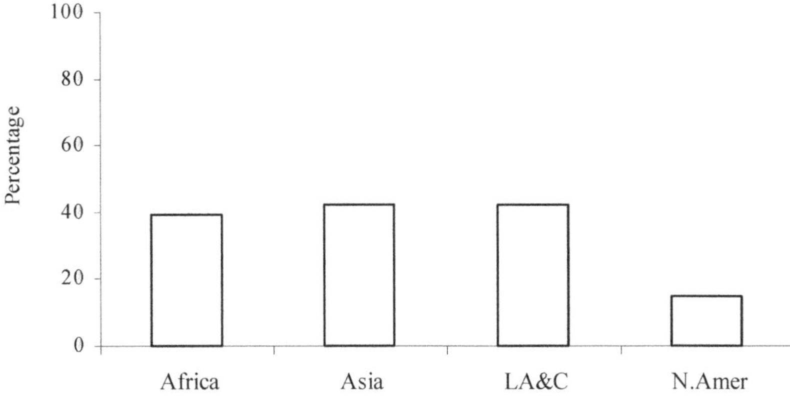

Figure 8.6 Mean unaccounted-for water in large cities (WHO, 2000).

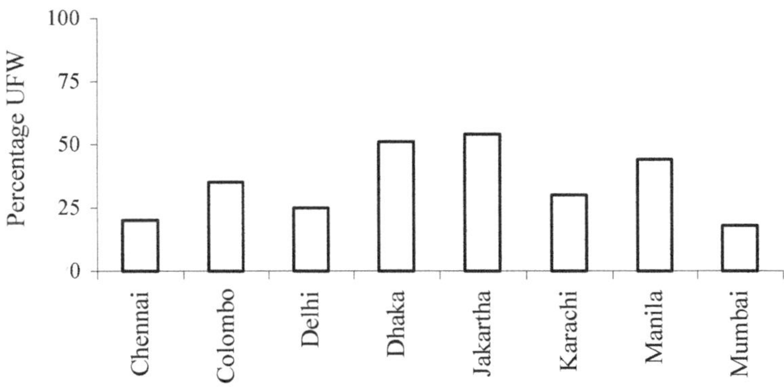

Figure 8.7 Mean unaccounted-for water in selected Asian cities (ADB, 1997).

8.4.2 Real losses

The major part of real losses is often due to leakage and is usually due to lack of maintenance or failure to renew and replace ageing systems. Successful leakage reduction programmes not only pave the way for an efficient and sustainable water supply system but also foster public support by providing affordable water to consumers. Reducing leakage minimises the difference between the available supply and projected demand, hence minimising the vulnerability of the system to supply shortages.

It should be noted that leakage reduction programmes follow the law of diminishing returns - the more effort to reduce leakage, the less will be the return in terms of water saved. Therefore for any distribution system one needs to define an economic level of leakage (ELL) - a target level of leakage below which the return on investment is not optimal.

ELL is defined as the 'level of leakage where the marginal cost of active leakage control equals the marginal cost of the leaking water' (WHO, 1985). Therefore the estimation of the ELL can only be understood by knowing how water is valued at various places.

In developed countries with abundant water resources, the unit cost of water is comparatively low. However, in developing countries where water is scarce, its unit cost is high. Leakage minimisation plays a crucial role in conserving the limited quantity of water and helps in extending existing resources to their maximum potential, brings down the operating costs and enables the deferment of capital works. In addition leakage minimisation improves the efficiency and reliability of the water supply system resulting in greater confidence of the system from the consumers. Ultimately this will result in an increase in the consumers' willingness to pay for the improved services. (Kusnur, 2000; Choe and Varley, 1997).

Leakage assessment, detection and repair

A leakage control programme consists of two components:
- Water audits
- Leak detection surveys.

Water auditing is the detailed accounting of flows into and out of the system and enables identification excessive leakages areas in the system. Detailed procedures for water auditing can be found in Chapter 7 of this book. In order to pinpoint the exact location of leaks, detection surveys need to be carried out.

A good description of the methods to undertake a detailed leakage assessment and detection programmes for continuous water supply systems can be found in WHO (1985) and also in Chapter-7 of this book (Trow and Farley). However as stated earlier, in most developing countries networks operate intermittently and the conventional approaches to leakage detection are not applicable (Farley and Trow, 2003). In this section only details of leakage assessment and detection for intermittent supply will be given (Kumar, 1991).

8.4.3 Leakage in intermittent systems

To measure leakage in a zone of an intermittent system all outlets should be plugged and the amount of water entering the zone during the assessment measured. This is a very laborious process. Typically a leakage assessment programme for an intermittent system is in two phases: a preparatory phase; and the leakage detection test.

For a particular zone or pilot area the preparatory phase involves checking mains records, identifying all service connections (including public stand posts), excavating trial holes to check for cross connections, and checking and repairing the boundary (zone) valves to ensure their water-tightness. If practicable, the normal operational pressures and flows into the zone are recorded prior to undertaking the test. In a typical zone these preparatory works could take up to two weeks to complete.

The test phase is carried out during non-supply hours and begins by isolating the test area. This is achieved by closing boundary valves and closing all service connections. The pipelines of the pilot area are then filled with water supplied by a series of tankers using small booster pumps. A mains pressure of approximately 10m is maintained and flow of water measured by means of water meters to give a direct measurement of leakage. A typical test area is usually 100 connections or 500 metre of pipe.

It is important to detect leaks during the test itself where artificially high pressures are maintained as most leakage detection equipment have limited use in low pressure systems that is the norm with intermittent supply. Farley and Trow (2003) reports that in Chennai, the total duration for the test is approximately 8 hours.

It is clear that leakage assessment and detection in intermittent systems is laborious and expensive. Hence there is need for research to develop more effective methods tailored specifically for such systems. Currently at Loughborough University (WEDC), decision trees are being developed to enable engineers to identify zones that are most vulnerable to leakage and select pilot zones for detailed leakage assessment. In addition models are being developed that establish optimal leakage control boundaries and optimal locations of flow meters (for leakage assessment). In addition statistical based methods are being developed to identify pilot zones for detailed leakage assessment with the objective that these zones are representative of leakage in the entire network. This will enable estimates of leakage to be made for the entire network with known degrees of uncertainty.

Rehabilitation, repair or replacement

After identifying defects in the system, decisions have to be made to determine the type of rehabilitation. A range of methods and techniques are available for pipeline rehabilitation but the selection of a suitable technique can be difficult. Decision support systems have been suggested to evaluate replacement and preventive maintenance priorities (Kulkarni & Reid 1991; Madiec *et al.*, 1996; Fenner & Sweeting 1999; Yan et. al., 2002; Yan & Vairavamoorthy, 2003a, 2003b).

8.4.4 Apparent losses

The main factors affecting 'apparent losses' are illegal connections (water theft), improper metering (including improper meter installation and inaccurate reading), billing anomalies and inefficient management.

Illegal connections

Illegal connections not only contribute to the loss of water but also play a major role in reducing the levels of service of the supply system. For example a recent study in Hyderabad, South India (Chary, 1997), found that 49% of houses have legal connections and the remaining ones are either dependent on public standpipes or have illegal connections.

Water scarcity, poor management and lack of awareness among general public are the factors behind Illegal connections. Often consumers resort to illegal connections if they are refused household connections, or if inappropriate tariffs are imposed that penalise the poor and lower middle classes. In developing countries it is not surprising to see consumers tapping into the unsecured and openly laid water mains with the assistance of the local plumber when there is a water shortage in the area.

In most developing countries handling and minimising illegal connections have left the utility staff with unpleasant memories and led to riots in communities. This collective opposition to prevent and minimise illegal connections arises from the view that water is a basic human need and charging for water is unacceptable. The situation is aggravated by the involvement of politicians who try to win the public support at the expense of sustainability. An example of such a situation is in Maharashtra, India where the water utility's efforts to detect and disconnect illegal connections resulted in staff being intimidated and this ultimately led to riots (Alwani, 2000).

More recently the move has been to encourage the regularisation of illegal connections by:

- Providing an amnesty for those with illegal or unregistered consumer connections, so that they may regularise their connections.
- Giving appeals and advertisements to the general public to regularise illegal or unregistered consumer connections.

- Giving 'on the spot' powers to the local managers and operators to regularise illegal and unregistered consumer connections.

However, in addition to encouraging regularisation, appropriate deterrents must be imposed such as legal intervention and fines, connections being disconnected etc., to discourage the proliferation of illegal connections. However, regularising connection should always be a preferred option, as this will ensure that consumers still receive the water they need.

Metering schemes

Although metering is discussed in more detail in Section 8.5.4, its relevance here is that accurate metering is essential for assessing authorised consumption. Therefore, it is essential that a water utility must have a programme to measure and calibrate the performances of the various types of domestic water meters. It is also important that meters are regularly maintained and replaced. For example in a recent study in Hyderabad, India it was found that (Chary, 1997):

"the accuracy of the present meters is not satisfactory ...... meters were found to be inaccurate, frequently going out of order due to locking of the gear mechanism and are susceptible to tampering ... owing to intermittent supply, it was observed that meters indicate wrong readings under air flow and back flow situations"

Figure 8.8 (WHO, 2000), shows the mean number of meters installed and replaced annually for large cities in Africa, Asia and Latin America. This figure indicates that meters are replaced on average every eight years or more and as meters tend to under-read with age it is likely that a considerable proportion of apparent losses are due to metering error.

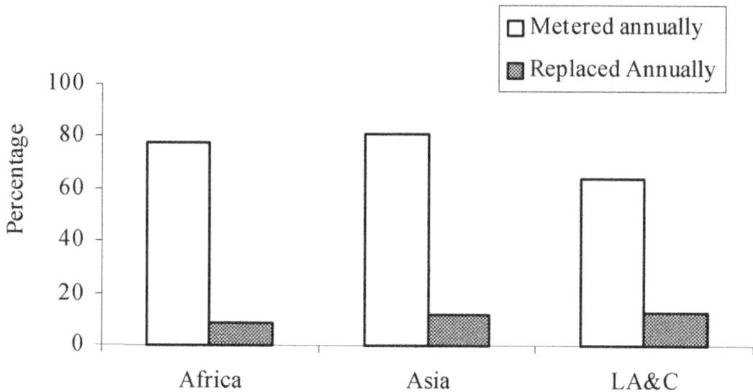

Figure 8.8 Regional means for largest cities: metered and meters replaced annually (WHO, 2000).

Billing aspects

It is important that the water utility has the manpower to carry out the meter readings and collections. In developing countries it is quite common to find discrepancies in meter reading and billing. This may be due to: failure to get access into a property to read the meter and hence the meter reader 'makes-up' a reading; meter readers recording inaccurate figures after obtaining financial favours from the consumers; and inefficiencies and flaws in the billing system especially in calculation and collection (Gokhale, 2000). Adequate staff training, automating the meter reading and billing system will reduce these problems greatly. However, in developing countries the applicability of the latter methods are questionable due to lack of financial resources.

8.5 WATER CHARGING, TARIFFS AND METERING

8.5.1 Water charging and subsidies

One of the most powerful methods to encourage DM is to charge for water on a per unit basis, with payment tied to volumetric use. A pricing system based solely on a flat rate tariff regardless of volume used or a system based on property value would not have any incentives to conserve water as once the flat rate is paid the price for the volume used is essentially zero (Arlosoroff, 1999).

When charging for water it should be recognised that DM is not the only objective of a charging policy and the other objectives such as improving the equity of supply of water (water to be provided to all at an affordable rate), and ensuring cost recovery (to make the water service sustainable), must also be considered (PPIAF, 2002). Therefore if we wish to make water affordable to all while ensuring cost recovery, then subsidies must be provided for low-income consumers.

The National Water Act of South Africa (RSA, 1998), takes equity further:

"... establish a pricing strategy which may differentiate among geographical areas, categories of water users or individual water users. The achievement of social equity is one of the considerations in setting differentiated charges. Water use charges are to be used to fund the direct and related costs of water resource management, development and use, and may also be used to achieve an equitable and efficient allocation of water"

8.5.2 'CAFES' Principal

The determination of tariff policies should seek to address both commercial and social welfare concerns .The simple but comprehensive 'CAFES' principles (Sansom *et al.*, 2002) are outlined below:

- *Conserving.* The tariff structure should influence consumption in such a way that customers are able to purchase enough water to meet their needs without being wasteful.
- *Fair.* The average tariff should be adequate for the utility to achieve financial sustainability. Allocation of required level of revenue must be fair and equitable between customer groups for both the poorer members of the community and the different levels of service options.
- *Adequate.* The tariff should produce a level of financial resources, which will enable the utility to meet its financial commitments with sufficient contribution towards future investment.
- *Enforceable.* The utility should have the capability of enforcing the tariff through viable sanctions such as court action, disconnections, etc. Tariffs that cannot be enforced are unlikely to be sustainable.
- *Simple.* The tariff structure should be simple for the utility to administer and easy for customers to understand. Customers usually display greater willingness to sustain payment of water bills when they understand the bills.

8.5.3 Block tariffs

Of all the available tariff structures, the most widely used method in developing countries is the increasing block tariff (IBT), where high-volume consumers can be used to subsidize the poor. With an IBT, the price per unit of water increases from one block of consumption to the next.

It is common with IBT's that the first block is charged at a low rate and covers the basic water needs of the consumers. Consecutive blocks can be charged at appropriate prices so that the consumers are discouraged from indulging in unnecessary use. Industrial, commercial and high volume users usually pay significantly more for their water under this system and are hence subsidising the low volume users.

A good description of the ideas behind the different blocks in IBT is given in (DWAF, 1999b) and summarised below:

Block 1-basic human needs

First block should be designed to cover the basic needs of the consumers and charged at a nominal rate. There are strong social and equity grounds for creating an affordable lifeline tariff to facilitate the consumption of a certain minimum amount of water.

Block 2-normal consumption

Normal consumption is defined as the average per capita consumption of a particular area. Tariffs in this block are designed to fully recover costs including the depreciation of the capital assets.

Block 3-luxury consumption

Luxury consumption is defined as the consumption above what is normally consumed. The availability of water supplies and the possible drought conditions are also taken into account in luxury consumption (additional incremental costs for the increased capacity are reflected in this block).

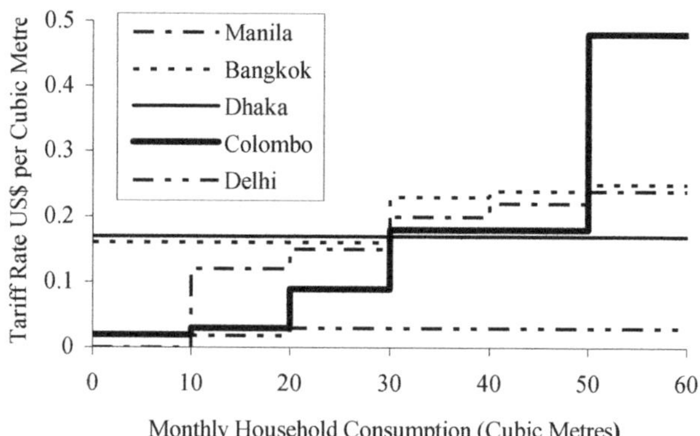

Figure 8.9 Tariff structures for selected Asian cities (ADB, 1997).

Figure 8.9 (ADB, 1997) shows different IBTs for 5 major cities in Asia. This figure indicates that for Manila, Colombo and Delhi the initial block is set at 10 cubic metres while for Bangkok it is set at 30 cubic metres. Also note that in Manila there is no charge for the first block and that Dhaka has a flat rate tariff.

Much has been reported about the disadvantages of IBT system and the way in which in some instances it may be disadvantageous to the poor. The main disadvantage for the poor concerns the practice of sharing household connections. This is where multiple families share a single water connection and hence their total water usage exceeds the lowest block of IBT (Whittington, 1992; Liu *et al.*, 2003). Hence these households pay a higher average cost for each unit of water than do households who do not share connections. Another disadvantage highlighted by Boland and Whittington (2000), is that the maximum subsidy available to poor households through an IBT is usually quite small. This is because to benefit from the maximum subsidy a household must use the entire first block. As the first block is often large and low-income households generally use less water, they do not receive the full subsidy. It is argued that the main reason for implementing IBT is that they are politically acceptable as they are perceived to be fair.

Alternative tariff systems have been proposed to overcome some of the limitations of the IBT. For example the increasing rate tariff (IRT) is a block tariff system that is based on per-capita consumption as opposed to a per-service connection consumption (Liu *et al.*, 2003). It is a tariff structure where the water price is dependent on both total household water consumption and household size. Although this type of tariff structure overcomes issues related to shared connections, its practical applicability is questionable. Boland and Whittington (2000), have also proposed an alternative approach to water pricing. Their proposal involves a two-part tariff consisting of a single volumetric charge equal to the marginal water cost coupled with a fixed monthly credit or rebate. Their scheme also incorporates a minimum monthly charge to avoid negative bills. Under this approach, they calculate that most poor households would receive a lower bill than the one they now receive under the IBT system. Meanwhile, a higher percentage of households would face the full marginal cost of the water. More radical views exist concerning water charging such as subsidizing access to water services (household connections, stand pipes etc.), and not subsidizing the consumption (PPIAF, 2002).

8.5.4 Metering

Water metering is essential for a successful DM programme as it is the basis to link consumption and price. Therefore before any DM programme is implemented unmetered connections must be reduced to an absolute minimum and metering coverage should be maximised in all sectors. In addition resources must be made available within the water utilities to undertake installation, calibration, maintenance and reading of a comprehensive water metering system.

Metering must be considered at all levels including within the main supply system (bulk metering, district metering), and individual consumer metering. For consumption a metering scheme can be implemented in stages starting from the larger water consumers like the industry, commercial users, higher income domestic users (including government offices, police, army, public taps) (Arlosoroff, 1999).

When implementing a metering programme the quality, precision, and repair of meters require thorough technical assessment and planning. This may involve considering the installation density (may not be in all households), their characteristics (precision, durability, ease to install and read etc), their installation programme (duration, procedure, equipment and personnel), their control and use (verification, repair, frequency of reading) (Buenfil, 1992). In addition several other important measures need to be considered such as: financial strategy to purchase and install meters (fee to consumers, loan, etc.), legal enforcement, motivation and information for consumers.

Before any new metering programme is undertaken, questionnaires and metering pilot areas should be prepared. This is important in order to: assess consumer cooperation for installation and behaviour with a new charging system; evaluate and classify sources of difficulty and find installation techniques suitable for different cases (inside the house, external); assess problems and solutions about reading, inspection and repair; and assess the unit costs and techniques for each task (Buenfil, 1992).

Recently studies have been undertaken in South Africa that consider unconventional metering or vending devices for shared connections (DWAF, 1997). Charging the shared connections is a complicated issue as there is no control over water use among different households. In South Africa, prepaid meters (electronic devices) for shared connections have been successfully tested. These prepaid meters have been found to be very effective devices for conserving water and recovering costs. The shortcomings of these systems are water theft at prepayment points and the high unit costs for low water consumption users.

Many problems have been reported in the literature mainly related to mis-measurement resulting from poor meter quality, poor water quality, use beyond useful life, improper sizing, improper installation (Dzikus, 2001; Farley and Trow, 2003).

Also as stated earlier most distribution systems in developing countries are intermittent and research has shown that meters function poorly in intermittent systems. Gokhale (2000) states the following problems associated with meters in intermittent systems:

- Start of the supply causes sudden acceleration of the meter impeller rotation from no flow to maximum flow rate (discharging capacity). Similarly sudden retardation is effected at the end of the supply. These sudden variations subject the meter mechanisms to undesirable strain
- Alternate drying and wetting of the parts of the meter, coming in contact with water is detrimental to continued satisfactory performance of the meters
- Air that enters the distribution system is forced out through the consumer connections at the start of the supply and that causes meters to run at excessive speeds affecting working of meters.

The frequent problems associated with meters means that it is essential to have a continuous monitoring and replacement programme for meters.

8.6. RETROFITTING

8.6.1 Potential for retrofitting in developing countries

Retrofitting provides one of the most effective short-term options for reducing water demand particularly in the high-income domestic and institutional sectors. Typical examples include: low-flush toilets, spring-activated/low flow/aerated faucets, and low-flow showerheads etc. For example in Mexico City 350,000 toilets were replaced with 6 litre models and this saved enough water to meet the household needs of 250,000 residents (Serageldin, 1995).

Many government buildings or institutions, which are state owned, do not pay for their water and hence consumers have no interest in conservation. Good examples are university campuses, ministry buildings, government hospitals etc. With very little capital investment, usually only a few dollars per fitting, water consumption in these buildings can be reduced by as much as 20% (Dziikus, 2001). All the cites involved in the 'Managing Water for African Cities' programme (see Section 8.10), are implementing retrofitting (particularly in government buildings etc.), as part of their DM programme. For example in Dakar Senegal (UN-HABITAT, 2003):

"Retrofitting of faucets, showerheads and toilet flushing will be tested using modern high-quality fittings which can be imported and later manufactured locally. These fittings will be installed in public/office/university buildings...."

The prospect of conserving water for industrial and commercial organisations is always an attractive proposition as it almost always results in a reduction of operation costs. Retrofitting also brings direct benefits to high–income domestic consumers whose spending on water can be reduced. Consumers with limited supplies can stretch their share of water to last for longer duration in order to avoid having to buy transported additional supplies at very high costs.

Although repairing and retro-fitting domestic house installations are classified as customer demand initiatives rather than distribution management initiatives, water services institutions should not ignore the potential costs savings by implementing such measures at their own cost. In order to monitor retrofitting programmes' effectiveness, especially when designed as a pilot demonstration, efficient metering is necessary. This can be for example, block metering for an entire institution of apartment block. Some type of incentive could be offered to those who retrofit including, payment for installation added to water bill to spread the cost, grants from local authorities.

8.6.2 Low-income communities

In general low-income communities are low water users (collect water from yard taps and public standpipes which have water saving valves (such as the waste not tap – Talbot (2003)), use low volume pour flush latrines (½ - 2litres per flush (Franceys, *et al.*, 1992), and hence retrofitting is not applicable. However, there have been many devices developed to provide a more equitable distribution of water for low-income households. For example the RFR (Lindeijer, 1986) is a device that has been developed to control flow rates in both standpipes and house connections of low-income communities. This device maintains a constant flow rate irrespective of pressure variation (within certain pressure limits). This ensures that pressure variations within the network do not affect the ability to collect water and hence provides a more equitable distribution. However, it is not clear from the literature the extent and success of such devices.

Another example of a system to conserve water is the Durban Tank system (DWAF, 1997; Macleod, 1997). This system has gained wide acceptance because of its flexibility and that it provides low-income communities with a reliable, affordable water supply at their houses. A 200-litre tank of water is provided at the front door of each house, which is filled once a day with clean drinking water. Within the community, a water bailiff is appointed who at a fixed time each day opens the main supply to each connection and thereby fill the tanks at each house. This process is relatively quick and is usually completed in less than one hour (for 200 customers). In addition the bailiff is entitled to install a standpipe on his property, which is metered, from which he may sell water to those residents of the area who are not able to afford the costs of the tank system. This method provides an acceptable quantity of clean drinking water at an affordable price, delivered directly to each house so as to do away with the need to carry water long distances.

8.7 WASTEWATER REUSE

Wastewater reuse contributes to DM in that it enhances the efficiency of a water supply system by supplying reused water to activities which otherwise would have utilised potable water from the distribution system. Potential applications of wastewater reuse in developing countries include centralised collection and reuse for landscaping and irrigation (parks, green belts, sports fields and cemeteries etc.); industrial recycling and reuse (cooling water, and where industries reuse water with various qualities as required for the various processes); and grey water reuse in households (mainly for gardening).

The most common form of wastewater reuse found in developing countries is effluent from wastewater treatment works used for agriculture. In India for example 55% of available wastewater is used for irrigation and in Mexico City this value is 100%. In Bulawaya, Zimbabwe the effluent from 5 out of 6 treatment works is used to water suburban parks, golf courses, nurseries, playing fields etc (Sibanda, 2002). However, to achieve this requires appropriate infrastructure facilities such as treatment plants, pumping stations etc.

Industries should also be encouraged to treat their effluents and recycle or reuse this water as it will provide them with an economic gain (as they abstract less water from the water system). For example in Beijing industrial water reuse has increased from 46% in 1978 to 72% in 1984 (Sibanda, 2002). Governments should put in place legislations to ensure industrial consumers maximise their recycle and reuse potential.

Wastewater reuse is also practised to a lesser extent in large households, hospitals, schools, and government buildings mainly for gardening purposes. For example in Botswana vegetable gardening area of 150 m² at a clinic in Lobatse was irrigated with wastewater from sinks and hand basins (CES, 2003). All wastewater was drained into drums dug into the ground and then used for gardening.

The important issues that need to be considered when practicing wastewater reuse are:
- Protection of water users: There is a danger of people accessing poor quality water sources and use for consumption which may have health implications
- Protection of the environment: Reuse water should not contain pollutants of significant concentrations that may lead to ground water pollution
- Social acceptability: Any reuse strategy should consider social preferences of the population as objections arise frequently due to aesthetic, religious and other reasons. Therefore it is important to embark on strong public education and awareness campaign.

8.8 INSTITUTIONAL CAPACITY

8.8.1 Institutional Capacity and Demand Management

In order successfully implement a DM programme, there needs to be sufficient capacity within the institution to implement this. Farley and Trow (2003) provide a good description of areas that needs to be strengthened within an institution to effectively implement a water loss programme (similar to what might be required to implement a broader DM programme).
These include:

- *Appropriate staffing level*: Sufficient numbers of skilled competent staff who will carry out the tasks.
- *Staff education and training:* Delivered through awareness seminars (for senior staff), training workshops (for engineers and technical staff) and continuous practical training (for operations staff).
- *Operation and maintenance (O&M):* Important as lack of O&M leads to inefficient practice, ineffective services and waste of resources.
- *Assessing and monitoring:* On going monitoring to maintain DM targets. This should be applied at three levels: *Strategic (*analysis of trends and projections); *Tactical (*maintain and periodic inspections of facilities that have been established during the DM programme); and *Operational (regular* monitoring of systems performance).

However, more fundamentally most water utilities in developing countries lack:

- An awareness of practical processes involved in a DM programme.
- An awareness of the potential financial and operating benefits of top management of a DM programme.
- A motivated staff at operation level to undertake a DM programme.
- Resources.

Therefore, it is important to raise the profile and change the perception of DM within the utility. In addition efforts must be made to show how actions can best be taken in logical steps to achieve effective control even when operational problems seem insurmountable. There is a need to maintain the motivation of staff, understanding of what's been achieved and the provisions to sustain DM.

8.8.2 Institutional development programme

Developing and managing institutional improvements is a difficult process. Edwards (1988) has developed a manual that provides practical and immediately useful information about developing and managing institutional change projects in the water supply and sanitation sector. He points out that institutional development projects should *"focus on the development of comprehensive organisational systems and the people within the system which make them work."* He goes on to say that *'the overall purpose of these projects is to achieve institutional learning or sustainability."*

More recently, Loughborough University (WEDC) have been involved in the development of the 'Change Management Forum' (CMF) in India[1]. One of the main areas of discussion within this forum has been water losses and the development of more efficient water supplies. The mission of this forum is to

"promote institutional and organisational development and support reform of the urban water and sanitation sector through capacity building, knowledge sharing and promotion of partnerships."

The CMF works through policy and decision makers from municipalities, water utilities and Pubic Health Departments to develop critical mass of change champions. Activities of the CMF include dissemination of information on best practices, knowledge resource products, introduction of performance indicators, and the development of a benchmarking database. This is achieved through seminars, theme specific study tours, and the 'Change management times' a newsletter (CMF, 2002).

8.9 PUBLIC AWARENESS

8.9.1 Importance of public awareness

One of the main aims of DM is to influence consumers to use water more efficiently. This may include the implementation of metering, changed tariff structure and retrofitting. Clearly, all these activities will directly or indirectly affect consumers and hence a public awareness campaign will play an important role in any DM programme. The focus of the campaign should be to raise the publics' awareness of the urgency to conserve water now to avert a crisis situation in the future. It should be noted that in the past implementation of programmes and initiatives were generally based on decisions of engineers and planners in isolation without consultation of the public and without public awareness campaigns. It has been widely recognized that this top-down approach has been the reason for the failure of many initiatives.

---

[1] www.cmfindia.org

8.9.2 Components of public awareness campaign

Public awareness campaigns generally consist of two distinct components: knowledge and information transfer; and education. Its major goals are to mobilise support for ongoing and future activities and ensure the sustainability of the programme (also see Chapter-12 of this book).

The dissemination of information requires appropriate "packaging" to suit the different target groups. To achieve this the publics involvement should be encouraged. Various different means are available to ensure appropriate information and knowledge transfer and adaptation. The use of mass media is probably the most cost effective in most cities as even the urban poor have access to such communication tools.

Apart from the dissemination of information materials to households, education is the second major cornerstone of an awareness raising campaign. These measures should address decision-makers and consumers at all levels. To ensure the sustainability of this campaign the consumer education should start with the young, and this could be achieved by trying to include water demand management in the school curriculum.

In Hyderabad, South India for example the implementation of a DM programme includes (Chary, 1997):

" ....*aggressively used audio and visual media to educate consumers on advantages of UFW management, water conservation, advantages of fixing ISO meters, regularization of illegal connection etc. Appeals on the above subjects have been made in English and vernacular languages, through posters, pamphlets, newspaper advertisements and public meetings.*"

Public awareness campaigns are an integral part of the 'Managing Water for African Cities' programme (see Section 8.10). For example in Dakar Senegal where an extensive retrofitting programmes is being initiated (UN-HABITAT, 2003):

"*A city-level awareness campaign will be developed and implemented, directed to all users in the pilot area before retrofitting and throughout the project duration.*"

8.9.3 Factors to consider for public awareness campaign

The extent of public involvement will depend on local conditions. For example in a small city, ad hoc advisory committees and close communication with elected officials can provide adequate input and feedback to formulate the program. In larger cities, a formal advisory group may be needed to develop a program that can be presented to the general public. As stated by Arlosoroff (1999):

"it is advisable to build into the planning process the time and techniques required to consult with community leaders and interest groups having major concerns about a water conservation effort. These steps are especially critical if a retrofitting program is planned. It is essential, as well, when implementing a comprehensive water metering installation, reduction of unaccounted for water program and pricing charges."

In addition Arlosoroff (1999) provides a useful checklist of groups that may have a role in DM planning, and this includes:
- Elected officials Staff persons from water companies, key personnel from local government agencies.
- Representatives of major local economic interest groups – major industries, community groups leaders, contractors associations, farmers, tourism-boards, parks and gardens contractors and others.
- Representatives of major community forces – school boards, local unions, religious groups, block leaders, local press and media owners.
- Representatives of local environmental interest groups, if they exist (they could be powerful support teams).
- Local professionals with credibility, such as economists, engineers etc.

8.10 DEMAND MANAGEMENT PROGRAMMES IN DEVELOPING COUNTRIES

As stated earlier the water sector in developing countries have now recognised the need to manage the available water resources more judiciously and with care and are embarking on DM programmes. Evidence of this can be seen from the 'Managing Water for African Cities' programme and the recently launched 'Managing Water for Asian Cities' programmes. In both these programmes DM is a key component and it is anticipated that the lessons learnt from these programmes will greatly influence the water management practices of other developing countries.

8.10.1 Managing water for African cities

The 'Managing Water for African Cities" initiative is a collaboration of the United Nations Human Settlements Programme (UN-HABITAT) and the United Nations Environment Programme (UNEP). This initiative is a direct follow-up of the Cape Town Declaration, adopted by African Ministers in December 1997.

The targeted beneficiaries of this programme include policy makers on water and the environment, city managers of water utilities, water consumers, school children for water education, and the mass media for awareness promotion.

The programme involves seven cities (Abidjan, Accra, Addis Ababa, Dakar, Johannesburg, Lusaka, and Nairobi), and focuses on three inter-linked priorities:

- Introducing effective urban water management strategies in African cities.
- Protecting freshwater resources from the growing volumes of urban wastes.
- Enhancing regional capacity for urban water management through information sharing, training, and public awareness and education campaigns.

This programme has generated wide political support with concrete commitments from African governments, and has already established a network of African policy makers, city managers, and professionals addressing urban water issues. The program's innovation and success has resulted in requests from other African countries to join the program. More specific details of the activities involved in this programme can be found at UN-HABITAT (2003).

8.10.2 Managing water for Asian cities

In March 2003 the Asian Development Bank signed a Memorandum of Understanding with UN-HABITAT to set up a Water for Asian Cities Program. The program aims to build the capacity of Asian cities to secure and manage pro-poor investments and to help the region meet the Millennium Development Goals (MDG) of "halving, by 2015, the proportion of people without safe drinking water and basic sanitation." To meet the MDG for water and sanitation in Asia, it is estimated that an additional 675 million people in Asia's cities need access to adequate sanitation, and 619 million need access to safe water.

The program will draw upon lessons learned in our successful Water for African Cities program and help ensure that pro-poor, sustainable water policies are implemented. As with the 'Managing Water for African Cities' programme one of the main aims of the programme will be promoting urban water DM by:

- Focusing on all aspects of DM, economical, social, technical, legal, administrative and institutional.
- Prioritising the reduction of unaccounted for water; pricing and public-private relationships; equity in distribution of services; regulation and reallocation of resources.

8.11 CONCLUSION

Most developing countries recognise that they need to manage their water resources more judiciously. It is also recognised that DM can postpone the need for large capital investment for expanding water resources. Hence there is now a drive in developing countries to initiate DM programmes and evidence of this can be seen with the establishment of the 'Managing Water for African Cities' and 'Managing Water for Asian Cities' programme. For both these programmes DM is central.

In developing countries DM provides opportunity of providing a more equitable distribution of water resulting in improved access and water availability to low-income consumers and this could go along way in improving their health status. Therefore in any DM programme, needs of the urban poor must be addressed.

There are many instruments available to implementing DM and these instruments are interdependent and mutually reinforcing and the most optimal way they are applied will depend on prevailing local conditions. In addition, these instruments must be tailored for the unique conditions that exist in developing countries, and it must be understood how these different instruments affect different types of consumers with varying income levels, who access water in different ways.

In developing countries, one of the main limitations to successfully implementing a DM programme is the lack of capacity within the institutions responsible for delivering water. This needs to be recognised at the outset and any programme should have a serious institutional development component.

Finally public awareness and participation is crucial for the success of DM. In developing countries the public receive a poor, inefficient and inequitable water service. It is likely that most consumers are already receiving limited quantities of water and already involved in their own water conservation and DM practices. By engaging with the public DM programme, it is possible to receive their support and hence improve the chances of success.

8.12 REFERENCES

ADB (1993) *Water Utilities Data Book: 1ˢᵗ Edition*, Asian and Pacific Region, Asian Development Bank, Philippines.

ADB (1997) *Water Utilities Data Book: 2ⁿᵈ Edition*, Asian and Pacific Region, Asian Development Bank, Philippines.

Alawni, A. (2000) Handling Unauthorised Connections Case Study of a City. In *Proc. International Seminar on Intermittent Water Supply System Management*, Indian Water Works Association, Mumbai, India, 49-59.

Arlosoroff, S. (1999) Water Demand Management. In *Proc. International Symposium on Efficient Water Use in Urban Areas*, IECT-WHO, Kobe, Japan.

Arti Karpi. (1997) Letter to the Editor. *Emerging Infectious Diseases* 3 (3), July-September.

AWWA (1990) *Water Audits and Leak Detection*, Manual of Water Supply Practices No. M36, American Water Works Association, Denver, USA.

Bhatia, R and Falkenmark, M. (1993) *Water Resources Policies and the Urban Poor: Innovative Approach and Policy Imperatives.* Water and Sanitation Currents, UNDP-World Bank Water and Sanitation Program, USA.

Boland, J. and Whittington, D. (2000) *The Political Economy of Increasing Block Water Tariffs in Developing Countries*, Oxford University Press. Oxford, UK.

Buenfil, M.O. (1992) *Water Metering Planning and Practices.* MSc thesis, Loughborough University, UK.

Chary, S. (1997) *Managing UFW-A Case of Hyderabad.* Internal Report, Administrative Staff College of India, Hyderabad, India.

CES (2003) Water Saving and Reuse of Water. *Rain Water Reservoirs Above Ground Structures for Roof Catchments.*
http://wgbis.ces.iisc.ernet.in/energy/water/paper/drinkingwater/rainwater/rainwater.html

Choe, K., Varley, R., and Bilani, H. (1996) *Coping with Intermittent Water Supply: Problems and Prospects, Environmental Health Project.* Activity Report No. 26 USAID, USA.

Choe, K., and Varley, R. (1997) Conservation and Pricing-Does Raising Tariffs to an Economic Price for Water Make the People Worse Off?. In Proc. Best Management Practice for Water Conservation Workshop, South Africa.

CMF (2002) *Change Management Times*, www.cmfindia.org, 1, Jan.

DWAF (1997) *Implementing Prepayment Water Metering Systems.* Department of Water Affairs and Forestry, South Africa

DWAF (1999a) *Water Conservation And Demand Management - National Strategy Framework.* Department of Water Affairs and Forestry, South Africa.

DWAF (1999b) *Draft Tariff Regulations for Water Service Tariffs.* Department of Water Affairs and Forestry, South Africa.

Dzikus, A. (2001) Managing Water for African Cities: An Introduction to Urban Water Demand. In *Proc. Regional Conference on the Reform of the Water Supply and Sanitation Sector in Africa – Enhancing Public-Private Partnership in the Context of the Africa Vision for Water (2025)*, Kampala, Uganda.

Edwards, D. (1988) *Managing Institutional Development Projects: Water and Sanitation Sector.* WASH Technical Report, No.49, Water and Sanitation for Health Project, USA.

Farley, M. and Trow, S. (2003) *Losses in Water Distribution Networks - A Practitioners Guide to Assessment, Monitoring and Control.* IWA Publishing, London.

Fenner, R. and Sweeting, L. (1999) A decision support model for the rehabilitation for the non-critical sewers. *Water Science and Technology* **39** (9) 193–200.

Franceys, R., Pickford, J. and Reed, R. (1992) *A Guide to the Development of On-Site Sanitation.* World Health Organisation, Geneva, Switzerland.

Gokhale, M. K. (2000) Metering in Intermittent Water Supply Systems. In *Proc. International Seminar on Intermittent Water Supply System Management*, Indian Water Works Association, Mumbai, India, 79-82.

Hardoy, J. E., Mitlin, D., & Satterthwaite, D. (2001) *Environmental Problems in a Urbanizing World: Finding Solutions for Cities in Africa, Asia and Latin America.* Earthscan, London.

Kulkarni, R. and Reid, F. (1991) Replacement/maintenance priorities for gas distribution system. In *Proc. Annual Conf. of the Urban & Regional Information Systems Assoc.*, Australia, 1, 156–66.

Kusnur, N. (2000) Leakage Control in Intermittent Water Supply systems-Experiences in Mumbai. In *Proc. International Seminar on Intermittent Water Supply System Management*, Indian Water Works Association, Mumbai, India, 95-100.

Madiec, C., Botzung, P., Bremond, B., Eisenbeis, P., Skarda, B.C., Ray, C.F. and Matthews, P. (1996) Implementation of a probability modal for renewal of drinking water networks. *Water Supply*, **14** (3–4), 347–350.

McIntosh, A. C. (2003). *Asian Water Supplies, Reaching the Urban Poor*. Asian Development Bank, Manila.

NEERI (1994) *Evaluation of Engineering, Economic, Health and Social Aspects of Intermittent vis-à-vis Continuous Water Supply Systems in Urban Areas*. National Environmental Engineering Research Institute, Nagpur, India.

PPIAF (2002) *New Designs for Water and Sanitation Transactions-Making Private Sector Participant Work for Poor*. PPIAF- Water and Sanitation Program, Washington D.C., USA.

RSA (1998) *National Water Act, 36 of 1998*. Pretoria, Government Printers, Republic of South Africa.

Rosegrant, M.W., Cai, X., Cline, S.A. (2002) *Averting an Impending Crisis, Global Water Outlook to 2025*. Food Policy Report, International Water Management Institute (IWMI), Colombo, Sri Lanka.

Sansom, K., Franceys, R., Njiru, C., Kayaga, S., Coates, S. and Chary, S. (2002) *Serving all urban consumers-A marketing approach to water services in low-and middle-income countries Vol. 2*. Water, Engineering and Development Centre (WEDC) Publishing, Loughborough University, UK. ISBN -1 84380 055 1.

Seckler, D., Molden, D. and Barker, R. (1998) *Water Scarcity in the Twenty First Century*. IWMI Water Brief 1, International Water Management Institute (IWMI), Colombo, Sri Lanka.

Lauria, D.T (1992) *Case Study: The Cost of an Unreliable Supply in Tegucigalpa*. Note prepared for the World Bank, June 1992, USA.

Lindeijer, E.W. (1986) How to Develop Distribution Control. In *Proc. 12th WEDC Conference: Water and Sanitation at Mid Decade*, Calcutta, India.

Liu, J., Saenije, H.H.G. and Xu, J. (2003) Water as an Economic Good and Water Tariff Design Comparison Between IBT-con and IRT-cap. *Physics and Chemistry of the Earth* **28**, 209-217.

Serageldin, I. (1995) *Towards Sustainable management of Water Resources*. Direction in Development Series, World Bank, Washington D.C., USA.

Sibanda, P. N. (2002) *Water and Sanitation: How Have African Cities Managed the Sector? What Are the Possible Options?* The Urban and City Management Course for Africa, Kampala, Uganda.

Singh, N. (2000) Tapping Traditional Systems of Resource Management, *Habitat Debate, UNCHS* **6** (3).

Talbot (2003). *"Valves and Stopcocks-The Talbot Talflo"*, www.talbot.co.uk/valves-stopcocks/valves-talflo.htm.

Thompson, J., Porras, I. T., Tumwine, J. K., Mujwahuzi, M. R., Katui-Katua, M., Johnstone, N. and Wood, L. (2001) *Drawers of Water II*. International Institute for Environment and Development, London, UK.

UN (2003) *Millennium Development Goals*. United Nations, New York, USA, http://www.developmentgoals.org/Education.htm.

UNDP (1999) *Water for India's Poor - Who Pays the Price for Broken Promises?* UNDP-World Bank Water and Sanitation Program – South Asia.

UNESCO (2003) *Water for People Water for Life* - The United Nations World Development Report, United Nations Educational, Scientific and Cultural Organisation, New York, USA.

UN-HABITAT (1999) *Managing Water for African cities - Developing a Strategy for Urban Water Demand Management: Background Paper No. 1.* Expert Group Meeting UNEP & UN-HABITAT.

UN-HABITAT (2003) *Managing Water for African Cities.* Demonstration Projects, UN-HABITAT & UNEP, www.un-urbanwater.net/cities.

Vairavamoorthy, K., Akinpelu, E., Ali, M., Anand, S., Elango, K., Harpham, T., Lin, Z., Patnaik, R. (2001) *Guidelines for the Design of Intermittent Water Distribution Systems.* Report submitted to Department of International Development (DFID), UK.

Vairavamoorthy, K., Akinpelu, E., and Lin, Z. (2001) Design of sustainable water distribution systems in developing countries. In *Proc. EWRI-World Water and Environmental Resource Congress,* May 2001.

Vairavamoorthy, K. and Elango, K. (2002) Guidelines for the design and control of intermittent water distribution systems. *Waterlines,* ITDG, **21** (1).

Vairavamoorthy, K., Yan, J., Galgale, H. and Mohan, S (2004) A GIS based spatial decision support system for modelling contaminant intrusion in to water distribution systems. In *Proc. 30th WEDC International Conference on People-Centred Approaches to Water and Environmental Sanitation,* Oct 2004, Laos PDR.

Weglin-Schuringa. (1999) Water demand management. In *Proc. International Symposium on Efficient Water Use in Urban Areas,* IECT-WHO, Kobe, Japan.

Whittington, D. (1992) Possible adverse effects of increasing block water tariffs in developing countries. *Economic Development and Cultural Change,* October.

WHO (1985) *Leakage Control, Source Material for a Training Package.* World Health Organisation, Geneva, Switzerland.

WHO (2000) *Global Water Supply and Sanitation Assessment Report:* World Health Organisation-United Nations Children Fund, Geneva, Switzerland.

Yan, J. M., and Vairavamoorthy, K. (2003) Prioritising water main rehabilitation under uncertainty. In *Proc. CCWI International Conference on Advances in Water Supply Management,* London, September.

Yan, J. M., and Vairavamoorthy, K. (2003) Fuzzy approach for pipe condition assessment. In *Proc. ASCE International Conference on Pipeline Engineering and Construction,* Baltimore, U.S.A., July.

Yan, J, Vairavamoorthy, K., and Lin, Z. (2002) Vulnerability assessment in intermittent water distribution systems – water quality aspect. In *Proc. International Conference on New Information Technologies for Decision Making in Civil Engineering,* London.

Yepes, G., Ringskog, K., and Sarkar, S. (2001) The high cost of intermittent water supplies. *Journal of Indian Water Works Association* **33** (2).

9

Drivers and barriers for water conservation and reuse in the UK

Susan Roaf

9.1 INTRODUCTION

'The centrality of water in our lives - social, economic, political and spiritual - cannot be overestimated. Water has been a major factor in the rise and fall of civilizations. It has been a source of conflict and tension between nations. Its quality reveals everything, right or wrong, that we do within our ecosystems. Water is an indicator of poverty and social development. Nearly every decision we make is directly linked to the use of our water resources. In short, water is life.

Unlike the energy crisis, the water crisis is life threatening. Unlike oil, fresh water has no viable substitute. Its depletion in quantity and quality has profound social, economic and ecological effects. Water is a particularly vital resource. Without water, ecosystems are destroyed.

Economic activities halt. People die.......

Water has to be used in an environmentally sustainable manner in order to maximize its economic and social benefits. It should not be used faster than it is replenished, nor should it be polluted.' [1]

The supply of sufficient clean water for populations around the world, even in 'rainy' Britain, has become an issue of not only quality of life, but of survival, as clean water becomes a scarcer and more precious resource year on year for a number of reasons including:

- Climate Change, with trends towards regional and seasonal droughts, extreme weather and unpredictability of weather systems. One third of the world's people already live in countries with moderate to high water stress. If current consumption patterns continue, two out of every three people on Earth will live in water-stressed conditions by 2025
- Demographics, with populations and households increasing annually, and more people living longer
- Increasing rates of consumption per capita due to increasing rate of attitudinal and lifestyle changes, including leisure pursuits, across all sectors of different economies and societies
- Increasing rates of ground water extraction and freshwater reserve depletion
- Increasing concentrations of chemical and organic pollutants in rivers and lake water from agriculture, pharmaceuticals and extreme weather events. Already, more than 5million people die every year as a result of poor water quality – ten times the number killed in wars. More than half the victims are children
- Increasing public antithesis, and practical barriers to the building of major new dams in the landscape

These reasons, alone, for water conservation and reuse in the UK, are compelling. Combine them with the fact that the UK water supply infrastructure is aging fast and was originally designed for very levels of use, and it becomes clear that there is a pressing need to implement effective water conservation, and water reuse, programmes in the UK.

'The potential for water reuse in the UK is considerable. The amount of water supplied for urban use in England and Wales is greater than 30% of the total potable water produced and so the UK situation affords great opportunities for sustainable water management. A decrease in urban water requirements will lead to a significant decrease in total water demand: far greater than in other EU countries.

---

[1] http://www.unep.or.jp/ietc/Focus/Water_in_21Century1.asp

Also, the UK has a high percentage of urban development compared to its European partners, with over 50% of the populace living in towns of greater than 100,000 inhabitants. This demographic trend provides the opportunity of creating water recycling schemes in cities and large towns, thus presenting the potential for large water savings' (WROCS, 2000).

9.2 THE THREE KEY DRIVERS

9.2.1 Climate change

Climate change predictions indicate that future climates will be more extreme than those of the past century, with wetter winters but significantly drier summers (Hulme *et al.,* 2002). Winter precipitation in the UK is predicted to increase for all periods and for all climate change scenarios. By the 2080s, this increase in predictions range from between 10% and 35%, depending on the emissions scenario chosen, ie. whether we increase, decrease, or keep the same levels of carbon dioxide emissions from our lifestyles. Conversely the summer pattern is reversed and almost the whole of the UK may become drier, with a decrease in rainfall of between 35%, and a staggering 50% or more. The highest changes in precipitation in both winter and summer are predicted for eastern and southern parts of England, while changes are smallest in the northwest of Scotland (Roaf, *et al.,* 2004a). What we looking at here is the increasing risk of summer droughts and winter flooding.

Rainfall is likely to also become more intense, and in both winter and summer the patterns of rainfall will change enormously, except perhaps for northwest Scotland. For the southeast in particular, in winter, there will be much more rain, falling more intensely, leading inevitable to much higher flood risks for the region, contrasting with the significantly drier summers and more frequent droughts (Hulme *et al.,* 2002b).

These changes will impact on water availability and quality in a number of ways:
- Less water available during drought periods, particularly in regions like the southeast of England
- Higher pollution levels of water sources in periods of flooding and droughts
- Long term depletion on ground water supplies
- More stress on the water supply networks, during periods of flood leading to failure of already stressed pipe systems and contamination of fresh water with sewage and other pollutants.
- Summer soil moisture by the 2050s may be reduced by about 30% over parts of England for the *High Emissions* scenario and by 40% or more by the 2080s. This has obvious implications for soil shrinkage and related damage to regional or local supply and drainage systems.

- Geo-structural failures: Drought could even increase the flooding risk: in Amsterdam in summer 2003, the drought conditions weakened a raised river embankment so much that it gave way, causing localised flooding. In Shetland in summer 2003, prolonged drought conditions weakened the peat covering a hillside, and when there was a heavy shower of rain it resulted in a peat slide which engulfed several houses. Such geo-structural impacts could affect regional supplies and significantly impact the cost of maintaining water supplies.

9.2.2 Demographics

The population of the UK is predicted to grow by 3.3 million people between 1996 and 2016, with a trend to more smaller households using more water per capita[2] and with more people living in urban environments. There were some 20 million households in England in 1996. This is projected to grow to 24 million by 2021, an increase of 3.8 million or 19%. By 2016 some 36% of all people will be living in single person households[3]. By 2025 over 20% of the population will be over 65. In 1950 0.4% of the UK population were over 85, in 2000 it was around 2% and by 2050 around 5% of the population will over 85%. With increasing trends towards reduced security in the pensions market, many of these retired people will be dependant on state aid. If we look at the basic requirements for decent living they include adequate drinking water, affordable food and fuel bills to provide adequate warmth and calorie intake to survive.

Over 4.7 million are already in debt to water companies today, that is one in five households with each owing an average of £166, or £781 nationally, a debt that is rising fast. Water, unlike gas or electricity, cannot be cut off, and the rising debt is leading to an increase in water costs across the board. The UK Water regulator warned the government that water bills were set to rise considerably over the next years from £234 average bill now by 31% to £306 a year by 2009 with one firm, United Utilities requesting increased bills by 71% from £243 to £416 for the same period. Climate change is already impacting the rapid rise in bills. In April 2004 millions of households faced big increases in water and sewage bills to help pay for flooding, consumer debts and build costs[4] Average bills were predicted to rise by as much as £75 to pay for the costs of renewing Victorian sewers and capacity building the water infra-structure. Anglian Water wanted to increase average bills of £279 by £101 between 2005 and 2010, and claims its 5 million customers must pay to fund sewer flooding

---

[2] http://www.ciria.org/suds/pdf/ciria_rp664%20_consultation_summary.pdf

[3] http://www.waterforesight.com/

[4] http://money.guardian.co.uk/utilities/story/0,11992,1211560,00.html

prevention and all utilities charge customers for the costs of repairing infrastructure damaged in storms.

The pressing need to reduce the cost of water for those in water poverty will provide an increasing driver to the move to reduce overall water consumption and look at imaginative ways in which adequate clean water supplies can be provided in a future when ground water and surface water supplies may be less dependable than they were in the 20th century.

9.2.3 Increasing rates of per capita consumption

The Environment Agency in 2004 published its advice to the Ministers in 'Securing Water Resources', in which they identified per capita consumption (pcc) is the amount of water used by each individual at home, as a key factor in the forward plans for the water companies. Water companies distinguish between consumption by people in households that are metered (measured pcc) and those in households without meters (unmeasured pcc), as households with metered supplies generally use less water. Both measured and unmeasured pcc data must be estimated from other water supply and demand information. As in previous years, the pattern for both measured and unmeasured pcc across England and Wales is that most of the high values are concentrated in the south-east.

Measured pcc in 2002/2003 ranged from 97 lpd in Yorkshire Water's East Surface Water Zone to 214 lpd in the South zone of South East Water. Average measured pcc across England and Wales in 2002/2003 was 134 lpd. This is an increase from the previous year's average of 132 lpd. It is possible that measured pcc has risen as a result of the small overall increase in metering. Households that opt for meters generally have low water consumption, as this means that it is financially beneficial for them to pay by the amount of water they use. As more households are metered, it is possible that households that opt for meters have slightly higher water use.

The value of 214 lpd for South East Water's South zone seems to be very high. The company explains that this increase is the result of revisions of Environment Agency Securing water supply population estimates as a result of the data from the 2001 census. In the third annual review we reported that the two lowest measured pcc values were in two of Yorkshire Water's zones. The values for these zones are higher this year, but are still very low. The South Meirionydd and Tywyn-Aberdyfi zones of Dw ˆ r Cymru Welsh Water also have very low measured pcc at 98 and 103 lpd respectively. The average unmeasured pcc for England and Wales for 2002/2003 was 145 lpd. This is about the same as for 2001/2002. The range of values for unmeasured pcc was between 108 and 194 lpd. Southern Water's Hampshire Kingsclere and Sussex Hastings zones also have high reported values of 194 lpd (previously 154 and 169 lpd). Both companies say that these changes are due to revised calculation methodologies. Water companies are aware of the significant variation in pcc and occupancy rate across England and Wales and that large variations in trends in consumption will require a range of DSM programmes for water conservation and reuse that respond to these regional variations[5].

The interaction between different drivers can exacerbate conditions of water stress, for instance where larger increases in water demand develop in areas where it is decreasingly available. There is a need, in the light of such drivers and their relationships, to strategically plan to match regional growth in population to available water resources (Roaf, et. al., 2004a). What has surprised not only the Insurance Industry but many politicians and Water Utility companies (DMB, 2003), is that they have not been consulted by the Government on the impacts of building millions of new houses in the regions of Britain where the worst water shortages are predicted this century, many of which, like the Thames Gateway development, have been planned in areas that will be prone not only to flooding, but to drought as well.

9.3 WATER CONSERVATION AND REUSE MEASURES

9.3.1 Water conservation measures

Water Conservation Measures can include[6, 7 & 8]:

- Setting minimum requirements for performance standards of products
- Water metering
- Water leak detection
- Conservation pricing - water rates and tariffs
- Landscaping
- Water Reuse
- Consumer Education
- Water conservation incentives eg. awards
- Emergency Water Use Restrictions

9.3.2 Greywater reuse systems

Greywater recycling is currently being developed to reduce:

---

[5] http://www.environment-agency.gov.uk/commondata/105385/
4_an_review_590262.pdf

[6] http://aggie-horticulture.tamu.edu/extension/xeriscape/xeriscape.html

[7] http://www.environment-agency.gov.uk/subjects/waterres/
286587/?version=1

[8] http://www.mrsc.org/Subjects/Environment/water/wc-measures.aspx

- The impact on the supply, private wells and wastewater treatment facilities
- The cost of water treatment
- Water bills
- The environmental impacts of water supply

Greywater recycling includes[9]:
- light grey - bathroom sink & shower
- dark grey - kitchen sink
- black water - includes toilets
- recycled water is used for toilet flushing and depending on treatment, sinks & showers

Reuse programmes in Australia also include[10]:
- Water reuse programmes for stormwater
- Re-use of effluent
- Water sensitive urban design
- Total urban water planning and management

9.4 BARRIERS AND DRIVERS FOR THE GOVERNMENT AND REGULATORS

9.4.1 Regulators and the Government: Drivers

The UK Government has a publicly stated commitment to Sustainable Development, in its many guises and an early manifestation of the application of environmental concern to the water sector was outlined in the 'Agenda for Action 1996 – Twin Track Approach' (DOE, 1996). Using the Twin Track Approach, the Government intends to work with stakeholders over the long term to encourage structural reform in the water industry, including a significant increase in the uptake of water conservation and reuse, whilst supporting practical short term local solutions on the ground, particularly in 'problem hot spots'. The latter builds trust and confidence, which can legitimise the further reaching structural reform of water consumption patterns and the water supply industry. The latest update of the UK water conservation duties has been published in the Statutory Instrument 2004 No. 641 (C.24) of the 2003 Water Act that came into force in March 2004. This Instrument had the effect of enabling Sections 81 and 83 of the Act on water conservation.

---

[9] http://www.awwa.org/Advocacy/govtaff/REUSEPAP.cfm
[10] http://www.clw.csiro.au/publications/consultancy/2004/national_issues_water_cons ervation_reuse.pdf

Short term issues have already prompted the government intervention, for instance, to minimise already extant environmental damage caused by excessive water extraction, cranked up to meet rising demand. Such actions are more apparent to the public in times of drought when supplies are threatened and information of the withdrawal of extraction permits by the Environment Agency are covered in the Press for in stretches of water where flows are at unacceptable low levels or ground water supplies are low. Hose pipe bans imposed by Local Water Companies as well are also a high profile incentives to reduce water use by homeowners and businesses.

As always the Government must balancing stakeholder requirements, including those of the environment, customers and the water industry. Concerns about climate change, particularly after the drought of 1995, have encourage a longer term structural approach to the introduction of Demand Side Management (DSM) strategies into the water supply industry, through the medium of legislation, including the following bills, the Water Resources Act, the Environment Act, the Water supply regulations 1999and the Water Bill.

Key players at the level of Government are Ofwat, the Water Industry Regulator, the Environment Agency, both of which are responsible for enforcing the legislation, and the DWI (Drinking Water Inspectorate). The remit of Ofwat includes the imperative to encourage water companies to use water efficiently to manage the water supply demand balance at least cost. The Environment Agency is the statutory body with a duty to secure the proper use of water resources in England and Wales.

The type of policy tools available to various stakeholders is partially a function of the legislative and financial relationship between government, the regulatory bodies and the commercial sector. Presently, a variety of responsibilities, liabilities and accountabilities are distributed amongst a number of stakeholders. Top down programmes to encourage less use of water by consumers are also developed and promoted by these bodies, as well at Water Companies, such as the 'Watermark' scheme aimed at enabling customers to minimise their water bills.

The government has also raised the profile of the Environment Agencies Water Efficiency Awards for different sectors ranging from education to research, and information and applications for this Award can be downloaded from the environment agency's website[11].

---

[11] www.environment-agency.gov.uk/savewater

9.4.2 Regulators and the Government: Barriers

Perhaps the most significant hindrance to more widespread adoption of water recycling in the UK is the lack of quality standards for reused water. In very simple terms, it is difficult to design recycling systems without some guidance on what is an acceptable water quality for specific applications such as toilet flushing or garden irrigation. Commercial concerns are generally, and understandably, not prepared to make estimations or conjecture on this point. In the absence of official guidelines or criteria, many organisations are waiting for the first precedent of a legal challenge to the use of recycled water before strategically committing themselves to water recycling. Similar barriers are found in many countries including Canada[12].

The development of reuse quality criteria for the UK is hampered by several issues:

- A lack of any empirical data upon which quantitative models of risk can be based
- An apparent unwillingness of any regulatory or governmental body to take responsibility for setting, and monitoring, standards

There are currently several sets of 'guidelines' available on water quality in the UK context. Perhaps the most widely referenced sets are those published by BSRIA (Brewer et. al, 2001) and WRAS (2000). Neither of these are based on any quantitative assessment of risk exposure and are broadly comparable with similar guidelines published in the USA. In addition, the European Commission are currently carrying out a project in collaboration with the World Health Organisation which aims to set standards for reused water.

Standards could relate to:

(a) the quality of the final effluent
(b) treatment processes which are to be applied to the source waters
(c) the design and operation of specific recycling systems

The risks associated with using recycled water are both context and scale specific. Appropriate standards might therefore also be made dependent upon the scale of application. We would caution that standards impact on several aspects of the potential for water recycling in the UK. For example, there is a clear relationship between the severity of standards for a particular application, and the cost of supplying water of appropriate quality. If set too severely, standards could effectively repress the financial motivation for recycling. It is also worth noting that standards aimed at prescribing particular treatment processes could serve to inhibit innovation in a field where there is still a great deal of opportunity for technology and technique development (Roaf et. al., 2004b).

---

[12] http://www.cmhc-schl.gc.ca/publications/en/rh-pr/tech/98101.htm

The most promising commercial option for recycling in single or multiple domestic premises is 'new build' developments, at least in terms of the economics. However, house builders and the manufacturers of 'in-building' recycling systems are still looking for guidance on the design, fitting and operation of recycling systems (BSRIA, 2003). Hence the need for acceptable and rigorous standards to be included in the legislation as standards or as quasi-legal guidelines.

In some quarters it is also considered a problem that the Government has, itself, lost control of the policy levers that are driving water conservation and reuse. With the dispersal of powers to regulate for, and oversee, water supply and consumption, to the regulators and the Utilities, the Government may well have instituted a real hurdle to the short term and longer term structural reform of the Industry. Where it may be sounder to invest in conservation schemes a Utility may well prefer a solution to supply problems that involve the building of new dams or other water storage capacity, and will have a leading say in where such future expenditure is directed.

The political acceptability of legislation itself may be in question, where stringent steps may have to be taken that are not popular with the public such at the curbing of the use of water profligate practices such as car washing or garden watering. This issue may well be one of the causes of the 'Cautious Approach' adopted by companies overseeing water efficiency where the curbs on high water using equipment and practices are reluctantly introduced.

A major barrier to more effective legislation is the paucity of research held at the centre of Government where legislation is drafted. At the same time Water Companies hold important studies and information but are reluctant to share them. Correct information is needed on which to base legislation. In addition the established research funding cycle, developed by bodies such as NERC and EPSRC, are out of synchronisation with the cycles of legislative drafting, so what legislation on the subject that does emerge is often several years behind the current developments in research.

Due to a lack of sharing and co-operative development of research on these issues there exists no consensus on the cost and impact of water saving options that would be sufficient to stimulate new resolve amongst the regulators and legislators to strengthen the arm of government in this field. This lack of resolve is manifested, for instance, in the paucity of the regulations on water efficient fittings to control water use in new developments. Regulations are introduced to prevent bad practice, not promote good practice generally within the industry.

For users the difficulty in obtaining information on current UK water use regulations and their application is also a major barrier to the uptake of water reuse and conservation technologies as exemplified by the following issues of water regulations and safety.

A further field that requires significant development is that of understanding, interpreting and applying the current Riparian and Water Extraction laws that can be extremely complex where related to particular localities and water sources. The local Environment Agency is a good place to start when investigating the statutory position in relation to a water source, but in extremis the Environmental Law Foundation[13] in London could also be contacted for advice on where to seek further legal advice on the subject.

9.5 LOCAL AUTHORITIES

9.5.1 Local authorities: Drivers

The key drivers, driving water conservation and management projects for Local Authorities (LAs) are often related to 'hotspot events', either during periods of flooding or droughts. Water recycling has obvious benefits during periods of water scarcity as does effective rainwater conservation, planning and management. For settlements low river flows can be hazardous, not only for its impact on water quality that can affect wild life and the viability of local ecosystems, but also have implications on the visual environment, more of a problem in tourist areas. Hence water flow enriching schemes may well be undertake by LAs, often in conjunction with local water Utility companies.

Flooding, as we have increasingly learnt this century, is even more of a problem. LAs had jurisdiction over only some stretches of water ways in their areas and will undertake water course management and control projects to ensure that local residents are a little affected by inundations during peak flows as possible. A major concern has been to eliminate or reduce overflows from combined sewers which may pose a significant health hazard for the local population. A number of LAs are investing in projects to enhance storm water retention rates through Sustainable Drainage Schemes (SUDS), the costs of which must be born by themselves and not central government. Such water management schemes may well actually save the costs of building storm water or combined sewers, that if fact would be borne by local Utilities or central government.

9.5.2 Local authorities: Barriers

The low level of use of water conservation, management and reuse systems by LAs must, in part, be caused by a lack of knowledge of subject by LA Managers, Council Members, Planners, Building Control officers and Environmental Health Officers. Poor communications between departments in

---

[13] http://www.elflaw.org/

LAs also exacerbates the lack of effective action. Well set out information and Guidance on the subject would facilitate its uptake in this sector just as was achieved for the SUDS systems.

Inevitably the unknowns concerning the costs involved in storage tanks, filtration systems, maintenance implications and water quality standards for grey water/rain water reuse schemes must also pose a significant barrier here.

9.6 WATER COMPANIES

9.6.1 Water companies: Drivers

Water companies are at the coal-face of water conservation. They are increasingly concerned about rising demand for water which, in some scenarios appears unsustainable. The impacts of Climate Change, hotter summers, and greater public demand for water has encouraged the need to adopt 'Twin Track' options to trouble shoot short term hotspot problems while putting into place medium and long term structural strategies to accommodate these trends is a real driver for the industry.

Increasingly stringent legislation to control water sales by Privatised Companies has been introduced over the years such as the Water supply regulations 1999 and 2002. Under this legislation the companies are placed under a Duty to promote water conservation to customers through legislation such as the Environment Act 1995, and Ofwat MD118.

Metering, promoted through the Utilities in domestic, industrial and commercial properties is also providing financial incentives to save water.

The limiting of ground and river water abstraction, through the licence system is also driving companies to optimise the efficiency of the use of available water.

Public opinion against the building of new dams places more onus on water management and conservation schemes for Utilities.

The Government's Sustainable Development Strategy, as manifested in the Water Summit's 10 point plan in 1997, is also placing the onus on Utilities to minimise waste through the greater uptake of water conservation schemes for Leak reduction, and the rebuilding of the water supply infrastructure.

The Competition Act 1998 allows water companies to operate outside traditional boundaries enabling Utilities to undertake new approaches to water conservation and waste reduction that can provide more water to sell to new customers.

9.6.2 Water companies: Barriers

A key barrier to the greater uptake of conservation and reuse schemes by the water companies is that they make their money in selling drinking-quality water from the tap, and that is what they have always done. Much in the field of water conservation and reuse can be done only at the scale of an individual building or development and such systems, by their very nature, will reduce the amount of drinking water sold to individual customers. So the profit motive for developing such systems is low.

Water companies are also impeded in their search for smaller building scale water supply solutions by the industry aim of driving down costs of supplying the water to individual customers. This aim is traditionally achieved by applying economies of scale at a macro-level, with the ultimate goal being the reduction of the cost of water at the taps rather than a sustainable supply over the next decades or even century. There is also a widespread lack of flexible cost tariffs that can encourage water conservation and supply management. The trade off between water conservation and the need to optimise returns to share holders is also a potent barrier to the introduction of water conservation measures.

Lack of good shared research means that often the cost benefits of water conservation still need to be quantified in relation to the environmental and societal impacts of related measures. For instance there is little publicly available research available on the demographics of the increasing population, affluence and ownership of water using appliances, although individual companies often have detailed research programmes covering such topics. Water conservation requires partnerships with customers which can be time intensive and costly to set up. Issues of the security of supply are often not factored into those studies that are done, and risk analyses are not applied to the industry in the same way as is currently done at all levels for the energy supply industry.

A less obvious constraint on more extensive use of water recycling is the existence of a deep-seated familiarity with the current water distribution arrangements. The population of the UK are accustomed to having access to as much high quality water as they require, when they require it. Furthermore, the UK water industry is accustomed to supplying the resource on demand. During the work on the WROCS (2000) project the research team noted the lack of imagination and creative thinking displayed by major water sector organisations. A shift from a single product production mindset to a multiple product service mind set would go a long way towards creating conditions conductive to the exploitation of recycling opportunities. In addition to conservation and recycling as demand management options open to the companies and agencies, leakage control, pricing, or tariff structuring can be used, as well, before recycling is seriously considered.

9.7 PRIVATE CONSULTANTS

9.7.1 Private consultants: Drivers

There appears to be a real, and potentially lucrative, role for private consultants to inform the public sector on ways to improve their water consumption, conservation and reuse in the climate of rising unti costs and environmental awareness. The UK Government's Sustainable Development Strategy places water conservation at the heart of the design of 'sustainable developments'. The opportunities for consultants to earn good money in building Capacity in terms of skills, knowledge and understanding of the issues may well prove a driver in the field.

9.7.2 Private consultants: Barriers

Water saving is seen as a poor relation of 'energy efficiency' and has less earning potential at the moment not least because the economics and paybacks of water conservation and reuse are poorly understood. Key barriers include the lack of clear standards for water conservation and greywater recycling and the lack of a developed market in the field, although this is changing with the lessons learnt from demonstration projects.

9.8 ARCHITECTS, DEVELOPERS AND PLANNERS

9.8.1 Architects, developers and planners: Drivers

The rise in general environmental awareness means that clients are increasingly willing to pay a 'green premium' for conservation and recycling schemes in their developments. Water conservation increasingly seen as part of the 'green credentials' of a development and greywater systems are considered trendy in some quarters. The desire for 'innovation', and new products, regardless whether they have been tried/tested, may be seen as a driver, or a barrier, to the uptake of related technologies as consumers are often risk averse (WROCS, 2000).

Flooding has heightened consumers concern about the role and nature of water use and availability. Flood Control may require water conservation and storage strategies for building on and off flood plains such as those promoted by the SUDS schemes.

SUDs systems are becoming better understood by designers, developers and planners alike. These requires water retention strategies that may involve attractive 'green building elements' such as green roofs and soft landscapes that appeal to designers and clients. Such designs also have other environmental selling features in that they can help to enhance local climate and absorbing the CO_2 over cities thereby reducing greenhouse effect. However it should be noted that green roofs are not optimal for water conservation as they provide poor water quality for recycling. Hence retention and recycling require different approaches.

Issues of soil shrinkage during extended droughts, leading to a rapid rise in related insurance claims is also raising public awareness of the environmental impacts of water shortages.

Security of supply issues, and climate change impacts, may influence developers to include additional water storage sources, particularly in areas where future droughts are predicted, such as the South East of England.

9.8.2 Architects, developers and planners: Barriers

Key barriers are the general lack of knowledge and information on the subject, and the relationship between water conservation and reuse and the more general application of 'Sustainability' principals to a development.

A front line barrier is the additional cost of construction and maintenance of such systems in relation to the currently perceived low price for water in general. Such systems are seen as an optional high cost extra in developments, unless they are integrated the whole philosophy of the scheme, as happened for instance at the Hockerton Housing Project[14]. Issues of who ultimately owns and maintains communally owned water resources is also a barrier to their wider use.

Concerns about the water quality in such systems and public health also are high on the list of reasons given by a range of potential stakeholders for not including innovative water supply and reuse schemes. This is understandable in relation to a number of project where grey water schemes have been shown to provide, at least at times, less than adequate standards of water quality (Roaf *et al.*, 2003). The lack of clearly defined and disseminated water quality standards for grey water and stored rain water form a key element in this barrier as they provide no recourse from liability claims by designers and clients alike by the lack of demonstrable performance standards. Previous bad experiences, such as the black frothing toilets at Linacre College (Roaf *et al.*, 2003) are often more widely advertised in the Press than the successes.

A lack of a common language with which engineers, planners and architects can discuss water systems is a problem with the lack of adequate educational programmes covering this field (Roaf *et al.*, 2004b). The integration of the subject into undergraduate and post-graduate courses for all related disciplines is a real barrier to their further uptake.

---

[14] www.hockerton.demon.co.uk

9.9 EDUCATION AND RESEARCH

9.9.1 Education and research: Drivers

The educational and research sectors may potentially earn income by acting as effective conduits for knowledge transfer between stakeholder groups and regions, countries and continents. Increasingly water availability is seen as a major global problem, particularly in the face of Climate Change.

Opportunities exist for interdisciplinary design and research including industry, architectural and engineering partners, and obtaining of new funds for interdisciplinary research and curriculum development. This is being encouraged by the trickle down impacts of the Government's Sustainable Development Strategy that eventually manifests itself in the research funding rounds in the UK.

The Regulatory regime actually support new scales of demand management and hence offers the opportunity for increased levels of related teaching and research. The desire for businesses and sectors to 'get their own house in order', eg. through the 'Green Campus Movement' is enabling an effective 'Learning by Doing' approach that should improve opportunities in the teaching and research fields.

In addition the design and property management professions, eg. Architects, Engineers, Surveyors and Facilities Managers are now requiring that 'Sustainability' to be included in teaching modules and courses and so steps must be made to ensure that water issues are included here (WROCS, 2000).

9.9.2 Education and research: Barriers

There are a number of fairly potent barriers to the growth of such studies in the education and research sectors. Related bodies are typically reactive, rather than proactive, in designing curricula and research projects, meaning that new fields are slow to develop. The relative lack of advanced skills and knowledge in the areas of environmental performance of such water systems is also a barrier hindering the teaching of, and research development in the field.

The lack of publications on the subject that are accessed by the relevant stakeholders is a barriers as is the lack of consistency in what constitutes Best Practice in the area, again due in part to the lack of established standards in the field. The lack of a widespread perception of need for such work impedes the development of the field.

Lack of co-ordination in what is taught in schools of architecture, planning and engineers removing 'common ground for understanding' limits the scale and impacts of related research and the feeling that this is an 'expert's subject' is enhanced by the competitive practices of a reluctance to share information between water companies, who have paid for results, and research bodies who wish to use them.

Water saving is seen as the poor relation in the family of environmental sciences with a low profile and a low research rating, enhanced by the consistent problem of relatively cheap mains water. Perhaps the scatological image of the subject, for instance of toilet performance, is not a glamorous field to the young researcher to move into.

9.10   MANUFACTURERS

9.10.1 Manufacturers: Drivers

Niche markets where conservation and reuse are already considered an appropriate solution to water supply problems include:
- Isolated buildings or remote communities
- Households or communities with private supplies
- Buildings located adjacent to sites which have additional recycled water use potential
- Building with un-adopted sewers

Business conditions under which water recycling becomes commercially attractive can change rapidly. Assumption should not be made that somehow water supply and sewage companies have a responsibility to initiate and operate water-recycling schemes. The market place provides a strong driver as is evidenced by the relaxation of the Water Supply Fitting Regulations in 1999 to permit the retrofitting of dual flush mechanisms in WCs installed before July 1999. This opened a new market for related water conservation products.

Some niche markets are already popular and booming. In terms of technology selection, rainwater systems are perhaps a more feasible and commercially viable option than grey water systems in the short term for several reasons:
- More equipment (often in the form of self-construct kits) is available
- The risks to public health are potentially lower
- Architects and builders are largely familiar with the concepts
- financial savings from demonstration schemes are well proven

9.10.2 Manufacturers: Barriers

Achieving and maintaining reliable water quality is a challenge that in some circumstances may prove to be a barrier for product development, particularly in the absence of well established standards for water quality. For grey and black water, the technology required to guarantee reliable effluent quality is not simple. Some biological process is desirable to ensure that the treated effluent can be used for most purposes. However, the types of technology which are suitable for these applications, such as ultra-membranes, are capable of treating black water to a similar degree. This raises the question of whether grey water only systems are making the most economical use of the resource. Installation and plumbing would be simpler and costs could be reduced if integrated grey/black treatment systems were installed in preference to grey water only applications.

Effluent quality, and subsequent risk to public health, in the vast majority of systems reviewed during the WROCS project was strongly influenced by performance reliability. Stable technology performance was an elusive target for many of these projects and in one case study, faecal coliforms (although sometimes only 1 or 2) were present in every grey water sample analysed. All water recycling systems should be fitted with some form of fail-safe device which both notifies a breakdown in processing and shuts off effluent supply to reduce the risk of system failure which can so badly impact on the public perception, and uptake of, such systems. However, the design of failsafe devices raises problems in its own right. They must themselves be reliable and effective. Furthermore, on-line monitoring of water quality is an expensive operation, although it is more economical at larger scales of application.

The lack of clear water quality standards remains a key barrier to the development of relevant technologies. The lack of established markets for such products, the life time in use and the pioneering status of much of the technology and the lack of a clear idea when the different systems will be required in the marketplace are all potent barriers to the development of such products. A lack of good research on such systems also impedes the development of water conservation and reuse products.

9.11   CUSTOMERS AND CONSUMERS

9.11.1 Customers and consumers: Drivers

In a number of building types where water supply is at a premium then water conservation and reuse systems are already employed. For example in remote houses and in hotels where water shortages in conventional systems may exists and the per capita demand for water is very high.

At the household level, the current payback time on commercially available systems makes them unattractive for all but a small number of households who have specific water demand and usage patterns. Clearly, variations in the price of delivered water or the cost (capital and operating) of the reuse system will change the balance of this equation. At particular times, such as in dry weather for garden watering, water conservation is already widely used in water butts and smaller stores. By using such simple systems customers can definitely save money over the year and many do.

Rising environmental awareness is a key driver for the further uptake of water conservation and reuse systems and grey water techniques are widely included in sustainability briefs, largely for new buildings. There is a changing perception of the value of water both in the wider, global environmental context but also closer to home as the unit costs of water supply steadily increases. Conservation is becoming trendy, but also the mind of the consumer is more clearly focused since the introduction of water meters for individual buildings. These enable customers to see what they are using, employ water management schemes and to reduce their overall consumption and water bill. New ways of saving money through water economy are thus of increasing interest to many consumers.

The cost of water will ultimately stimulate the uptake of local conservation and recycling schemes. There were in mid-2004 around 4.7 million people in Britain who could not pay their water bills and companies were threatening significant increases in annual bills to pay for infra-structural repairs and capacity building. While considerable attention is paid to 'Fuel Poverty' (only around 1.4 million people cannot pay their electricity bills in the UK) increasingly 'Water Poverty' will rise up the political agenda, providing an increasing strong driver in the field.

9.11.2 Consumers: Barriers

Customers tend to be reactive in their habits, whether through apathy, laziness or a real reluctance to invest in what is perceived as untested, and therefore high risk, products. There is not an established culture of water conservation and reuse in the UK because typically there was never a need for it in this rainy climate.

The paucity of good information on the subject also increases the consumer's anxiety as to the reliability of such systems and products. In addition customers feel disenfranchised from the water companies, because they presume they cannot live without clean water and they have no say in its supply. This is reinforced by the lack of regulations that link performance to accountable standards for the water industry, it is all a black box experience where customers take for granted they will have a guaranteed supply of clean abundant water, for a reasonable price.

Fear of unknown is always a barrier to innovation and the development of new product ranges and when the commodity being experimented with is as fundamental as drinking water then the health implications of making the wrong choice in this field are a worry for the public.

The poor availability of reliable products is the field is also a real draw back for the wider adoption of these practices which provides rather a Catch 22 situation.

However again we fall back on a key driver to the wider scale uptake of conservation and reuse practices being the lack of water quality standards and the associated consumer perception of risk related to such products. These strong barriers are exacerbated by the perception of the low value of water due to its apparent abundance in the UK. This may well change fast in the UK with Climate Change and a couple of dry years, as happened in 1995.

9.12   CONCLUSIONS

What is clear is that the problem of providing adequate clean water to the increasing numbers of UK citizens, with their increasing per capita demands, in a changing climate is a challenging one. This is particularly so at a time when we take it for granted that we can wash our cars and water our gardens with drinking water, ad infinitum. Our water supply industries must adapt to survive, and water conservation and reuse provide core tool for that adaptation.

The safe and economic exploitation of water the water conservation and recycling potential in the UK could be promoted with more inter-disciplinary, and inter-agency, collaboration, building on the drivers outlined and above, and working together to reduce the barriers at the same time. The WATERSAVE Network, developed to provide a forum for debate and leadership on water conservation and reuse, with membership drawn from commercial, regulatory, academic and governmental interests, has been instrumental in identifying the above drivers and barriers. However in the words of the Water Regulation Advisory Scheme:

'Due to a lack of straightforward, factual information about costs, reliability and control of hazards there has been little enthusiasm to install reclaimed water systems. The adoption of reclaimed water systems is strongly influenced by user perceptions and economic benefit, but once installed, reclaimed water systems should not be allowed to fall into a state of disrepair or disuse and attention is drawn to the importance of prompt investigation and resolution of problems.' (WRAS, 2000).

The above paper is by no means exhaustive or comprehensive but it does outline the need for joined up thinking in the water industry and market place. It is apparent from this study that the Barriers are numerous, but that the Drivers for water conservation and reuse in Britain are increasingly powerful in the changing climate of the water demand and supply industry of the 21st century.

9.13 REFERENCES

Brewer, D., Brown, R. and Stanfield, G. (2001) *Rainwater and Greywater in Buildings: Project report and case studies.* Technical Note TN 7/2001, BSRIA.

BSRIA (2003) S*afety and performance of grey water systems.* Building Services Research and Information Association

DOE (1996) *Agenda for Action.* Department of Environment, London.

DMB (2003) *Demand Management Bulletin.* Environment Agency. No.59, June.

Hulme, M., J. Turnpenny, and G. Jenkins (2002) *Climate Change Scenarios for the United Kingdom: The UKCIP02 Briefing Report,* Tyndall Centre for Climate Change Research, School of Environmental Sciences, University of East Anglia, Norwich, UK.

Ofwat (2001) *Leakage and the efficient use of water 2000-200.* Office of Water Services.

Roaf, S., Horsley, A. and Gupta, R.(2004a) Flooding. *Closing the Loop: Benchmarks for Sustainable Buildings.* RIBA Publications, London.

Roaf, S., Crichton, D. and Nicol, F. (2004b) *Adapting Buildings and Cities for Climate Change.* Architectural Press, Oxford.

Roaf, S., Fuentes, M. and Thomas, S. (2003) *Ecohouse 2: A Design Guide.* Architectural Press, Oxford.

WRAS (2000) Water Regulation Advisory Scheme. HMSO.

WROCS (2000) *Final Report of the Water Recycling Opportunities for City Sustainability (WROCS) project.* Cranfield University, Imperial College London and The Nottingham Trent University.

10

The economics of water demand management

Paul R. Herrington

10.1 INTRODUCTION: SETTING THE STAGE

10.1.1 Definitions and categories

It is now received wisdom that reconciling present or future water supply-demand imbalances by augmenting supplies – e.g., damming, diverting or pumping the basic resource – may in certain circumstances lead to significant environmental damage as well as economic waste. Imbalance resolution may in fact be achieved through (i) supply expansion, (ii) supply-management through enhanced use of supply sources (e.g., conjunctive use), (iii) operational control by authorities responsible for water services supplies and (iv) the management of demands exercised by one or more user sectors (households, industry, agriculture, etc.). It is then convenient to simplify the taxonomy and classify all such reconciliation measures as *either* supply-management *or* demand-management, and consequently to define water demand management[1] as:

those policies, measures or other initiatives which serve to control or restrict the demand for, use of or waste of water supplies or other water services.

Note that there is no restriction here of demand-management to initiatives either promoting the *efficient* allocation of water (as in the definition offered by Renzetti (2002)) or taking into account the *value of water in relation to its cost of provision* (as in those of Baumann *et al.* (1980) and Winpenny (1994)). Thus a proposed demand-management initiative may be economically good or bad. Indeed, it is one of the central purposes of this chapter to describe appraisal techniques which can be used to distinguish the better policies and options from the not so good, the bad and the ugly. The *economics* of demand-management in the chapter title makes explicit our central concern with *economic* appraisal and viability, and therefore, inevitably, with the identification, quantification and evaluation of the benefits and costs of demand-management options and programmes. Detailed consideration of the above definition suggests that measures and policies may be usefully classified as in Table 10.1.

Table 10.1 Categories of demand-management measures

Category	Types of Demand-Management
i. Economic (or Incentive-based)	Pricing structures and levels; subsidies, taxes and tax credits; rebates and buy-backs; low-interest loans; fines for regulatory non-compliance
ii. Educational	Education, information provision and campaigns, water audits
iii. Regulatory	Use and consumption regulations; building, plumbing and landscaping codes
iv. Restrictive	Rationing[2], moral suasion and voluntary use reductions
v. Operational control	Leakage detection and alleviation, pressure control and sewer infiltration control

In principle all the elements in Table 10.1 may be applied to the different *stages* of the water cycle (e.g. abstractions, storage, water service users' own facilities) *and* to different water-using *sectors*, e.g. households,

---

[1] The terms *water demand management, water conservation* and *water use efficiency* are often used interchangeably in the literature. For this study of policies and actions at the consumer end, however, the term *demand management* is preferred since the former of the two alternatives emphasises preservation of only the *natural* resource and thus draws attention towards abstractions and discharges, while the latter is too technical and runs the risk of being identified with water efficiency in just 'productive' sectors – agriculture, industry, etc.

[2] Through supply-side rationing, demand is clearly 'managed' in an *ex post* but not an *ex ante* sense.

offices/commercial, industry (EEA, 2001). Because of (i) the homogeneity of much domestic-type water usage across these sectors and (ii) the current stress of demand-management policy on such use, this survey is restricted to households' public water supply demands and piped wastewater services and conventional domestic-type use in other settings (offices, hotels, factories, schools, etc.). No consideration is therefore given to operational control, non-domestic usage of the public piped supply systems and abstractions/discharges.

10.1.2 Scope for demand-management

Per capita household consumption data for 21 out of 30 OECD member states covering various periods over 1975-2000 show recent estimates stretching from Germany's 116 litres/person/day (lpd) for 2000 to the US's 382 lpd in 1995 (Herrington 1999, updated). Four countries reveal increasing trends (Canada, Slovakia, UK [unmetered households], US), five an average of over 200 lpd (Canada, US, Australia, Italy and Japan) and three an average of 150-200 lpd (England & Wales [unmetered], Switzerland and Sweden). On these simplistic criteria alone, nine out of 21 states might be expected to have a serious interest in demand management. In fact, only five of these nine have such concerns: Australia, Canada, the UK, Japan and the US. Others embrace the interest for different reasons, e.g. Denmark (supplies affected by groundwater pollution), Netherlands, Belgium and Germany (other environmental issues), and Spain and Portugal (with often acute development/environment trade-offs).

Table 10.2 Importance of domestic-type PWS use in England and Wales (Ofwat 1996, BSRIA 1998 and Ofwat 2003b)

Consumption (Ml/d)	1995-96	2002-03
1. Households (unmeasured)	7736.2	6364.9
2. Households (measured)	146.0	1325.1
3. Domestic-type use in non-h/holds	2502.2	2423.5
(as a percentage of 4.)	*(60.9%)*	*(60.9%)*
4. Total non-household use	(4108.7)	(3979.5)
5. Total domestic-type use of PWS	10384.2	10113.5
6. Total useful consumption of PWS	11990.9	11770.3
7. Importance of domestic-type use	86.6%	85.9%
(5. as a percentage of 6.)		

More detailed information for England & Wales shows the extent to which 'useful' consumption of the public water supply (PWS) – i.e., that excluding leakage – is dominated by domestic-type usage. An investigation estimated that in 1995-96 domestic-type usage (WCs, urinals, washrooms, showering, garden watering) constituted 61% of 'useful' non-household consumption in the aggregate of the following building types: factories, hospitals, hotels/motels, leisure centres, nursing homes, offices, schools, universities and retail stores (BSRIA, 1998). Using recent water company returns to Ofwat, Table 10.2 shows how domestic-type use has remained at 86% of 'useful' PWS consumption over the 1995-2002 period.

Recent studies detail the range of domestic demand management measures: Baker *et al.* (1996a) list 97 and Pinnie *et al.* (1998) 33. After Tate's (1990) earlier Canadian survey, White (1998c) has edited an excellent Australian manual stressing the importance of economic evaluation through a 'total resource cost test', while Grant (2003) adopts a practical 'household economics' approach to assessing the viability of many UK measures on offer, using water savings and tariff levels to generate estimated net benefits expressed variously as annual savings (£/year), net present value (£) or payback periods (see 10.2.3, below).

However, the most comprehensive treatment of current demand-management measures in all major sectors – from an exclusively United States viewpoint – is the monumental work by Vickers (2001). Historical tables for all the major domestic (and other) types of water use and appliance, including typical water use volumes and likely savings for model vintages back to the pre-1980 period, show the enormous scope for demand management in the U.S. water context. There is no examination, however, of the *economics* of the measures described.

10.2   ECONOMIC APPRAISAL

10.2.1 Different types of appraisal

Any proposed water demand management measure, policy or programme should be subject to systematic appraisal, which will normally have *technical*, *economic, environmental* and *social* dimensions. We outline these in turn, before turning to a more detailed investigation of the techniques of economic and financial appraisal.

10.2.1.1 Technical appraisal

To be seriously considered for adoption, a water demand measure must be technically effective, meaning it should function satisfactorily, have a relatively low failure rate and result in a reduction in water use or wastewater disposal that is significant. But what is 'satisfactory', 'relatively low' and 'significant' will partly depend on the measure's costs and also its benefits – the economic, environmental and other social costs of water service provision that are avoided as a result of its implementation. Thus *technical* effectiveness may in part be subsumed under the *economic* and *environmental* dimensions. However, it is still worth separating out the technical criterion, since a proposed measure may clearly be found to be technically ineffective without recourse to detailed economic analysis. Sometimes decisions on technical effectiveness are the responsibility of a national 'standards' organisation, and such a body should not be immune to the economic and environmental consequences of the minimum standards (and thus the technical effectiveness) that it may be required to determine. If it is, standards may be set too high.

10.2.1.2 Economic appraisal

Economic appraisal is an essential element in the evaluation of all types of demand-management practices, measures and programmes, as well as for supply-management and supply-enhancement schemes. At a fundamental level, economic appraisal is indifferent to actual cash flow. The aim of the exercise should be to identify, quantify, evaluate (in monetary terms) and compare the economic benefits and costs of a measure, programme or policy change, timing them according to their impact on the community. For some such benefits or costs no cash may flow. It therefore amounts to an assessment of the economic implications of using scarce resources in a particular manner.

A related *financial* appraisal examines the financial effects of the introduction of a demand-management option or programme on one or more stakeholder groups (e.g., customers in general or a single sector like households, the water utility itself, any national environmental protection agency, 'users of the environment'). It is sometimes instructive to present a social cost-benefit appraisal (embracing all socio-economic effects) so that the financial and other gains and losses to each of the stakeholder groups are separately distinguished. In that situation the aggregate net gain of all the interest groups (comprising 'society') is necessarily equal to the net social benefit that emerges from the cost-benefit analysis, if all the relevant cash flows have been accounted for (see 10.2.3.5, below).

10.2.1.3 Environmental appraisal

Some practitioners believe that if any environmental and other non-economic social consequences (gains, losses) of a proposed measure could be satisfactorily reduced to monetary equivalents, then all such consequences should be subsumed with economic elements under a single test comparing 'social benefits' and 'social costs'. Others are more sceptical, believing with Bowers (1997) that "....with many environmental assets no meaningful monetary valuations can be derived."

In the latter situation, there are two ways forward. First, adverse environmental effects may be dealt with by the prior erection of a set of *environmental standards*, implicitly or explicitly reflecting society's chosen sustainability preferences. Then environmental assets do not have to be valued; instead the costs of meeting the environmental standards are included in the calculus, if either standards might otherwise fall (e.g., through lower river flows in a supply response to an excess abstractions demand situation) or if it is decided that standards should be increased (Bowers, 1997)[3].

Alternatively, recourse may be had to a range of distinct environmental appraisal techniques, such as Environmental Impact Assessment (EIA), Strategic Environmental Assessment (SEA), Risk Assessment (RA), Life Cycle Analysis (LCA), as discussed in chapter 6, and a range of Multi-Criteria Analysis procedures (MCA) (DETR, 1999). LCA and its associated 'Life Cycle Costing' (Tucker *et al.*, 2000) helpfully stress asset disposal costs in what is really a cradle-to-grave environmental cost-benefit analysis (CBA), although they can also lead to confused analysis due to the possible double-counting of resource scarcity in both higher market prices *and* in the LCA (Pearce and Barbier, 2000). MCA formally recognises a multiplicity of decision-taking objectives (e.g., distributional impact, environmental sustainability, economic efficiency) and measures the extent to which any measure/ programme can achieve each of these objectives. Attaching weights or scores to each goal enables individual options or programmes to be scored and therefore compared. Typically, alternative sets of weights are considered, to check the sensitivity of the results. Note that any score (in points) given to economic efficiency (in £) implies a valuation (£) wrapped around the other goals, so CBA returns to the scene.

10.2.1.4 Social appraisal

The term 'social appraisal' is used here as a shorthand for appraising a conservation option or programme in terms of its *social, political* and *institutional* feasibility – referring respectively to the consensus of public opinion, the legislative position and the institutional capacity to implement and deliver (Flack, 1982; Herrington, 1987a). The term 'feasibility' suggests overcoming a hurdle, perhaps more appropriate for these social dimensions than the implied quantification of an 'appraisal'.

10.2.2 Economic appraisal: from micro to macro

There have emerged in the last 20-30 years two distinct but related literatures dealing with the economic assessment of water demand-management activities: one to be applied at the micro level, and the other at the macro level in the establishment of broad programmes.

---

[3]The strict corollary appears to be that increases in environmental standards above the required minima are not granted any monetary value.

Micro-appraisal

The micro approach recognises that engineers, water utility decision-takers, government policy advisers and others often need to be able to forecast the economic implications of a demand management policy or measure that is either (i) proposed, in the process of development or just formulated, or (ii) already developed and trialled and now being considered for application in a new context. Clearly capital and operating/maintenance costs of the activity will somehow have to be balanced against expected benefits. Various techniques for making the comparison and arriving at a decision-rule have been used in the water industry, and these will be examined in Sections 10.2.3 (the underlying theory) and 10.3.2 (dealing with applications). 10.2.3 will discuss in turn:

- cost-benefit analysis
- internal rate of return
- cost-effectiveness analysis
- payback period
- the total resource cost test

White (1998a) has argued that the total resource cost test is the most appropriate test for determining the net economic benefits of a demand management measure when undertaken from the perspective of the whole community.

Macro-appraisal

At an altogether different scale, the macro approach reflects the fact that increasingly water utilities in their strategic planning exercises are obliged (for a whole range of technical, financial, regulatory and political reasons) to make transparently rational and coherent decisions about how to reconcile future demands and supplies for core water services. In principle reconciliation embraces both quantitative and qualitative dimensions of the services, but the practice is usually to take the quality (both the quality of service and intrinsic quality) as given and then concentrate on the resolution of any quantitative problem that emerges. Any postulated quality increases – or even decreases - in future years (e.g., as a result of government legislation or EU directive) can be built into the demand forecasts, which complicates but does not change the essence of the reconciliation problem. Just how the optimal balance of policies, schemes and options should be decided upon amounts to fixing an optimal investment programme, and Section 10.2.4 outlines the methodology for dealing with this problem for the core water services.

10.2.3 Economic appraisal: micro

10.2.3.1 Cost-benefit analysis (CBA), discounting and net present value (NPV)

Introduction

CBA is a procedure for assessing an individual project, policy or policy change by reference to the difference between the monetary values of the social benefits and costs it is expected to generate over its likely lifetime (or some agreed shorter period). A social benefit (or cost) may in turn be defined as anything with an economic and/or environmental dimension which increases (or decreases) human welfare. Human welfare is assumed to be determined by what individuals prefer (their wants) and what they are prepared to pay to satisfy those wants. It is thus individually based and in its most familiar variant indifferent as to *who* secures the benefits and incurs the costs[4].

Theory

Individuals' preferences are revealed through 'direct' market behaviour and outcomes, surrogate markets (e.g., housing market prices can reveal information about the value put on air pollution reductions) and market research investigations (e.g., asking people about their willingness to pay – WTP – for defined environmental outcomes). Typically social benefits take the form of increased output, savings in economic resources or time, and increased quantities or qualities of environmental services. Social costs are typically represented by the use of economic resources for a particular purpose (thus ruling them out for other uses), or by environmental or social welfare reductions.

In this way it is possible to estimate a demand-management measure's aggregate net impact upon a community (nation, region, water utility area, etc.). Suppose the monetary values of a measure's (or project's) social costs and benefits are defined over its lifetime of n years by:

(i)   the capital costs (K_0), assumedly borne in year 0,

(ii)   a steam of annual social benefits B_t arising in years $t = 1,.....n$,

(iii)   a stream of annual social costs C_t incurred over years $t = 1,.....n$.

People's preferences (ceteris paribus) for experiencing benefits sooner rather than later – and incurring costs later rather than sooner – are summed up in the idea of *time preference*. This concept is normally reflected in CBA in an annual *discount rate*, through which the present value of a benefit or cost (say £X_t)

---

[4] CBA may be adjusted to reflect distributional objectives, but this has been seldom used practically in the last 20 years. For a brief discussion, see DETR (1999).

arising or being incurred t years into the future is estimated in present value (PV) terms as

$$PV(X_t) = \frac{X_t}{(1+r)^t} \qquad (10.1)$$

where r, the annual discount rate (expressed as a proportion), is assumed to represent the rate at which *either* 'society' (viewed as a collective of consumers) trades off benefits (or costs) received (or incurred) in future years, giving rise to a 'social time preference' rate, *or* the water utility/agency has to borrow or otherwise raise funds in order to finance the measure (giving rise to the idea of the 'opportunity cost of capital')[5]. The discount rate is generally (although not invariably[6]) assumed to be constant over future time.

The net present value (or present value of the net benefits) of the measure is then given by

$$NPV = K_O + \sum_{t=1}^{n} \frac{(B_t - C_t)}{(1+r)^t} \qquad (10.2)$$

The Decision-Rule
The decision-rule for the project in the 'simplest' situation is:

$$\textit{accept} \text{ if } NPV > 0 \text{ and } \textit{reject} \text{ if } NPV < 0 \qquad (10.3)$$

Complications alter this rule. First, if the costs and benefits are uncertain, so is the *NPV*. If only possible ranges for the costs and benefits are known, the

---

[5] Guidance issued in 2003 to the water companies in preparation for the 2004 price review (PR04) incorporated both approaches to the discount rate (Environment Agency 2003). In assessing candidate schemes for the PR04 Environment Programme and thus for framing advice to Government Ministers about the extent of the Programme, Ofwat and the Environment Agency would use a social time preference annual discount rate of 3.5%. Once the overall Programme had been decided on and an "efficient level of regulatory control…thereby identified", water companies would be expected to use a cost of capital discount rate to arrive at the least-cost compliance option and decide on financing matters. For a statement of the theory and numbers behind the derivation of the central government determined social time preference rate at 3.5%/year, see Her Majesty's Treasury (2003).
[6] The Treasury (Her Majesty's Treasury, 2003) recently recommended that in public services economic appraisal exercises costs and benefits accruing more than 30 years into the future should be discounted at a gradually decreasing rate, from 3%/year for those 31-75 years ahead to 1%/year for those over 300 years ahead. This deals to some degree with the 'tyranny of discounting' claim, which maintains that discounting – and particularly higher discount rates – militates against the environmental interests of future generations. See Turner *et al.* (1994) for a useful discussion of this issue.

possible range of values for the *NPV* can be presented to the decision-taker. However, if the uncertainty of the 'inputs' can be characterised through probability functions, the use of Monte Carlo methods will enable a probability function for *NPV* to be estimated. Second, if there is a fixed budget for capital or any other input for year 0 which cannot be exceeded, then it may be that not all projects with *NPV* > 0 can be afforded. In that case, overall programme *NPV* would be maximised by ranking projects by the project *NPV* per unit of capital (or other input) and working down the list until the budget was exhausted.

The relatively few published direct applications of the *NPV* criterion set out above to the appraisal of water demand-management measures and policies have dealt mostly with the metering of unmetered homes and the introduction of marginal cost pricing in metered situations. These will be reviewed in Sections 10.3.3.10 and 10.3.3.11. The number of studies is small because *as the decision-rule stands* any particular estimated monetary value for *NPV* is difficult to assess. Certainly, a positive *NPV* is 'good' from an economy efficiency viewpoint – but how much more impressed should the decision-taker be if *NPV* has the value £5000 rather than £500, or £500,000 rather than £50,000? Presumably the answer depends on the size of the measure or policy change – a 'small' positive *NPV* for a 'large' demand-management initiative might not be viewed well, whereas a 'large' *NPV* for a 'small' initiative would impress.

It is thus clear that the result needs to be *scaled*, by somehow relating the *NPV* outcome to the 'size' of the endeavour – perhaps in terms of an input used (for example, the size of the capital budget involved or the funds incurred at the outset) or of the 'output' of the measure (by for example measuring the size of the water savings that result). Such corrections for scale provide the basis of the techniques that will be discussed in the next two sub-sections.

10.2.3.2 Internal rate of return (IRR)

Theory

The *IRR* is formally defined as that rate of discount which would make the present value of a demand-management measure's benefit stream equal to the present value of its costs stream, i.e. it is the discount rate which would make the measure's *NPV*, as defined above in equation (10.2), equal to zero. The relationship between the *NPV* and *IRR* of a measure is easily illustrated if we assume (i) the capital cost is wholly borne in year zero, (ii) the net benefit stream is positive in every year up to and including the end of its life and (iii) the *NPV* is positive if the discount rate is zero. For in that situation if the discount rate increases the *NPV* must decline, and eventually it must equal zero and then become negative, as is shown by the tracing out the line NPV_A in Figure 10.1, where the *IRR* of a measure A is seen to be r_A (for the moment, ignore the measure B and its *NPV* profile in the figure).

The advantage of the IRR measure is that it is in a language that is easily understood – it is an annual proportional return (or, multiplying by 100, a percentage return). Such a figure measures the overall annual average 'return' to the initial capital expenditure which is represented by the net benefit stream. Suppose a demand-management measure costs £100 in year 0 and has a two-year life, generating net social benefits worth £55 in year 1 and £60.60 in year 2. The annual discount rate which makes the *NPV* zero is readily seen to be 0.1, i.e. 10%. This is the annual *internal rate of return* of the option since £100 invested for a year at 10% would yield £110 at end-year. If we took £55 out, and invested the remaining £55 for a further year at 10%, this would yield £60.60 at the end of the year. So the net benefits ensure that the 'investment' just earns 10% per annum on the initial expenditure.

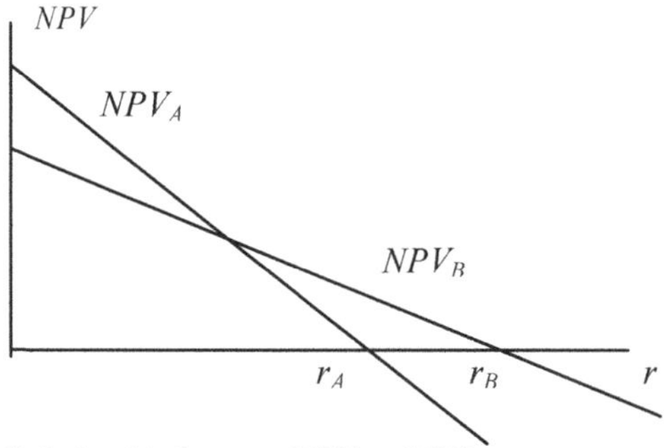

Figure 10.1.Relationship between NPV and IRR.

A Decision-Rule

It is also seen that with the assumptions (i) to (iii) listed at the start of 10.2.3.2, a decision-rule can be derived which will ensure that *NPV* and *IRR* measures can be used to give identical decisions. If the cost of capital for measure A has a value less than r_A, it is seen in Figure 10.1 that NPV_A is positive, so the decision is: go ahead with the option. But the same decision will of course emerge if the *IRR* of the project (r_A) is compared with its cost of capital, since the latter is less than r_A. So the decision-rule becomes: accept the project if $r_A >$ cost of capital, but reject if it is less.

Problems with IRR

With apparent 'equivalence' of the two decision-rules and the much greater familiarity of the 'percentage per annum' dimensions of the *IRR*, is the *IRR* therefore generally to be preferred? No, because it is only in 'simple' situations ((i) to (iii) above) that equivalence holds. If, for example, there are changes in sign of the net benefit stream over the life of the project after year 0, multiple solutions for *IRR* may result (caused, for example, by decommissioning/disposal costs at the end of a project's life). This could give ambiguity in the *IRR* decision-rule, whereas the *NPV* remains single-valued for a given cost of capital. Second, if there is capital or other input rationing, and ranking is required, then ranking measures by *IRR* may well result in a non-optimal selection of measures from the point of view of maximising overall *NPV*.

This may occur when it is necessary to choose between a number of options or projects representing alternative ways of achieving the same objective, and Figure 10.1 illustrates the problem. Suppose the questions are: which to prefer of options A and B, to achieve a given objective, and whether in absolute terms the preferred option should be proceeded with? If an *IRR* rule is utilised for preference determination, B would be preferred, since $r_B > r_A$, and B will be recommended for action so long as the cost of capital is $< r_B$. But it is seen that for lower values of r the *NPV* of A is the higher, so that for a range of values of the cost of capital option A is to be preferred. For higher values of r, B would be chosen, however. So it depends on the cost of capital, and the preferred general procedure is that *NPV* should be used to maximise net social benefits. We conclude that only in the 'simple' situation defined above can *IRR* be used to generate correct go-ahead-or-not decisions, and even then ranking options by IRR can give wrong option selection.

10.2.3.3 Cost-effectiveness analysis (CEA)

CEA associates non-monetary indicators of the scale or 'achievement' of a measure/policy with the monetary valuation of its costs. CEA therefore provides a way of comparing and ranking measures of different shapes and sizes, since scale is accounted for[7].

Consider a demand-management option, and re-write its *NPV* expression in equation (10.2) above as:

---

[7] Further, recognising that the *benefits* of a demand-management measure will often find expression in the consequent water service *supply* costs avoided (economic, environmental and other social costs), it is possible to conceive of what could be termed *benefits-effectiveness analysis* (BEA), associating the same 'achievement' indicator of a demand-management measure with monetary valuation of its *benefits*. By comparing the resulting CEA and BEA ratios, a formal decision-rule on whether to proceed with an option or not can thus be derived, and this *CBEA* procedure (as it could be called) is obviously precisely – and rather trivially – equivalent to the NPV decision-rule in expression (10.3) (see Section 10.2.3.1, above].

$$NPV = \left[\sum_{t=1}^{n}\frac{B_t}{(1+r)^t}\right] - \left[K_0 + \sum_{t=1}^{n}\frac{C_t}{(1+r)^t}\right]$$

i.e. if PW = present worth, $NPV = PWB - PWC$ (10.4)

Assume now that the *water savings over the lifetime of the measure* have a present value of PVW, where

$$PVW = \sum_{t=1}^{n}\frac{W_t}{(1+r)^t}$$ (10.5)

where W_t is the water saved (in m^3) in year t.

Now divide each term in expression (10.4) by the total water savings generated over the life of the measure, as given by (10.5), to produce:

$$\frac{NPV}{PVW} = \frac{PWB}{PVW} - \frac{PWV}{PVW}$$ (10.6)

and call the three ratios, in turn, the measure's average incremental net social benefit ($AINSB$), average incremental social Benefit (AISB) and average incremental social cost $(AISC)$[8]. Thus:

$$AINSB = AISB - AISC$$ (10.7)

The measure's economic acceptance test may now thus be stated in three equivalent versions:

$$
\begin{array}{lll}
\text{(i)} & NPV > 0 \ ? &) \\
\text{(ii)} & AINSB > 0 \ ? &) \\
\text{(iii)} & AISB > AISC \ ? &)
\end{array}
$$ (10.8)

Numerous uses have been reported of variants (ii) and (iii), and of the ranking of options by $AISC$ ratios, in the appraisal of resource-expansion and

---

[8] The $AISC$ measure may in fact be applied to any supply-expansion, supply-management or demand-management option with a defined expected lifetime and progress towards full utilization. It divides the present worth of social costs incurred in the option's construction, operation and maintenance by the present worth of the total water delivered to consumers or saved. In Australia the same measure applying to a demand- or supply-side option has come to be known as the Levelised Unit Cost (LUC).

demand-management measures in the last 30 years; these will be discussed in Section 10.3.2.3.

10.2.3.4 Payback period

The payback period is usually defined as the time it takes an 'investment' in a given measure to generate sufficient financial savings to repay the capital and operating/maintenance costs incurred. In the past it was a commonly used but crude method for analysing the financial attractiveness of capital projects in the commercial private sector. It is still used occasionally in the appraisal of water demand-management measures, usually when the emphasis is on the financial return to a water customer from investment in a water-saving measure. Its main defects are that it (i) takes no account of the distribution of costs and savings over time, (ii) ignores the total savings generated over the whole lifetime of the measure and (iii) provides in itself no obvious decision-rule for acceptance/rejection of a proposed measure. Ranking of options follows the simple (and misleading) rule that the shorter the payback period, the better.

Defect (i) can be corrected by profiling the *NPV* of a 'simple' project (as defined in 10.2.3.1, above), taking account of, successively, only the first 0, 1, 2, 3,.......*n* years of its life (and thus of its benefits and costs in those periods), in order to identify how long it takes before the *NPV* turns from a negative into a positive figure (for at that point in time the *PV* of the benefits is just sufficient to 'pay for' the PV of the costs). This is shown in Figure 10.2, which illustrates the NPV_t/time relationship where NPV_t is the net present value of a measure including only those benefits received and costs incurred up to the end of year *t*.

The problem with payback can be seen here. Consider two alternative options, A and B. A, with low capital cost and low annual net benefits, would be preferred to B (higher capital cost and annual net benefits) since the payback period (t_A) is shorter than that for B (t_B). However, considering all costs and benefits up to the end of the projects' lives (*n* years), B should be preferred under the *NPV* rule since it has a higher NPV *for* t = n. The payback period rule can therefore easily generate wrong decisions if maximising *NPV* is the aim of project selection.

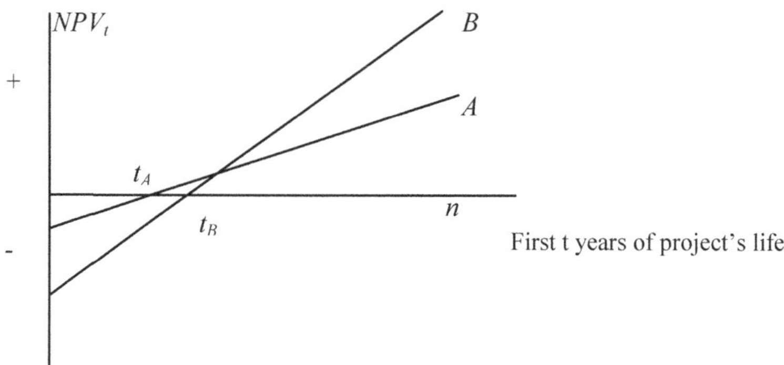

Figure 10.2 *NPV* and Payback Period.

10.2.3.5 Total resource cost test

In the late 1990s Stuart White confronted in the Australian context the issue of the forgone revenue and surplus that a water utility might suffer as a result of the success of demand-management measures (1998a, [with Howe] 1998b, 1998c). Using the Total Resource Cost test originally suggested by Dziegielewski *et al.* (1993), he presented a simple matrix identifying in present value terms all the financial and other economic benefits and costs accruing to (and incurred by) the water service provider and the customer sector following the introduction of a measure or programme. So long as the utility and the customer sector comprised the whole of 'society', cash flow changes were seen as pure transfer payments, affecting the welfare of each of the two sectors but cancelling out overall and thus playing no part in the aggregate net benefit total (which can therefore be seen as equivalent to the *NPV* expression, (10.2), above).

Table 10.3 sets out an elaborated version of the matrix introducing, first, an 'environmental sector' in order to identify a separate environmental balance sheet and, second, a brief taxonomy of the different types of real costs[9] that customers may choose or be obliged to bear as a result of demand-management initiatives in both the price and non-price categories (see Table 10.1, above). For convenience, 'the environment' reflects the activities of both the water resource

---

[9] Those familiar with the language of economics will note no 'loss of consumer surplus' term in the Customers/Costs cell. This is because any consumer surplus loss is already accounted for in the combination of the CC (or RWC) entry in the Costs row and the reduction in water bills under Benefits.

agency (the EA in England & Wales) and organisations responsible for energy provision[10] (other environmental consequences could also be located here).

As is seen in the bottom right-hand cell, the overall economic-environmental test, embracing welfare gains and resource costs (the latter of course referring to all economic resources and not just water), is that the overall net benefits resulting from any proposed change (e.g. from a specified new additional demand-management programme) must be greater than zero, i.e.

$$OCS + CCS + EG[W] + EG[E] - PC - CC - RWC - RQS > 0 \qquad (10.9)$$

Put another way,

$$OCS + CCS + EG[W] + EG[E] > PC + CC + RWC + RQS \qquad (10.10)$$

To induce all stakeholder groups to agree to the programme would probably require the 'Net Benefits' total in *each* column to be positive, which by definition is theoretically possible (through income redistribution, e.g.) so long as the *overall* 'Net Benefits' total is positive. Indeed, the customer and water utility perspectives (columns, in Table 10.3) are essential for addressing the issue of the appropriate customer incentives for participation in water conservation programmes (Fiske & Weiner, 1994). More broadly, the art of gaining acceptance for a new policy or policy change that is clearly beneficial in overall economic terms may often turn on 'teasing out' the net gains so that they are shared in an acceptable manner among all those with an interest. Note also the need to take account of the economic and other social benefits of lower sewage flows and planned sewerage capacity when appraising demand-management of public water supplies. White *et al.* (1999) maintain that sewerage-infrastructure-derived benefits are often larger than those associated with lower water supply costs.

---

[10] The institutions involved could, however, easily be separated from the 'environmental public' and placed in a distinct column, following the United States Environmental Protection Agency (EPA) recommendation that the interests of various 'program participants' be distinguished in any CEA of conservation programmes (Fiske and Weiner, 1994).

Table 10.3 Assessment of a Proposed Water Demand-Management Measure/Programme[11]

	Water Service Utility	Customers	'The Environment' [incl. water resource + energy institutions]	Net Total Resource Cost Test
	(1)	(2)	(3)	(4 = 1+2+3)
Costs	programme costs (*PC*); customers' reduced water bills (*RWB*)	customers' resource costs (*CC*)[a]; reduction in value of water consumed (*RWC*)[a]; reduction in quality of service (*RQS*)[a]	reduced abstraction bills (*RAB*)[b] reduced energy bills (*REB*)	$-PC-CC$ $-RWB-RQS$ $-RWC-RAB$ $-REB$
Benefits	operating cost savings(*OCS*); capital cost savings(*CCS*); reduced abstraction bills (*RAB*)	reduced water bills (*RWB*); reduced energy bills (*REB*)	environmental gain (water; *EG[W]*)[b]; environmental gain (energy; *EG[E]*)	$OCS + CCS$ $+ RAB + RWB$ $+ REB + EG[W]$ $+ EG [E]$
Net Benefits	$OCS + CCS$ $+ RAB - PC$ $- RWB$	$RWB + REB$ $- CC - RWC$ $-RQS$	$EG[W] + EG[E]$ $- RAB - REB$	$OCS + CCS - PC$ $- CC - RWC$ $- RQS$ $+EG[W]+ EG[E]$

Notes for Table 10.3

a. Consider the effects on households of 3 different demand-management measures. If metering induced consumers to spend time/resources on finding and alleviating in-house wastage, *CC*>0 but *RWC=RQS*=0. If metering led to the entirely voluntary forgoing of some sprinkling, *RWC>0 and CC=RQS*=0; the same result would occur from a restriction on ex-house consumption for unmetered households. Voluntary acquisition by a metered household of a low-flush WC might lead to *CC, RWC* and *RQS* all >0, but *RWB* would be expected to be > *CC+RWC+RQS*. Valuation of *CC, RWC* and *RQS* would in principle be possible via contingent valuation or information about the position and shape of the house- hold sector's demand curve. See (Hanke, 1980a) for early mention of *PC, CC* and *RWC*.

b. Under some circumstances a reduction in the EPA's revenues (following a utility's reduced abstractions because of a demand-management programme) may result in a cutback in its environmental protection work, which will have to be offset against the environmental benefit (higher river flows, etc.).

---

[11] (Source: based on White, 1998a).

10.2.4 Economic appraisal: macro

10.2.4.1 Introduction

As indicated in Section 10.2.2, the appraisal of demand-management occurs at very different levels, stretching from a new toilet cistern economy measure being trialled in one resource zone of a small utility to the long-term strategic planning of a large utility serving 5-10 million people and required to include, alongside supply-based measures, a comprehensive and transparent appraisal of the full range of demand-management options. The latter planning exercises go by a variety of names – Least Cost Planning (LCP), first mooted in the United States energy industry in the1980s; Integrated Resource Planning (IRP), applied to the US water industry in the mid-1990s and in Australian water planning a few years later; and , from 1995 onwards, the development of LCP for the UK water industry as, first, the Economics of Demand Management (EDM) and, in 2002, the Economics of Balancing Supply and Demand (EBSD)[12]. The following sections deal with these developments in turn.

10.2.4.2 Least cost planning (LCP) and integrated resource planning (IRP)

The origins of LCP are found in the Office of Conservation at the United States Department of Energy in the mid-1980s (Mieir *et al.*1983; Sant *et al.* 1984). The starting-point for Sant's analysis is that consumers require energy *services* and not energy as a product *per se*; i.e. the demand is not for kilowatt-hours or therms but heat, light, mechanical drive, etc. Increases in the demand for energy services can obviously be met by energy supply increases, enhanced efficiency of energy-using equipment or some combination of the two. The least cost energy strategy is then defined as the mix of energy inputs and energy efficiency improvements having the lowest overall cost to society. LCP is the process needed to identify and achieve this lowest cost (Brown, 1990). Very restricted LCP solutions to future expected deficits were briefly considered (examining

---

[12] LCP, IRP, EDM and EBSD are all examples of strategic planning, normally pursued within the following framework:
1. Construct supply and demand forecasts (with present demand policies unchanged)
2. Identify and quantify future supply-demand balance problems (averages, peaks, zones)
3. Screen supply- and demand-based options to identify feasible list
4. Evaluate cost-effectiveness of feasible options
5. Undertake option selection routine (CEA or other)
6. Allow for tariff and demand feedbacks, risk, environment and equity, as required
7. Identify preferred plan.

supply-side options only) by the UK Water Resources Board in the early 1970s (see 10.3.2.3, below), and the Office of Water Services commended 'proper' LCP, minimising the present worth of costs drawn from supply- *and* demand-side options, for a period in the mid-1990s (see 10.3.2.1). Section 10.3.3.12 includes a summary of the results of two recent LCP exercises undertaken with reference to large water utilities in England and Australia.

IRP can be seen as a broader framework into which LCP may be fitted, with many US utilities preferring the former term because of the implication of LCP that there is only *one* optimal solution to the planning problem. Indeed, White and Fane (2001) argue that an IRP process would include the iterative reapplication of LCP as part of a cycle of evaluating and assessing options and also – and perhaps more important – take account of a broader range of issues including equity, the inherent uncertainties of supply side and conservation options, implementation timeframes and the potential for changes in tariff structures.

In the United States IRP principles have been transferred to the water sector by Dziegielewski *et al.* (1993) and Beecher, the latter stressing an "open and participatory decision making process….the construction of alternative planning scenarios….and recognition of the multiple institutions concerned with water resources and the competing goals among them." (Beecher, 1995; 1996).

10.2.4.3 Economics of demand management (EDM) and economics of balancing supply and demand (EBSD)

Increasing interest in the UK in these matters found practical expression in a contract let out by UK Water Industry Research and the Environment Agency (UKWIR & EA) to NERA (London) in 1995. The objective was to develop a framework, to be used by the public water supply industry in England & Wales, for "balancing future water supplies and demands with an optimal mix of initiatives or schemes for water resource and production management, water distribution management and customer-side management." (Baker *et al.* 1996a; see also Baker *et al.* 1996b, the 'Practical Guidelines' report). The EDM framework had to be acceptable to the economic and environmental regulators – the Office of Water Services and the Environment Agency, respectively.

The report put into practice least-cost planning, the agreed formal planning objective being to minimise the net social cost of balancing the supply and demand for water into the future[13]. This could be achieved by calculating the *AISC* for each of the supply and demand options, which then need to be

---

[13] 'Net social cost' was defined to cover the sum of (i) investment and operating costs of service provider and recipients, (ii) net environmental costs (+ or -) resulting from water abstraction and return changes and (iii) the value of net welfare losses (+ or -) arising from service delivery changes.

"... selected and timed in increasing order of AISC to form an initial solution: a programme of options which balances supply and demand over the horizon" (Baker et al. 1996a).

The initial solution would then be amended until internally consistent (adjusting for option links and indivisibilities), and improved as demanded by risk and equity (or fairness) considerations.

Within four years it was claimed in a second NERA report commissioned by UKWIR/EA that there was (i) significant variability in the use made of the 1996 report and Guidelines, (ii) some misinterpretation of the suggested approaches and (iii) key outstanding methodological issues (Atkinson & Jones, 2001). This 2001 report recommended that revised guidance should increase the conceptual emphasis on total cost minimisation relative to that on AISC-based option ordering, and also provide further explanation of how risk, uncertainty, critical periods, scheme interaction and scheme lumpiness should be taken into account.

The subsequent third set of NERA reports (EBSD; see Atkinson & Buckland, 2002a and Atkinson & Buckland, 2002b for the Guidelines) confronted these issues and also (i) incorporated headroom and reliability in modelling the supply-demand balance, (ii) allowed a planning solution to be modified by testing it against a stochastic representation of supply and demand and (iii) presented and discussed more sophisticated algorithms (than had been used in the 1996 reports) that minimised costs directly.

These reports emphasised two key elements of the planning process that planners would need to choose. First, the complexity of the *Modelling Framework* would have to be determined. Should it be:

- the Current Framework (i.e. as in the 1996 report: with an exogeneous required service level);
- the Intermediate Framework (using Monte Carlo analysis to predict the level of reliability that will be provided by any given solution set); or
- the Advanced Framework (as for Intermediate, but also allowing the   level of reliability to be set optimally on the basis of customer preferences, ie willingness-to-pay) (Atkinson & Buckland, 2002b, stage 6) ?

Second, NERA authors argued that the technique to select options – i.e. the *Selection Routine* – needed to be explicitly selected by planners, suggesting three possible approaches, in increasing order of complexity:

- the AISC approach (as spelt out in the 1996 report);
- linear/integer programming-based formulations; and
- stochastic-programming model formulations.

The available approaches would have to be assessed against the following criteria: accuracy, likely success, costs, and benefits relative to AISCs (Atkinson

& Buckland, 2002b, Stage 7). The chosen approach would also have to be acceptable to economic and environmental regulators.

10.3 APPLICATIONS IN PRACTICE

10.3.1 Introduction

Section 10.3 turns away from the 'principles' of economic appraisal as outlined and discussed in Section 10.2 and concentrates on the documented use of those principles in the water sectors of three countries: Australia, the United Kingdom and the United States. 10.3.2 follows the historical development of the major micro appraisal techniques introduced in Section 10.2.3, from *NPV* through to *AISC* and *LUC*, and 10.3.3 reports results from applications of those techniques to the major components of household water consumption and of sanitation use in non-household sectors.

10.3.2 Application of major micro-appraisal techniques

10.3.2.1 Cost benefit analysis and net present value

Over 40 years ago Hirshleifer, De Haven and Milliman (1960) criticised the then prevalent US practice of comparing water projects by measuring the *ratio* of the present worth (PW) of benefits (PWB) to the PW of costs (PWC)[14]. They argued strongly and correctly that it was NPV, the *absolute difference* between the two, that should (i) determine whether a given one-off water scheme should go ahead or not and (ii) form the basis of scheme ranking.

In a critique of New York's then "water crisis", the authors put these ideas to work in a comparison of various suggested options for the "improvements of existing supplies" (leakage reduction, domestic metering extension and a price increase for metered consumers) with a new (Cannonsville) dam supply project already begun. This led the authors to a comparison, based on present worth calculations, of the costs of the three demand-management options (respectively $1, $148 and $267 per million US gallons saved, using an annual discount rate of 6%) with the estimated Cannonsville cost of $459 per million US gallons supplied. The cases for the demand-management options were concluded to be, respectively, "overwhelming", "impressive" and "arguable".

Nearly ten years later Steve Hanke spent a hectic decade publicising (1981) the application of CBA (via *NPV*) to improve decision-taking in the US water industry, covering both the formulation of marginal cost based pricing techniques and the assessment of a wide range of demand-management measures. Meanwhile, in the UK Warford had briefly applied CBA to the domestic metering decision in a wholly

---

[14] A practice which had been popularised by the US government's so-called *Green Book* (Subcommittee on Benefits and Costs of the Federal Inter-Agency River Basin Committee, 1950)

academic context (1968). This, like other occasional and uncoordinated economic forays into the UK metering issue, had little practical effect, and economic analysis in Britain thus began to concentrate its weaponry in the 1970s on non-price demand management and the intellectual poverty of extrapolative water demand forecasting methods.

More recently, over 1993-97, the Office of Water Services (1993) required that water companies in prospective deficit situations should demonstrate that compulsory household metering and leakage control had been properly costed and compared with the economics of resource developments, through calculations of the present worth of costs, before new resources could be funded via increased price caps. At that stage, however, there was no expectation that environmental costs would be included.

10.3.2.2 Internal rate of return

In the late 1970s Martin Rump at the Building Research Establishment (1978) added a powerful economic dimension to the pioneering trials of domestic and other personal water economy measures then being undertaken at the BRE. Covering WCs, urinals, showers, washing-machines and domestic recycling systems, he reported *IRRs* for demand-management measures ranging from <1%/year (domestic recirculation and installing a shower over an existing bath) to 58%/year (a low-volume flush WC in a new installation) – in terms of additions to net economic welfare, from the ridiculous to the sublime.

10.3.2.3 Cost-effectiveness analysis

Discounted Unit Cost

As noted, Hirshleifer, De Haven and Milliman (1960), without using the term *cost-effectiveness*, were already applying CEA to New York water problems in 1960. In the UK, the Water Resources Board began to embrace CEA in the early 1970s, but only in its comparisons of *supply-side* schemes. The use of present value (PV) techniques arose because of the enormous range of types and sizes of supply-side options (desalination, reservoirs, barrages, groundwater) then being considered in that predict-and-provide/supply-fixated era in Britain; with only a gradual take-up of the yield of some options, the need was for a unique measure of the incremental cost of a given scheme, so that some options could be filtered out, and the remainder properly compared, in the determination of the large regional supply programmes then being formulated. As made clear in expressions (10.6) and (10.7), above, the *discounted unit cost* (DUC)[15] of a supply-side option can be defined as the PV of supply costs over a suitable horizon (say, 30 years) divided by the present worth (PW) of water actually

---

[15] For a supply-side option the DUC can also be interpreted as that uniform price which, if charged each year on each unit of delivered output, would ensure that the PV of income equalled the PV of costs incurred.

delivered[16] to meet a deficit over that time (see Herrington 1979, for a detailed explanation). Use of the DUC measure was crucial in the report on desalination (Water Resources Board 1972) and in the integration of the WRB's major regional exercises in its national strategy report (Water Resources Board 1973).

AISCs and other Unit Cost Measures
CEA and CBEA continued to be used intermittently in the UK in analyses of the economics of domestic metering (e.g., National Water Council, 1976), sometimes cloaked in the language of long-run marginal cost (LRMC; e.g., DoE 1985a and 1985b). By 1995 the National Rivers Authority (1995) was publicising in *Saving Water* its own estimates of what it termed the *Demand Management Costs* of numerous demand-management options (in pence/m^3), and then comparing these figures with its estimates of a national range of resource development costs. A year later NERA reported on its EDM contract (see 10.2.4.3, above), using average incremental social costs (AISCs) for both supply and demand options as the main vehicle for generating a least-net-social-cost solution to the planning problem.

In Australia White's work on least cost planning in the late 1990s, similarly rooted in CEA, made use of what he has termed *Levelised Unit Cost* (LUC), defined as:

$$LUC = [PV(\text{costs to water service provider}) + PV (\text{costs to customers})] \text{ divided by PV (water saved or supplied)}$$

and where the *costs to customers* are the CC, RWS and RQS terms in Table 10.3, Section 10.2.3.5, above (White and Howe, 1998b). In the United States, a similar but less sophisticated measure has been used in California for the costing of the replacement of old domestic water-using fixtures by economy variants, the cost[17] being divided by the annual water saving that results (Gleick *et al.* 2003).

10.3.2.4 Payback period

The Building Services Research and Information Association (BSRIA, 1998) reported to the EA on its Water Consumption and Conservation in Buildings Research Project, which sought "to identify the scope for reducing domestic water consumption" (in buildings of all kinds). It added a section showing the estimated indicative financial payback times of low and high conservation scenarios for each of the 11 building types examined

---

[16] Although both obtained through discounting, PV is expressed in monetary units while the PW of water increments has a physical dimension (eg, kilolitres).
[17] The Californian analysis defines costs as the sum of the annual expenditure to amortize the capital investment required to capture water savings *plus* the net increase (+ or −) in annual operation and maintenance costs that results from implementing the replacement measure.

In four of the building types – factories, offices, schools and retail stores – the high conservation scenario, with higher capital cost, revealed a longer payback period than a low conservation/low cost variant; but for all these types except offices, calculations assuming an annual discount rate no higher than 6% and an economy appliance life of at least ten years showed the high scenario to have much the higher *NPV*, thus illustrating the potential pitfalls of the payback period shown in Figure 10.2 (Section 10.2.3.4, above). In the first set of the *Conserving Water in Buildings* fact cards which resulted from the project, the Environment Agency (1999) reported the 'maximum net payback time' for each product considered, but a later re-issue of the cards by the Agency contains no detailed economic or financial appraisal (2001).

10.3.3 Micro-appraisal in practice

10.3.3.1 Introduction

In this section the results of actual economic appraisals of domestic demand-management measures are summarised, first by appliance-specific type and then through the economics of tariff level and structure alterations. The following will thus be covered, in turn:

- WCs and urinals
- taps and washroom controls
- showers and bathing
- washing-machines and dishwashers
- gardens and gardening
- rainwater collection and greywater recycling
- water use restrictions on households
- demand-management and wastewater flows
- the metering decision
- econometric studies of demand-management

10.3.3.2 WCs and urinals

In the late 1970s Rump (1978) suggested the conversion of an existing WC installation (then generally 9.5 l. single-flush in the UK), would give an IRR of 32% p.a., equivalent to an AIC of 4 p/m^3 (in 1976 prices; water charges were then 11p/m^3 and LRMC estimated as 16.5 p/m^3) – in theory a good investment, financially (for metered consumers), economically and environmentally. Other WC measures were still under development – low-flush in a new installation and a user-controlled flush conversion of an 'old' WC – and Rump speculated they would give IRRs in the 50%-60% p.a. range (AICs 2p–2.5p/m^3). Four years later Flack (1982) was estimating US payback periods for new shallow trap toilets and dual-flush conversion as 1-3 years, depending on the water charge assumed.

In 1981 dual-flush (9l./5l.) became compulsory for new UK installations, but only until 1993, when the maximum single-flush volume was decreased to 7.5 l. and dual-flush was banned (at least, for new installations) following unpublished reports that it had resulted in such double-flushing that water use had actually increased.

Appraisal of economy variants continued, however, and 1995 conversions of 'original' WCs to low-flush (7.5 l.) and dual-flush (9l./5l.) had increased in cost to £30 each (Rump had assumed £5); the resulting AICs were 28 p/m^3 and 18p/m^3 respectively, still significantly less than the assumed national resource-expansion cost range of 33-66 p/m^3 (National Rivers Authority, 1995). A Southern Water/Environment Agency study on retrofitting dual-flush toilets (Keating & Lawson 2000) reported, in an eight month trial of five retrofit devices in 33 customers' homes, an average saving of WC use of 27% and that double-flushing had not significantly increased. Metered households would retrieve the capital costs of the kits (£20) in about a year, and CEA compared them favourably with Southern Water's resource-development schemes, but the likely life of the devices was unknown. Cistern displacement devices like Hippos are generally thought to be very cost-effective (Environment Agency, 1999; NWDMC, 1999), although water companies in southern England were, worryingly, reporting savings varying from 2 l./*property*/day to 21 l./*person*/day in their assessments for the 1999 Ofwat price review as well as sometimes seriously misunderstanding the calculation of AISCs (Howarth, 1999b).

Following the 2001 Regulations change to a maximum flush of 6 l. and the restoration of dual-flush (6l./max.of 4l.), Grant (2003) showed that dual-flush retrofit in the UK (6l./4l. or 6/3), costing about £40/unit, will often generate net benefits (no environmental costs included), although region (and thus resource costs saved) and household size assumed are important determinants of overall economic viability. Benefits from new ultra-low flush models (4.5 l. rather than 6l.) may also be economically attractive (same determining factors) – and thus permit an incentive to purchase to be offered – but extra uncertainty is created by the lack of evidence of overall water savings in the UK, stemming from the possibility of valve leakage and sticking under the new (2001) regulations which allow valve flush mechanisms. Grant also found that WC appliance *replacement* before end-life, with higher capital costs, has potentially higher savings, but longer life assumptions (for the replacement model) are necessary to generate any given NPV figure per appliance.

White (1994, 1998c) reports successful WC rebate programmes in the last 15 years in Massachusetts (public buildings), Auckland, Goleta and San Jose (California), Lismore (NSW), and a gift programme in Streaky Bay (South Australia). Mostly these have offered up to half the purchase price of a new low/dual-flush WC or cistern displacement bag. Another approach to water saving has been to evaluate economically, financially and in terms of consumers' preferences the provision of low-use appliances (and domestic recycling systems) in new homes as they are occupied. In a housing association development in Heybridge, Essex, dual-flush WCs downstairs (6/4l.) and 6l. flush models upstairs both proved very cost-effective, ranking second and third out of six innovations in NPV terms (Smith & Shouler 2001). Recently Gleick *et al.* (2003) in California have found that ultra-low flow toilets are very cost-effective under both 'natural' and accelerated replacement scenarios. Waterless WCs – composting or incineration – turn out to be economic only when water is very scarce and/or extremely highly-priced (NWDMC 1999).

Urinal controls (through time clocks, infra-red or microwave sensors, magnetic door switches, tap use or pull cord/button operation) continue to show very high rates of return compared with uncontrolled 24/7 cyclic flushing in offices. Payback periods on low capital cost controllers often look especially attractive, but high capital cost models – with consequently higher payback periods – may generate even higher water savings, with significantly higher NPV values after a couple of years (see also Section 10.3.3.3). The exception is for some waterless urinal systems – which can be retrofitted into bowls for about £10 for capital and £35/year maintenance (adoption by the National Trust is described in NWDMC 1999) – which also have very high *NPVs*, and therefore score highly on both counts. White (1998c) has a useful guide to the delicate issue of urinal economics and technology in the Australian market.

10.3.3.3 Taps and washroom controls

Washroom controls and devices divide into those that are clearly not designed for use in homes and those that might work in that environment. The former group embraces taps that are battery, infra-red and push-top operated or incorporate flow-restriction through valves, aeration or sprays (costing between £10 and £200). Widespread use in commercial washrooms attest to their financial and economic attractiveness (Environment Agency, 1999; NWDMC, 1999; White, 1998c). There are also available complete washroom control systems that limit hot and cold water supply (through solenoid valves), lighting and ventilation. High capital costs (£500 + installation) mean that the economics depends crucially on how many people use the facility (Environment Agency, 1999).

Because homes often require taps that work for both flow and volume requirements (in the UK about 8% of household use is at the washbasin), required specifications are more complex. Recent innovations include an insert which ensures a spray pattern at low flows (teeth, hands) but opens up to give unrestricted flows, and a cartridge for single-lever mixer taps which generates resistance once a flow of 5-10 l./minute has been reached but allows the lever to be pushed past to give full flow (Environment Agency 2001). The economics looks marginal, however, with reported average savings of only 2l./person/day (Smith & Shouler 2001).

10.3.3.4 Showers and bathing

In the UK there were for some years meaningless claims that replacement of baths by showers could mean lower water use (e.g., DoE, 1992). These ignored earlier information on the secular trend towards shower *and* bath ownership in the home, the increasing popularity of power showers, and UK and overseas data on showering frequency (Herrington, 1996). Figures prepared by the Building Research Establishment for Essex and Suffolk Water (1997) on the basis of trials suggested that users preferred high pressure mixer and power showers (with mean preferred use times and flows giving water uses of 79 l. and 84 l. respectively) rather than low pressure mixer (41 l.) and electric showers (44 l.). The former figures are similar to average recorded bath volumes, but average showering frequencies are thought to be up to around twice those of bathing.

In countries with traditions of higher domestic shower flow rates (> 11 l/min., e.g., Australia, US) large savings (water use, financial, economic) have been reported from flow regulators (threaded in, or a disc) and replacement of the existing head by a water-efficient shower-head. White (1994) reports a trial in Lismore, NSW, in which about 200 water-efficient showerheads were fitted with the aid of a \$A30 rebate each, generating savings at a cost of 14c./m^3 compared to a LRMC of 80c/m^3 for expanding supplies with a new dam; and also a mid-1980s retrofitting programme in San Jose, California (shower-heads and WC displacement bags), which suggested a net saving of \$US400 per residential customer in operating costs and deferred capital works as against one-off programme costs per customer of \$15 (White 1998c). Similar very large returns to natural and accelerated replacement programmes for older showers in California are predicted by Gleick *et al.* (2003).

In the UK the main concern in recent years has been the high present and forecast use of high-flow power-showers using 12-20 l./min. (Herrington 1996; Environment Agency, 2001a). Management of mains-pressure shower use could in principle occur through the atomiser effect of showerheads on high pressure flows, with US experience suggesting that enhanced wetting can thereby be provided from flows of only 4-10 l/min, but Grant (2003) draws attention to a 'cold feet' effect, and noise and moisture problems in bathrooms. He notes that it remains the case that research shows that flow rates and perceived comfort are correlated. Graded efficiency or eco-labelling has also been discussed, but a 'water-efficient' shower with a low water flow may be seen as a negative factor for sales. Additionally, because a 'shower' is made up of a number of elements, each of which may be more or less relatively efficient, labelling may become a complex exercise, perhaps based on a number of comfort dimensions (eg, pressure on skin, head to foot temperature drop, temperature stability, rinse efficiency). It might be simpler simply to try to dissuade consumers from purchasing flows of > 10 l/min. (say).

Domestic baths are not an obvious candidate for saving water, but in the Heybridge trial Smith and Shouler (2001) showed capital *and* water and energy savings through a reduced size (shallower) bath, with a ranking of six demand-management measures "by CBA" placing reduced bath volume at the top of the list. Researchers have speculated that tapered or peanut-shaped baths may give more space for bathing with less water (e.g., Environment Agency 2001).

10.3.3.5 Washing machines and dishwashers

Washing machines were thought to account for about 21 lpd (14%) of domestic use in the south-eastern half of England & Wales in 2001 (85% household ownership) of all households' use), with dishwashers responsible for 2% and concentrated in 30% of homes (Herrington, 1996). New washing machines use about half the water and energy of an average ten year old machine (over 75% of current models./cycle). Dishwashers are also now much more energy and water-efficient; the fact that in 2001 new machines were using only 16l. per cycle (12 place settings), 40% less than the new 1991 models[18], means average use will continue to decline as households renew machines. Probably only in the most water-scarce areas could an economic rationale be made for rebates to encourage people to trade in old models early. All appliances are now common to the European market, and a uniform EU labelling system for energy efficiency is in operation (A, most energy-efficient, to G, least).

Indeed, water use differences between 'more' and 'less' economical machines are now so small – probably due in large part to the effect of the A to G grade EU energy label on competitiveness[19] – that incentives to purchase the former are unlikely to be worthwhile.

Similar relative efficiencies of average and new machines are reported from California, where Gleick *et al.* (2003) have shown that accelerated replacement of 'old' washing machines through rebates up to $500 is cost-effective for water utilities as compared with the supply-expansion alternative. Savings in higher-efficiency dishwashers would not justify an accelerated replacement programme, however. In Australia top-loaders still had 90% of the washing machine market in 1998, for price reasons (White, 1998c)[20], but direct incentives for consumers to choose more efficient front-loaders would probably be cost-effective only where water charges and/or energy savings were very high (which would be only with a preponderance of hot washes). Australian dishwasher water use, however, seems to be following the efficiency trend in Europe.

---

[18] In 1991 electrical appliance manufacturers contacted in connection with research into the effects on demands of climate change (Herrington, 1996) saw no technological scope for further reductions below the 27l./cycle then being reported. In 1976 the standard use for new machines had been 50 l./cycle.

[19] An EU 'EcoLabel' award scheme, initiated in 1992, aims to identify the (environmentally) 'top' 30% of models in a given EU product market. At present this scheme covers washing-machines and dishwashers (as well as numerous non-water-using goods and services), but display of the label (a 'Euro-flower') is voluntary for manufacturers. Nationally, many countries also have their own environmental labelling systems. In the UK the Energy Saving Trust administers a voluntary scheme which allows manufacturers of energy-using products to display an EST-recommended label if the EU Energy Label has the best (A) grade.

[20] White reported (1998c) that whereas European manufacturers offered the convenience of a top loader with the (water) efficiency of the horizontal axis drum, such a machine had not been developed in Australasia. Australian top loaders apparently used 50% more water but were $A100 cheaper.

10.3.3.6 Gardens and gardening

Outside-the-home household use varies enormously across the developed world, due to climate, social history, lifestyle and affluence. In Europe, England & Wales probably has the highest per capita total (an estimated 9 lpd in the south-eastern half, equal to 6% of household consumption), whereas garden use in Australia and North America, dominated by sprinkling, is reported as frequently comprising 30% to 50% of annual household consumption in single family houses, the latter being between 80% and 160% higher than the UK's 150 lpd.

Relevant household demand-management efforts in these other areas are concentrated on the occasional use of seasonal charges and a range of educational and information initiatives. Sample surveys suggest that about 10% of all US water utilities use peak/seasonal pricing, but for those with populations served of more than 0.5 million the figure is 20% (Raftelis, 2002). The most common initiatives are leaflets, free advice lines, demonstration gardens, xeriscape promotion, free checking and adjustment of garden reticulation systems. Other economic incentives have been reported in Mesa (Arizona), North Marin County (California) and Kalgoorlie-Boulder (W. Australia) and have included discounted tap timers, cash rebates and discounted brick paving for agreed lawn area reduction, free mulch, low water use plants and water saving landscape rebates in developer charges (White, 1998c).

The importance of garden watering in England & Wales is that on a hot, dry summer evening up to 50% of the public water supply may be used for garden watering (Environment Agency, 1999). Three types of demand-management strategies are employed, involving *tariffs*, *storage* and *irrigation technology*. Concerning *pricing*, apart from the volumetric unit price message provided to metered households (26% of all households in 2004-05), levies are used by some of the 26 water supply utilities for unmetered households. One imposes an annual charge for hosepipe use, six on sprinkler users (nearly all 26 insist additionally on a metered supply) and eleven for swimming pool filling (Ofwat, 2003a). No UK water utility yet levies seasonal tariffs on metered households.

Concerning *storage*, water butts for collecting rainwater from downpipes – usually from roof gutters – have been promoted or provided free by some water companies. More recent innovations include (i) preventing a water butt from flooding when it is full and permit it to be placed away from the downpipe and (ii) incorporating a valve on a downpipe which diverts greywater (baths, showers, wash handbasins) into a water butt on demand. Low-water garden *irrigation* systems have proliferated in DiY stores in recent years, using time switches, moisture-sensing devices or even a small PC to minimise water loss through evaporation or run-off. *Seep hoses* – buried or threaded through plants – allow a pipe to leak along its length, but are classified as sprinklers; *trickle hoses* have small holes allowing water to seep out; and *drippers* let water drip into the soil at a controlled rate.

10.3.3.7 Rainwater collection and greywater recycling

Rainwater can also be collected on a significantly larger scale from the roofs and other surfaces of individual buildings and across developed areas. To make use of it for watering gardens, parks, leisure and sports areas etc., little or no treatment is required; for WC and urinal flushing and general washing, some treatment; and for drinking and related purposes, local treatment to a potable standard (see Chapters 2 and 3). At the same time greywater collected from homes and washrooms (*excluding* washing machine, dishwasher and kitchen sink waste) can be recycled for a variety of non-potable uses in residential and other buildings, with a number of systems already available (Environment Agency; 1999 and 2001).

The attractiveness of such developments is their large potential scope: in theory, up to 60% of daily water use in England & Wales could be saved through rainwater and greywater reuse. Peak (garden) demands can be reduced, and intercepting rainwater also reduces the storm-water loadings of combined sewer and wastewater systems. Even direct potable reuse (of *all* sewage effluent) is possible given advanced treatment (in Windhoek, Namibia, a plant has been in operation since 1969, reused water making up 40% of the potable water supply (White, 1998c); in developed countries dilution permits this to occur through continuing urban reuse of the same river). Against this are the diseconomies of small scale (attempting to emulate the orthodox PWS and wastewater systems at a sometimes very localised level), perhaps visual intrusion from rain and treated water storage tanks of sufficient size to cope with climatic variation, and possible contamination of roofs through atmospheric particles and bird droppings and of water by the material of the storage tank.

NPV and cost-effectiveness analyses of rainwater and greywater reuse systems (separate or combined) depend upon assumptions made about the system (and its costs), water availability, water consumption and the price of mains water. Results suggest that while greywater reuse systems (e.g., for WC use) at the level of the individual house in the UK remain uneconomic, rainwater collection systems are at or near breakeven for new developments, as witnessed by the proprietary brands for individual homes now on the market (Smith and Shouler, 2001; Environment Agency, 1999 and 2001; CIRIA, 2001). An Australian study with rainwater tank lives of only 15-20 years showed tank retrofitting to be uneconomic (White, 1998c). On a larger scale, greywater systems would be unlikely to have a large enough water supply to match the need for toilet flushing in non-household buildings (offices, factories, etc.), although a detailed Australian desk study compared the costs of five scales of greywater system (including advanced treatment) from 12 to 120,000 household connections and found that the cost-minimising scale (between 1,200 and 12,000 connections) could deliver water at A$0.5/m3, less than the potable water cost in most Australian cities (Booker, 2000).

10.3.3.8 Water use restrictions on households

Hanke (1980a) used NPV/CBA analysis to value the potential net economic impact of "light restrictions"[21] on water use in Perth, Australia, over the 3 months December 1975-February 1976. Transforming Hanke's analysis using the symbols of Table 10.3 and expression (10.9) in Section 10.2.3.5, above, he (i) implicitly assumed no environmental gain and zero reduction in customers' quality of service ($EG[W]$ = $EG[E]$ = RQS = 0), (ii) estimated the value of water saved (OCS + CCS) as the reduction of water use times the estimated long-run marginal cost, (iii) in the absence of other information assumed programme (administrative) costs (PC) and customers' resource costs (CC, in responding to the restrictions) were both zero, while agreeing both would have been positive – but small – in practice, and (iv) used an earlier study of the impact of Perth water use restrictions (Hanke & Mehrez, 1979) and the assumption of a price elasticity of -0.4 to value water consumption forgone (RWC)[22].

The result for Perth was that if light water use restrictions had been imposed for the period under consideration, the estimated social gain through savings in the operating and capital costs of water services (OCS + CCS) would have been \$1.9m and the estimated loss (RWC) \$1.4m. There would thus have been a net social gain valued at \$0.5m. Note that a social *gain* is an "inevitable" result if prices are originally set below marginal costs, programme and customer costs are assumed to be zero and the restrictions affect only the lowest-valued demands. Additionally, no account has been taken here of equity considerations (who gains? who loses? although that would have been comparatively easy), nor of the social feasibility of introducing restrictions at that time.

10.3.3.9 Demand management and wastewater flows

Blanksby (in chapter 5) has summarised the opposing effects of water demand management on UK wastewater systems. The reduction of water in *foul* sewers means less energy to drive solids through the system; they would thus be more likely to be "stranded" if grey and blackwater flows decreased too much and as the distance from the household source of the water increased (Littlewood and Butler, 2002). If, however, households are connected at regular intervals, this should not be an important factor. On the other hand, there could be significant reductions of pipe sizes and tank capacities in sewage treatment works. With *combined* sewers, the potential problems caused by demand-management (during dry weather lower sewage velocities giving

---

[21] A ban on outside sprinklers and/or a limit on the number of hours when outside sprinklers can be used.

[22] In a separate note Hanke (1980b) recognised that the 1979 study assumption that restrictions would "take out" the lower-value uses in sequence (lowest value uses first and then higher- and higher-value uses) was in error, since system output was rationed by *non-price* means. He claimed, however, that the loss of some 'out of sequence' higher-valued sprinkling demands would have only a "small" effect.

rise to more deposition and sedimentation) could be compounded if larger sewers were installed to cope with the higher peak rainfall expected under climate change, for even lower sewage velocities would result (see the extended discussion of these possibilities in chapter 5 of this volume).

While not reporting the adverse effects, Australian case studies confirm the economic gains (White and Robinson, 1999). Byron Bay (north coast of NSW) has been experiencing a rapidly growing residential and tourist population, leading to serious overloading of the sewage system. A LCP approach to the issue suggested that retrofitting water efficient showerheads, toilets and taps would cost about $3m., whereas savings would occur in deferring both STWs ($5m.) and new water supply infrastructure ($5m. to $15m.). Increased concentration of suspended solids, BOD and nutrients would not constrain operation at STWs. At Clunes, a small unsewered NSW village, a preliminary study showed that efficient showerhead installation could reduce water use by 25 kl. per year, also suggesting potential for cost-saving wastewater management options.

Cartwright, White and Carew (1999) examined the problems of Mount Victoria, a village of 500 homes (most with potable water and half sewered) in the Blue Mountains. Sydney Water's earlier look at options had resulted in a preference for enlargement of the sewerage system, but it asked the University of Technology's Institute for Sustainable Futures to re-examine the option in a LCP framework. Ten demand-management programmes were considered (some combinations of others), and the new preference was for an option that maximised the reductions in water demands and sewage flows while providing a net overall economic benefit and positive gains to customers (see the framework of Table 10.3, above). This included water efficient showerheads and taps and 6l./3l. dual-flush toilets in all dwellings over 1999-2001, non-residential audits with similar appliance refitting, a 3-month rebate offer of A$200 at point-of-sale for a front-loader if replacing a top loading washing machine and discounted phosphorus-free washing powders for households. Sydney Water anticipated that there would result a reduction of 15% in water demands and dry weather sewage flows and of 25%-50% in the phosphorus loading.

10.3.3.10 The metering decision

The formal analyses of (i) the decision whether to meter a group of consumers in a situation of static demands and (ii) the timing of the decision in the expectation of growing demands need not be repeated here (see, e.g., respectively, Herrington (1987b) and Warford (1968) for exposition of the theory). The major benefits are seen as the saving or deferment of economic resource use resulting from reduced demand, while costs are those associated with the installation, maintenance and operation of the meters, plus resource costs to the consumer and the value of any useful consumption forgone.

Warford showed that if short-run marginal costs were actually or approximately zero it would never be economically beneficial to introduce meters at once if demand was expected to remain static, but if supply costs are large and demand is responsive to price then metering might be justified even if demand was not increasing; while the former reference drew attention to both the variation of demands and costs within the year and also the gains to be had from improved waste detection (and reduction) and demand forecasting which would be expected to stem from a metering programme. It is also obvious that selective metering where costs are lower (in new houses, e.g.) and potential demand reductions are higher (e.g., with sprinkler-owners) will reveal lower costs per cubic metre saved than would a universal metering programme.

Numerous analyses of domestic metering have been undertaken, with varying degrees of sophistication. All measure the economic gains either via *NPV* estimation (10.2.3.1, above) or through CEA, facilitating comparison with other demand-management (and supply-based) options (10.2.3.3). Hanke's short but classic CBA, applied to Perth, Australia (1982), showed through NPV estimation and a two-season model that the completion of metering in 1976-77 (18,000 extra households) would have been economically justified in terms of a positive net benefit. Examples of domestic metering CEAs for England & Wales may be found in the work of the National Water Council (1976), the more sophisticated work undertaken for the Watts Committee (DoE, 1985b) and various comparative studies undertaken over 1995-2000 and referred to in Table 10.4. Demand effects resulting from decisions in OECD countries to switch a whole or part of a residential community from an unmeasured to metered charging, or extend metering to individual apartments, are surveyed in Herrington (1987b) and Herrington (1999), and in EEA (2001; see Case Study 5, dealing with the 1997 metering of apartments in Seville, and Case Study 30, concerning sprinkler users in Cambridge in the 1990s).

There is also some evidence of further savings, in particular through peak demand reductions, as a result of the introduction of more sophisticated tariffs (increasing-block, seasonal) for already-metered householders. For example, four US studies in the 1970s showed the introduction of summer premia into tariff schedules to be associated with reductions in peak day ratios of 10% to 15% (Herrington, 1987b; 1999). No full analyses of the economics of such changes, comparing monetary costs and benefits, have been reported, however, although as real supply-costs rise (higher-cost sources being brought into use, increasing environmental costs) it would be expected that both the economic case for metering hitherto unmetered homes and that for introducing more sophisticated (but acceptable) tariffs would become more compelling.

10.3.3.11 Econometric studies of demand-management

Some studies have combined econometric estimates of demand and cost functions for specific utilities to predict the economic implications of urban pricing reform. Simulations for Washington, D.C. (Davis and Hanke, 1971) and Victoria, Canada (Sewell and Roueche, 1974) showed how seasonal pricing could have deferred system expansion, leading to a predicted 8% peak demand fall (6% in Victoria). Hypothetically reforming water prices on a LRMC basis in Vancouver led Renzetti

(1992) to estimate a 4.5% increase in aggregate net benefits (see Table 10.3), while Russell & Shin (1996), in a similar exercise for Phoenix, Arizona, predicted an increase in the consumer surplus enjoyed from public water supplies of 7%-11%. These increases seem small, but remember that prior net benefits will have been high because of the large consumer surpluses received on higher-valued units of water.

Other studies have used econometric methods to attempt to assess utilities' experiences of demand-management. Cameron & Wright (1990) investigated the factors influencing Los Angeles households in decisions to install two efficiency measures, finding the adoption of shower flow restrictors linked to the potential savings on energy costs while WC retrofits were determined more by "conservation mindedness". Renwick and Archibald (1998) simultaneously estimated the factors governing the adoption of water-efficient devices/practices and the structure of household water demands. Interviews and analysis of monthly data over 1985-90 for 119 households in two Californian communities generated (i) an average price elasticity of -0.3 (-0.5 for poorer households – < $20K/year income – and -0.1 for the more affluent – > $100K/year), (ii) estimates that observed real price increases had reduced demands over the period by 9% (Santa Barbara) and 26% (Goleta), and (iii) an estimate that adoption of low-flow WCs and showerheads had reduced household demands in Goleta by 28% while low irrigation practices in Santa Barbara had caused a fall of 16%.

Michelsen *et al.* (1998) used maximum likelihood regression to test three models of residential water demand with a pooled annual database over 1984-95 covering seven utilities in the south-western U.S. Estimated elasticities were low – annual: -0.1 and summer: -0.2 – and non-price conservation programmes appeared to be effective only if a utility achieved a critical mass of programmes undertaken over a longer period of time. For cities with few programmes or limited experience with conservation, non-price measures had no observable impact.

The objective of Renwick and Green (2000), analysing monthly consumption data over 1989-96 for each of eight Californian water supply utilities covering 7 million people (24% of the state's population), was to assess the relative performance of price and non-price demand-management policies in a period overlapping with statewide drought (1985-92). A complex econometric model was used to identify price, climate and water demand equations, with the former – following the work of Nieswiadomy & Molina (1989) – explicitly including the influence of endogeneous price effects on demand under block rate schedules (since the marginal price depends upon the quantity demanded as well as vice-versa). The estimated year-round own-price elasticity of demand was low at –0.16, as was the summer-quarter estimate of –0.2. Dummy variable analysis suggested that there were small effects on households from utility policies on information (7 utilities; demands cut by 8%) and retrofit subsidies (7; 9% cut), whereas water rationing (2; 19% cut) and restrictions (2; 29% cut) had larger impacts. However, ultra-low-flow toilet rebate programmes (7) and the Los Angeles water efficient device affidavit policy (if not filed, households faced higher prices) appeared statistically insignificant[23]. The conclusion was that the menu of options for Californian utility policymakers included moderate (5-15%) reductions in *aggregate* demand through modest price increases and 'voluntary' demand-management policies, and larger effects (> 15%) from relatively large price increases and/or more stringent policy instruments.

10.3.3.12 Comparative cost-effectiveness analyses over 1995-2000

Table 10.4 summarises the results of Rump's IRR estimates and five sets of recent CEA estimates for a range of demand-management measures (four from the UK, one Australian). The CEA analyses vary greatly, with the National Rivers Authority (1995) and Howarth (1998) data sets drawn from measure-specific desk studies undertaken at the EA's National Water Demand Management Centre while the Howarth (1999a) column is a summary of AISC values provided by varying numbers of water companies in water resources plans submitted in connection with the 1999 Periodic Review of water charges in England & Wales (the lowest, and first and third quartile, values are listed).

On the other hand, the *AISC*-type estimates of Foxon *et al.* (2000) and White & Howe (1998b) both derive from LCP system planning exercises attempting to resolve current and future imbalances in the areas served respectively by Thames Water (1996-2016) and Sydney Water (1990-91 to 2000-01). Unusually, operating savings from avoided supply-side expansion are credited to demand management options in the Foxon *et al.* analysis, not affecting comparisons with resource expansion (the extra costs of which are consequently restricted to capital elements) but making comparisons with other *AISC* estimates difficult.

It should be emphasised that the prime purpose of regional strategy studies (like those for Thames Water and Sydney Water) is to identify an optimal programme of investment and other options (measures, policy changes, etc.), which can be identified with the *least-cost programme* as long as all relevant costs have been included and risk and equity satisfactorily handled (see 10.2.4.3, above). In this type of study the relative sizes of *AISCs* do not in themselves determine whether an option is in or out of that optimal programme.

---

[23] The authors were unconvinced by this result as a result of aggregation across utilities, problems with the definition of when the policy went into effect and varying emphasis within a policy-on period.

Table 10.4 Comparing IRR and Unit Costs from 1978-2000 Studies

Study	Rump (1978)	National Rivers Authority (1995)	Howarth (1998)	Howarth (1999a)	Foxon et al., (2000)	White & Howe (1998b)
Costs apply to ? year	end-1976	1994	1998	1998	1996	1998
Location covered	U.K.	E & W	E & W	WCos'WRPs[1]	Thames Water	Sydney Water
Measure used	IRR (% p.a.) (+ cap. cost)	AIC (p/m^3) (+ cap. cost)	AIC (p/m^3)	AISC (p/m^3)	Net AISC (p/m^3)	Levelised Cost (Aust.c./m^3)
WCs low-vol/user-control (new or conversion)	32%–58% (£5–£7)	18 (£30)	16		7*	
conversion to low-flush (9.8l. → 7.5l.)		28 (£30)			16*	
replacement: low-flush (9.8l. → 6l.)	3.5%–6.5% (£50–52)	172 (£300)		156 (167-194)	137*	
Hippo bags			0.5	0.5 (3-13)		
Urinals Controllers		9 (£200)		4 (32-64)		
Showers Atomiser: gas/elec.	<1%/14% (£115)					
Conventional: replacement for bath	<1%/11% (£115)	94 (£200)				
Low-volume. s/heads		102	33	78 (159-243)		
S/head rebate						14
Rainwater Butts			34	263		211[2]
Washing M/Cs Efficient M/C/standards		0				4.1
Rebate programme						70

Table 10.4 (continued)

Study	Rump (1978)	National Rivers Authority (1995)	Howarth (1998)	Howarth (1999a)	Foxon et al., (2000)	White & Howe (1998b)
Water Audits Audits (+education)				(0-64)		19/25

Metering
 General
 Compulsory
 Targetted/Spklr users
  optants
  free
		89	142
			
			 85
				(100-763)
				 88 (200-338)
				 50 (156-520)
				 (160-187)
				 (105-609)
					47*
					
					 63*	

Domestic Recycling greywater, for WCs general greywater		321–493 (£1000)	 448	 22 (178-400)	 192*	 244
Industrial Reuse						53/65
Resource Costs	MC est. as 16.5 p./m^3	33/56 p./m^3	35-70 p./m^3		51 p./m^3	
Notes	IRRs embrace program costs +water/energy savings benefits, but no other environmental benefits	Costs to utility only; 6% discount rate and 20-year implementation; no account taken of environmental benefits	Probably costs to utility only; 6% discount rate and 10-year implementation assumed	[1]Nos. are WCos own ests. as submitted to EA - they give min. value and 1st/3rd quartile values; may be costs to utilities only; see below	*Costs low since they have operating savings in PWS & S+ST netted out, as are some environmental *benefits*.	Resource costs only included, as in bottom right hand cell of Table 10.3; uncertainty about environmental benefits [2]Rainwater *tanks'* LUC

Note: In Howarth (1999a) cell entry 'x (y-z)' means x is the minimum value of the set submitted to the EA by different WCos. in their Water Resource Plans, and y and z are the 1st and 3rd quartile values. Between 1 and 8 WCos. submitted costs for each measure, and between 1 and 5 (respectively) of these included a socio-environmental cost element (usually small).

In the first four columns of the table, however, where there was no particular regional or other area study in mind, the *AISCs* (low values preferred) (or *IRRs*; high values better) have a definite role in guiding policy-makers, strategists and planners as to the likely desirability of taking a given measure forward to the next stage of the planning and programme formulation process.

Such a comparative table is also helpful in identifying areas or sectors where there is little dispute about economic viability (e.g., some WC options and urinal controls are clearly inexpensive, while recycling options and 'unnatural' WC replacement programmes are the opposite) and to measures and 'directions' where there is no apparent consensus and therefore further analysis may be required (e.g., shower and water butt options and metering).

10.4   CONCLUSIONS AND RECOMMENDATIONS

- Precisely what counts and what does not count as demand-management does not really matter. In deciding on operating and investment strategies, it is much more important to appraise all options – supply-expansion, leakage control, and both tariff-based and non-price demand-management – on a level playing field, no matter how they are defined.

- Calculations show that domestic-type usage of the public water supply – including that in offices, factories, hotels, schools, etc, as well as homes – accounts for about 85% of overall 'useful' consumption in England and Wales.

- Complete appraisal of a demand-management measure, policy, policy amendment or programme will usually have four dimensions – technical, economic, environmental and social. It is desirable wherever practicable to bring together the economic and environmental dimensions in the form of a cost-benefit or cost-effectiveness analysis. Financial appraisal should be seen as deriving from the economic dimension, the introduction of cash flows permitting estimation of the extent to which individual stakeholder groups gain or lose from a measure.

- The best way of assessing the overall economic implications of a single demand-management option is through net present value estimation. If a comparison of different demand- and/or supply-based options is being sought, estimate and then compare the average incremental social cost (AISC) of each option, calculated as the present worth of the costs incurred over its lifetime divided by the present worth of the quantity of water it will save for or deliver to consumers.

- Under some circumstances the internal rate of return measure – using familiar percentage per annum units – gives as good a decision-rule as the net present value expression. Under other circumstances it may give wrong or misleading guidance. The payback period, on the other hand, is a crude and potentially more misleading way of assessing the attractiveness of an option. It usually ignores the time value of money and takes no account of what happens after payback is achieved.

- The total resource cost test is very similar to net present value estimation, but allocates the present-value-based net benefits deriving from an option or programme to the various stakeholder groups and interests involved. Redistribution of those net benefits from a measure or policy among the stakeholders may be crucial to getting it accepted.

- Least cost planning (LCP) and integrated resource planning (IRP) treat appraisal in a macro framework, since they aim to establish optimal strategies and programmes for resolving medium- and long-term supply-demand imbalances at utility, regional and national levels. Recent variants of LCP incorporate headroom, stochastic representations of supply and demand and reliability; while IRP stresses more issues of equity and democratic participation in decision-taking as well as the different interests of the various agencies often concerned with the water cycle.

- Empirical work undertaken on the economic appraisal of demand-management suggests a wide range of viable measures for reducing domestic-type usage in residential and other settings in developed economies. Unsurprisingly, retrofitting water-saving components (e.g., WC adjustments, new shower-heads) are more likely – given the socio-economic costs of supply-side alternatives – to satisfy economic criteria than the accelerated replacement of white goods with new economy models (e.g., WCs, dishwashers and washing machines).

- Significant rainwater harvesting and greywater recycling for potable use in individual homes are shown as unlikely to be economic at present, although evidence from the UK and abroad shows systems to satisfy non-potable demands as sometimes viable if little or no treatment is required.

- The economic analysis of the metering decision is well-known; examples of its use in the UK and Australia are reported. Often selective household metering generates higher economic gains than universal metered charging, and as the real costs of supply provision rise over time (scarcity, environmental factors) the balance will shift further towards more metering, for example of individual apartments.

- A table compared *IRRs* and *AISCs* over a range of demand-management and supply-expansion options in one Australian and five UK studies. Such comparisons may serve to assist policy-makers and planners in deciding which measures to take forward in the planning process and also point to those options where cost or net benefit variations across studies suggest that further research and/or analysis is required.

This chapter has attempted to identify best practice in the appraisal of demand-management measures, both for individual options and in area planning studies which set out to reconcile future supplies and demands. It is clear from the survey of practical case studies that significant scope remains for the furtherance of beneficial demand-management in domestic-type water use settings. The translation of this scope into practical policies is essential for the achievement of sustainable development in water – for the resource, for the industry and for the sector as a whole.

10.5 REFERENCES

Atkinson, J. and Buckland, M. (2002a) *The Economics of Balancing Supply & Demand (EBSD) - Main Report*. Report 02/WR/27/3, UK Water Industry Research Limited, London.

Atkinson, J. and Buckland, M. (2002b) *The Economics of Balancing Supply & Demand (EBSD) Guidelines*. Report 02/WR/27/4, UK Water Industry Research Limited, London.

Atkinson, J. and Jones, S. (2001) *Economics of Demand Management – Phases I & II*. UKWIR Report 01/WR/03/3, UK Water Industry Research Limited, London.

Baker, W., Reehal, R., Kretzer, U., Jones, S. and Herrington, P. (1996a). *Economics of Demand Management – Main Report*, UKWIR Report WR-03. UK Water Industry Research Limited: London

Baker, W., Reehal, R., Kretzer, U., Jones, S. and Herrington, P. (1996b) *Economics of Demand Management – Practical Guidelines*. UKWIR Report WR-03. UK Water Industry Research Limited, London.

Baumann, D. D., Boland, J. J. and Sims, J. H. (1980) *The Problem of Defining Water Conservation. Cornett Papers*. University of Victoria, Victoria, British Columbia.

Beecher, J. A. (1995) Integrated Resource Planning – Fundamentals. *AWWA Journal*, June, pp. 33-44.

Beecher, J. A. (1996) Integrated Resource Planning for Water Utilities. *Water Resources Update*, Issue Number 104, Summer.

Booker, N. (2000) Economic Scale of Greywater Reuse Systems. Commonwealth Scientific and Research Organisation Email Innovation Online No. 16, December, available online at www.cmit.csiro.au/innovation/2000-12/economic_scale.

Bowers, J. (1997) *Sustainability and Environmental Economics*. Longman: Harlow.

Brown, I. (1990) *Least-Cost Planning in the Gas Industry*. Office of Gas Supply, London.

BSRIA (1998) *Water Consumption and Conservation in Buildings: Potential for Water Conservation*. BSRIA Report 12586B/3. The Building Services Research and Information Association, Bracknell.

Cameron, T. and Wright, M. (1990) Determinants of Household Water Conservation Retrofit Activity: A Discrete Choice Model Using Survey Data. *Water Resources Research* **26** (2), 179-88.

Cartwright, T., White, S. and Carew, A. (1999) Rigorously Reducing Sewage Flows – Case Study of Water Conservation in Mount Victoria. Conference paper presented at *Water Down the Track – Victoria and NSW branches of AWWA Joint Regional Conference*, Albury, NSW, October.

CIRIA (2001) *Rainwater and Greywater Use in Buildings: Decision-making for Water Conservation*. Report PR80. Construction Industry Research and Information Association, London.

Davis, R. K. and Hanke, S.H. (1971) *Planning and Management of Water Resources in Metropolitan Environments*. George Washington University Natural Resources Centre, Washington, D.C.

DETR (1999) *Review of Technical Guidance on Environmental Appraisal*. A Report by EFTEC. DETR, London. Available online at: www.defra.gov.uk/environment/economics/rtgea.

DoE (1985a) *Joint Study of Water Metering – Report of the Steering Group*. HMSO, London.

DoE (1985b) *Water Metering – Main Report of Coopers & Lybrand Associates*. Department of the Environment, London.

DoE (1992) *Using Water Wisely*. Department of the Environment and Welsh Office, London.

Dziegielewski, B., Opitz, E., Kiefer, J., and Baumann, D. (1993) *Evaluating Urban Water Conservation Programs: A Procedures Manual*. American Water Works Association, Carbondale, Illinois.

EEA (2001) *Sustainable Water Use in Europe Part 2: Demand Management*. European Environment Agency, Copenhagen.

Environment Agency (1999) *Conserving Water in Buildings – factcards*. National Water Demand Management Centre, Worthing.

Environment Agency (2001) *Conserving Water in Buildings – factcards*, revised set. National Water Demand Management Centre, Worthing.

Environment Agency (2001a) *A Scenario Approach to Water Demand Forecasting*. National Water Demand Management Centre, Worthing.

Environment Agency (2003) *Water Resources Planning Guideline* Version 3.0 February 2003 and Version 3.1 April 2003. Environment Agency, Bristol. Available on line via www.environment-agency.gov.uk.

Essex and Suffolk Water (1997) *Water Conservation Shower Evaluation*. Report prepared by Building Research Establishment Ltd.

Fiske, G.S. and Weiner, R. A. (1994) *A Guide to Customer Incentives for Water Conservation*. US Environment Protection *Agency*, Washington, D. C.

Flack (1982) Urban Water Conservation: Increasing Efficiency-in-Use Residential Water Demand. *American Society of Civil Engineers*, New York.

Foxon, T.J., Butler, D., Dawes, J. K., Hutchinson, D., Leach, M. A., Pearson, P. J. G., and Rose, D. (2000) An assessment of water demand management options from a systems approach. *Journal of the Charted Institution of Water and Environmental Management*, 14, June, 171-78.

Gleick, P., Haasz, D., Henges-Jeck, C., Srinivasan, V., Wolff, G., Kao Cushing, K. and Mann, A. (2003) *Waste Not, Want Not: the Potential for Urban Water Conservation in California*. Pacific Institute: Oakland, California.

Grant, N. (2003) *The Economics of Water Efficient Products in the Household.* Environment Agency Water Demand Management, Worthing.

Hanke, S. and Mehrez, A. (1979) The relationship between water use restrictions and water use. *Water Supply & Management* Vol. 3, 315-21.

Hanke (1980a) A cost-benefit analysis of water use restrictions. *Water Supply & Management* Vol. 4, 269-74.

Hanke (1980b) Additional comments on cost-benefit analysis of water use restrictions. *Water Supply & Management* Vol. 4, 297-98.

Hanke, S. (1981) *Studies in Water and Wastewater Economics.* Report No. 3046. Department of Water Resources Engineering, Lund Institute of Technology, University of Lund, Lund, Sweden.

Hanke, S. (1982) On Turvey's Benefit-Cost "Short-Cut": A study of water meters. *Land Economics* **58** (1), 144-46, February.

Her Majesty's Treasury (2003) *The Green Book, Appraisal and Evaluation in Central Government*, The Stationery Office, London.

Herrington, P. (1979) *Nor Any Drop to Drink? The Economics of Water.* Occasional Paper, Economics Association: Sutton, UK.

Herrington, P. (1987a) *Improved Water Demand Management: State of the Art Report.* ENV/NRM/87.2, OECD Environment Directorate, Paris.

Herrington, P. (1987b) *Pricing of Water Services.* Organisation for Economic Cooperation and Development: Paris.

Herrington, P. (1996) *Climate Change and the Demand for Water.* Her Majesty's Stationery Office, London.

Herrington, P. (1999) *Household Water Pricing in OECD Countries.* Report ENV/EPOC/ GEEI(98)12/FINAL, Organisation for Economic Cooperation and Development, Paris (*note that most of this text is reproduced with minimal amendments in the more accessible publication by the OECD (1999)*).

Hirshleifer, J., De Haven, J. C., and Milliman, J. W. (1960) *Water Supply: Economics, Technology and Policy.* University of Chicago Press, Chicago.

Howarth, D. (1998) Progress on Demand Management. In *Proc. IBC Conference on Water Resources Management*, London, 25/26 November.

Howarth, D. (1999a) The Economics of Demand Management Options. Paper delivered at *AWWA conference* (and explanatory letter to Paul Herrington dated 31 December 2001).

Howarth, D. (1999b) Email to Paul Herrington concerning Water Company Supply/Demand Balance Submissions sent to Ofwat in mid-1998 (June 1999).

Keating, T. and Lawson, R. (2000). *The Water Efficiency of Retrofit Dual Flush Toilets.* Southern Water and Environment Agency, Worthing and Bristol.

Littlewood, K. and Butler, D. (2002). Influence of diameter on the movement of gross solids in small pipes. In *Proceedings of the International Conference on Sewers Operation and Maintenance*, Bradford University, November.

Michelsen, A. M., McGuckin, J. T. and Stumpf, D. M. (1998) *Effectiveness of Residential Water Conservation Price and Nonprice Programs.* AWWA Research Foundation: Denver.

Mieir, A. K., Wright, J. and Rosenfeld, A. H. (1983) *Supplying Energy through Greater Efficiency.* University of California Press, Berkeley.

National Rivers Authority (1995) *Saving Water.* NRA, Bristol.

National Water Council (1976) *Paying for Water.* National Water Council, London.

Nieswiadomy, M. L. and Molina, D. J. (1989) Comparing residential water demand estimates under decreasing and increasing block rates using household demand data. *Land Economics* 65, 280-89.

NWDMC (1999) *Saving Water – On the Right Track 2*. Environment Agency National Water Demand Management Centre, Worthing.

OECD (1999) *The Price of Water: Trends in OECD Countries*. Organisation for Economic Co-operation and Development, Paris.

Office of Water Services (1993) *Paying for Growth*. Ofwat, Birmingham.

Ofwat (1996). *1995-96 Report on the Cost of Water Delivered and Sewage Collected.* Office of Water Services, Birmingham.

Ofwat (2003a) *Tariff Structure and Charges: 2003-2004 Report.* Office of Water Services, Birmingham.

Ofwat (2003b) *Security of Supply, Leakage and the Efficient Use of Water 2002-2003 Report.* Office of Water Services, Birmingham.

Pearce, D. and Barbier, E. (2000) *Blueprint for a Sustainable Economy*. Earthscan, London.

Pinney, C., Waggett, R., Mustow, S. and Smerdon, T. (1998) *Water Consumption and Conservation in Buildings: Review of Water Conservation Measures*, BSRIA Report 12586B/1. The Building Services Research and Information Association, Bracknell.

Raftelis Financial Consulting Group (2002) *2002 Water and Wastewater Rate Survey*. Raftelis Finacial Consulting, PA: Charlotte, North Carolina.

Renwick, M. E. and Archibald, S. O. (1998) Demand side management policies for residential water use: who bears the conservation burden? *Land Economics* **74**(3), 343-59.

Renwick, M. E. and Green, R. D. (2000) Do residential water demand side management policies measure up? an analysis of eight california water agencies. *Journal of Environmental Economics and Management* 40, 37-55.

Renzetti, S. (1992) Evaluating the welfare effects of reforming municipal water prices. *Journal of Environmental Economics and Management* **22** (2), 147-63

Renzetti, S. (2002). *The Economics of Water Demands*. Kluwer, Boston.

Russell, C.S. and Shin, B. (1996) Public Utility Pricing: Theory and Practical Limitations. In Darwin Hall (ed.), *Marginal Cost rate Design and Wholesale Water Markets, Advances in the Economics of Environmental Resources*, Vol. 1. JAI Press, Greenwich, Connecticut.

Rump, M. (1978) *Potential Water Economy Measures in Dwellings: Their Feasibility and Economics*. Paper CP 65/78. Building Research Establishment, Watford.

Sant, R.W., Bakke, D., Naill, R.F. and Bishop, J. (1984) *Creating Abundance: America's Least-Cost Energy Strategy*. McGraw-Hill, New York.

Sewell, W. R. D. and Roueche, L. (1974) Peak Load Pricing and Urban Water Management, *Natural Resources Journal*, 13 (3).

Smith, S. and Shouler, M. (2001) Sustainable New Homes, Heybridge, Essex. *Powerpoint presentation*. Essex and Suffolk Water, Chelmsford, 8 March.

Subcommittee on Benefits and Costs of the Federal Inter-Agency River Basin Committee (1950) *Proposed Practices for Economic Analaysis of River Basin Projects: Report to the Federal Inter-Agency River Basin Committee*. Government Printing Office, Washington, D.C.

Tucker, S. N., Mitchell, V. G. and Burn, L. S. (2000) Life cycle costing of urban water systems. Commonwealth Scientific and Research Organisation Email Innovation

Online No. 16, December, available online at www.cmit.csiro.au/innovation/2000-12/economic_scale.

Tate, D. M. (1990) Water *Demand Management in Canada: A State-of-the-Art Review.* Inland Waters Directorate Social Science Series Paper No. 23. Environment Canada, Ottawa.

Turner, R. K., Pearce, D. and Bateman, I. (1994) *Environmental Economics: an Elementary Introduction.* Harvester Wheatsheaf, Hemel Hempstead.

Vickers, A. (2001) *Handbook of Water Use and Conservation.* WaterPlow Press, Amherst, Mass.

Warford. J. (1968) Water Supply, Chapter 6 in (ed.) R. Turvey, *Public Enterprise.* Penguin, London.

Water Resources Board (1972) *Desalination 1972.* HMSO, London.

Water Resources Board (1973) *Water Resources in England and Wales,* Volume 2: Appendices. HMSO, London.

White, S. (1994) Preferred Options. In T. Fiander & Associates, *Report on the Lismore Water Efficient Hardware Incentives Trial,* Lismore City Council, New South Wales.

White (1998a) Regulating for Economic Water Efficiency. Discussion paper presented to National Working Group on Water Conservation, Canberra, May. Mimeo.

White, S.B. and Howe, C. (1998b) Water efficiency and reuse: a least cost planning approach. In *Proceedings of the 6th NSW Recycled Water Seminar.*

White, S. (ed.) (1998c) *Wise Water Management: A Demand Management Manual for Water Utilities.* Water Services Association of Australia: Sydney.

White, S., Dupont, P. and Robinson, D. (1999) Water demand management and conservation including water losses. *International Report. IWSA Congress 1999,* Paper IR-5.

White, S.B. and Fane, S.A. (2001) Designing cost effective water demand management programs in Australia. Paper presented at *IWA 2002 Berlin World Water Congress,* 15-19 October.

White, S. and Robinson, D. (1999) Costs and benefits of reducing wastewater flows through improving the efficiency of water-using applainces. In *Proceedings of the 18th Federal Convention of the Australian Water and Wastewater Association,* Adelaide, April.

Winpenny, J. (1994) *Managing Water as an Economic Resource.* Routledge, London.

11

Legislation and regulation mandating and influencing the efficient use of water in England and Wales

David Howarth

11.1 INTRODUCTION

In 1989 the public water authorities of England and Wales were privatised. The privatisation consisted of the assets and the management of those assets passing into private ownership. The privatisation was accompanied by a strong regulatory system to protect the public interest. The current 22 water companies are regulated by the Office of Water Services (OFWAT – the economic regulator), the Environment Agency (the environmental regulator) and the

Drinking Water Inspectorate (for drinking water quality). The Secretaries of State for the Environment, Food and Rural Affairs and for Wales have a wider role in developing policy and the legislative framework, often driven by compliance with EU Directives, as depicted in Figure 11.1.

Of the four regulators shown in Figure 11.1, only the Drinking Water Inspectorate has no locus on water efficiency matters.

The drought of 1988-92, which resulted in southern and eastern England being short of approximately ten months rainfall, was largely responsible for stimulating government interest in the conservation of water. Using Water Wisely (DoE, 1992) was a consultation paper that considered options for managing demand with an intention to initiate a wide and constructive debate on those options. In the last ten years the amount of regulatory activity has been considerable – this activity has been consistent with the existing legislative framework and new legislation has been developed or is in the process of being developed where needed.

Rather than an historical record of legislation and regulatory initiatives this paper considers how regulation and legislation applies to different water use components in the supply chain.

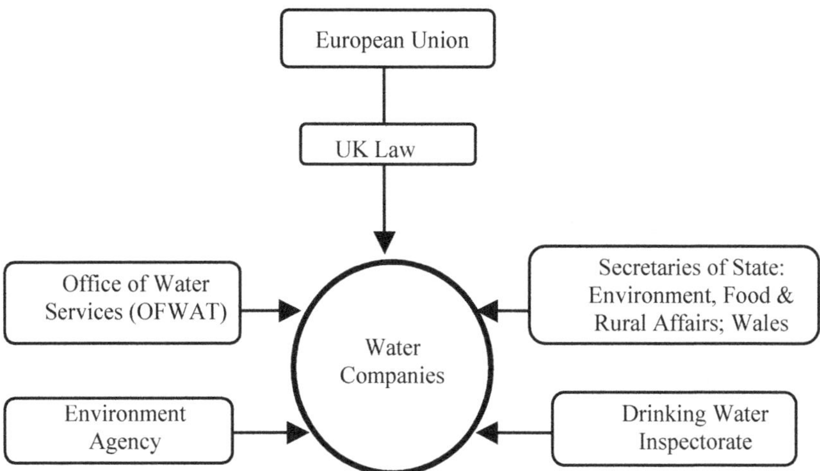

Figure 11.1 The regulation of the water industry.

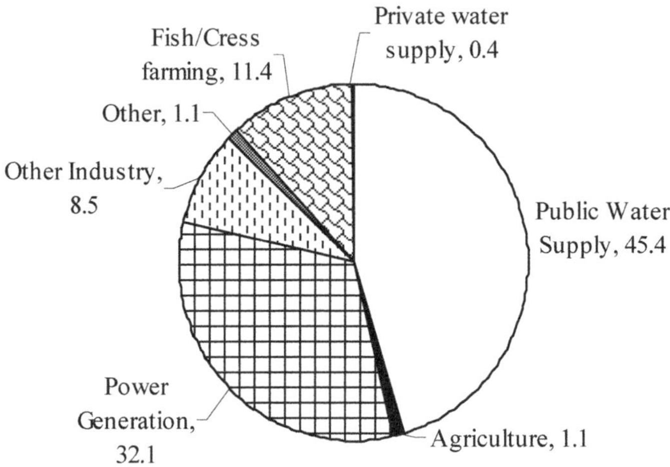

Figure 11.2 Water use by abstraction.

11.2 WATER USE BY ABSTRACTION

A breakdown of water use by abstraction (volume licensed) is shown in Figure 11.2.

11.2.1 Legislation/Regulation applying to all abstractions

Under Section 19 of the Water Resources Act 1991 the Environment Agency has a duty to *secure the proper use of water of resources* and under section 16 of the same act to *further the conservation and enhancement of the environment.* Under Section 4 of the Environment Act 1995 the Agency is given a principal aim to make a *contribution towards attaining the objective of sustainable development* that includes *discharging its functions so as to protect and enhance the environment.*

The Agency's vision for water resources for the next twenty-five years is *'abstraction of water that is environmentally and economically sustainable. Providing the right amount of water for people, agriculture, commerce and industry, and an improved water-related environment'.* (EA, 2001).

The Agency's principal vehicle for complying with these legal duties is the abstraction licensing system. The abstraction licensing system was introduced in

1965 and abstractions existing prior to this date are effectively authorised by 'licences of right' where no conditions (for example maintaining a river flow or level) can be applied retrospectively without liability for payment of compensation for loss of rights. The conversion of all licences to time limited status (whereby the Agency can withdraw all or part of the licence at the end of the specified time period) is one of the issues currently being debated in the Water Bill.

The Environment Agency administers around 50,000 abstraction licences and receives some 1500 to 2000 new applications each year. The majority of these licences are not held by water companies, but by industrial and agricultural abstractors. The Agency can refuse licence applications if the need is not considered reasonable.

11.2.2 Industrial and agricultural abstraction

11.2.2.1 Abstraction licensing

The Environment Agency has developed a range of water use benchmarks for industries that can be used for assessing water use requirements in licence applications. Current legislation requires that charges should be sufficient only to recover the Agency's costs in performing its water resources functions. The charge for a particular abstraction is calculated by multiplying the authorised quantity by a standard unit charge and by factors for 'season', 'loss' and 'support'. Abstraction charges for typical usage vary between $0.8p/m^3$ to $1.96p/m^3$ across the country Government sponsored research (DETR, 2000) concluded that 'abstractors on the whole are unlikely to respond to increases in charges unless those increases are significantly higher than current rates'. Government concluded that increasing charges beyond cost-recovery level was unlikely to be the best way to reduce abstraction.

All new abstraction licences are issued with a time limit. This allows the Agency to review the need for the licence and the efficiency of its use, and to take account of changes in the flow regime, the environment, or improved understanding of the impact of abstraction.

Water rights trading gives a realisable value to the rights contained in abstraction licences. There are limited opportunities to trade water rights under current legislation, where the Environment Agency acts as the mediator, ensuring the environment is protected. The Agency does not get involved in negotiating the price of the trade.

11.2.2.2 Discharge consenting

A second regulatory instrument available to the Agency to protect and enhance the environment is discharge consenting. Under the Water Resources Act 1991 the Agency is responsible for the protection of 'controlled waters' where controlled waters are defined as all watercourses and groundwater. From 1 July 1991 the National Rivers Authority (a predecessor body of the Environment Agency) introduced charges to recover related costs in respect of applications and consents for discharges to controlled waters in England and Wales. The application charge and the volume charge take into account the volume and content of the discharge as well as the quality of the receiving water. As well as volume reduction being a cost driver the Agency also has the power to refuse consent.

All industries, commerce, institutions and householders not connected to sewers that discharge into watercourses require discharge consent from the Environment Agency to do so. The Agency has issued approximately 100,000 discharge consents. The efficient use of water is not a direct regulatory consideration when the Agency considers an application but it can be a by-product. Refusal of discharge consents because of possible contributions to flooding and worsening water quality has in some cases made industries and building owners opt for more sustainable solutions such as rainwater collection for toilet flushing. Water company wastewater treatment plants require discharge consents but at present it is not considered that this has influenced water company demand management policy.

11.2.2.3 Integrated pollution prevention and control (IPPC)

Integrated Pollution Prevention and Control (IPPC) is an integrated regulatory approach to control the environmental impacts of certain activities, by determining controls for industry to protect the environment through a single permitting process. IPPC applies an integrated environmental approach to the regulation of certain major industries, requiring emissions to air, water, including discharges to sewer, and land, and a range of environmental effects, to be considered together. The IPPC regulations have been made under the IPPC Act 1999 that implemented European Commission Directive 96/61. The Directive covers only specified industries, but some of those use large volumes of water. Industries are required to carry out water audits within 2-4 years of the permit being issued and to present proposals for improving their efficiency of use. IPPC Guidance for the Food and Drink, Paper and Pulp, and Pigs and Poultry industries set out benchmarks and best practice water use for compliance with the Directive. Full implementation will be across 30 industry sectors and 6000 installations on a phased basis to 2007.

11.2.3 The public water supply

11.2.3.1 Water resources plans

In 1996, as a policy response to the 1995 drought, the Government (DoE, 1996) stated that the Environment Agency should be fully involved with water companies' new resource development plans. The Environment Agency has produced Water Resources Planning Guidelines (EA, 2003) for plans that have to be submitted to the Environment Agency every five years. Water Resources Plans have been submitted to the Environment Agency in 1999 and 2004. In 2004 the plans have been more closely integrated into the Periodic Review process: an example of the environmental and economic regulators working closely together. Plans are updated and reviewed annually. The guidance is based on collaborative projects involving the water companies and the regulators (UKWIR, OFWAT, DEFRA and EA, 2002) outlining economic methodologies for selecting options (demand and supply side) to manage the supply-demand balance. The methodologies are essentially cost-benefit analyses of a range of options that take full account of social and environmental costs and benefits to produce a least cost plan. The Agency expects a full consideration of demand management options. The Water Resources Plans also require the water companies to produce a demand forecast, allowing the Agency to challenge assumptions made about demand growth in addition to the selection of options to manage it.

The Agency's water resources planning guideline (EA, 2003) states:

'The plan must consider options available across the full range of 'total water management' options. These include:

- Customer side management (policies affecting customer use and supply pipe losses).
- Distribution management (policies targeted at activities between distribution input and the point of delivery, e.g. leakage control).
- Production management (policies targeted at activities between abstraction and distribution input, e.g. recycling of filter backwash).
- Resource management (policies affecting deployable output, such as new reservoirs or resource transfers)'.

The Water Act 2003 has placed water resources plans, and drought plans, on a statutory footing.

11.2.3.2 The periodic review process

The Director General of Water Services (of OFWAT) has a primary duty under the Water Industry Act 1991 to secure that:

- Water and sewerage functions are properly carried out through England and Wales.
- Water companies are able (in particular, by securing reasonable returns on their capital) to finance the proper carrying out of these functions.

The Director General is also required by Section 2 of the Water Industry Act 1991 to use his powers in a manner which he considers best calculated to promote economy and efficiency on the part of water and sewerage undertakers.

The Director General uses the Periodic Review process as the means to fulfil these duties.

Every five years a periodic review process takes place whereby OFWAT set price limits that enable the water companies to finance the delivery of services, in accordance with the relevant standards and requirements. OFWAT uses the RPI±K approach to price setting where the change in prices is inflation (as per the retail price index, or RPI) ± K, where K is a factor that, if positive, recognises that the companies are required to deliver a large capital investment programme to finance obligatory environmental and quality improvements. If K is negative this recognises that savings from expected efficiency gains have been predicted to be greater than obligatory financial commitments.

The price limits are set on the basis of assumptions about company performance in the following elements of the companies' business (OFWAT 2002):

- Future efficiency gains – assumptions are made about future efficiency gains.
- Enhancements to environmental and drinking water quality – reasonable expenditure is allowed to meet new quality standards.
- Enhancements to security of supply – if security of supply is an issue reasonable expenditure is allowed.
- Enhanced service levels – where service standards need to be improved, for example by solving problems of sewer flooding.
- Financing functions – a check is made that efficient companies can finance their functions and maintain a sound financial position.
- Past performance – companies are allowed to retain the benefits of out performance for up to five years, after which the benefits are passed back to customers.

The next Price Review will take place in 2004, where price limits will be set for the period 2005-10. Following the last Periodic Review (1999) the costs of maintaining the balance between supply and demand amounted to 1% of customer's bills (OFWAT, 2002). The price limits for maintaining the balance between supply and demand at PR04 will be determined following acceptance by the regulators of the water company Water Resources Plans. OFWAT expects water companies to consider options to manage demand across the public water supply (leakage control, household demand, non-household demand and operational use).

In their approach to the fourth Periodic Review (OFWAT, 2002) stated:

There is a general consensus that a planning approach should aim to identify the optimal way of delivering adequate security of supply to customers. The options include demand management, development of new water resources, and reducing leakage. We expect companies to present robust plans which have given full consideration to the options available. The companies should demonstrate that they have chosen the optimal strategy.

In the same document it also states:

If the cost of saving water by adopting a demand management measure is less than the cost of delivering additional water, then it is economic for a company to carry it out. We expect companies to assess the role of demand management within a long-term plan to balance supply and demand.

The statement promotes the idea that demand and supply side options should be treated equally, consistent with the approach set out in Economics of Balancing Supply and Demand (UKWIR, OFWAT, DEFRA and EA, 2002).

11.3 THE CONSTITUENTS OF THE PUBLIC WATER SUPPLY

Figure 11.3 shows the breakdown of the public water supply by volume. The 16% losses are on the distribution system only – additional losses are present in the household and non-household components.

In addition to the regulatory demands placed upon them by the environmental and economic regulators, the water companies have their own legislative requirements to comply with. The main ones are covered in sub-sections 3.1 and 3.2, but in addition the companies have a duty, under section 3 of the Water Industry Act 1991 *to further the conservation and enhancement of natural beauty.*

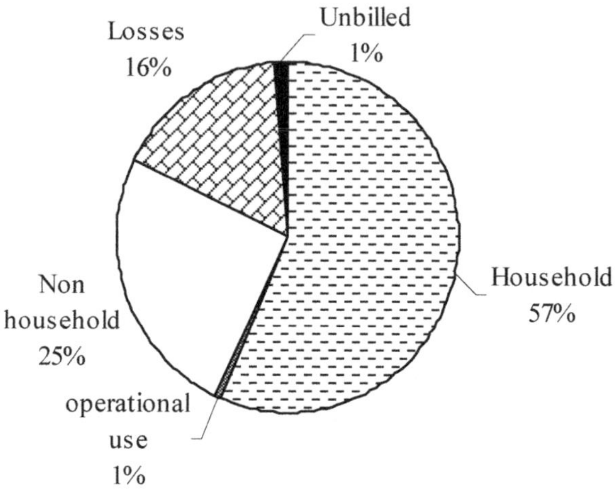

Figure 11.3 The public water supply by volume.

11.3.1 Leakage

Under Section 37 of the Water Industry Act 1991 water companies have a duty to *develop and maintain an efficient and economical supply system within its area.* In spite of this most companies' leakage increased between privatisation in 1989 and 1995. In 1997 the Government held a Water Summit and launched a 10-point plan for a better water industry as a response to some of the criticism targeted at water companies in the way they dealt with the 1995 drought. The first two points of that plan were as follows:

- OFWAT to set tough mandatory leakage targets.
- All water companies are expected to provide free leak detection and repair service to household customers' supply pipes.

The mandatory leakage targets, initially derived 'pragmatically' and subsequently on the companies' own economic analyses have been responsible for a reduction in leakage of 29% from its peak in 1995/96 (OFWAT, 2003). Leakage control expenditure has not historically been allowed for at price setting as it was believed to be in the companies own interests to operate at their economic level of leakage. In some years companies have failed to meet targets and OFWAT's response to this has been to place them on quarterly reporting where the company's plan for remedying the situation is closely scrutinised. In extreme circumstances a water company faces the loss of its operating licence for poor performance. The consistent failure of Thames Water to achieve leakage targets is currently challenging the regulatory system.

Since the inception of leakage targets in 1997 they have been set on an annual basis. When water companies submitted their Water Resources Plans in 2004, part of that submission was an assessment of their economic level of leakage for the period 2004-09. The resources required to achieve or maintain this level of leakage will form part of the price-setting consideration. There is then an incentive for companies to achieve this level of leakage in the most efficient way, which could be due to technological development or improved management practices, because any savings are deemed 'legitimate profit'. At most companies will reap a five-year benefit: at the next Periodic Review (2009) the improved technology and management practices would feature in the new ELL calculation and subsequent target.

11.3.2 Public water supply to buildings (households and non-households)

11.3.2.1 Duty to promote the efficient use of water

In 1996 the Water Industry Act 1991 was amended by the addition of section 93A giving the water companies a duty to *promote the efficient use of water by their customers*. It is the role of OFWAT to ensure that the water companies are exercising this duty. In 1996 water companies were required to submit a water efficiency plan to OFWAT setting out how they intended to fulfil their duty. A revision to the original water efficiency plan was requested in 2001. Water companies report annually to OFWAT on water efficiency activity and since 1996 the following inputs have been recorded (OFWAT, 2003):

- 428,621 supply pipes repaired
- 58,413 supply pipes replaced
- Over 7m cistern devices distributed
- Over 13m water audit packs distributed to households
- 102,572 household water audits carried out by water companies (or their agents)
- 136,718 water audit packs distributed to institutions and commerce
- 31,985 institutional and commercial water audits carried out by water companies (or their agents)

(For reference purposes there are approximately 21m households and 1.6m non-households in England and Wales.)

Water company estimates of the savings resulting from these activities totals 483 Ml/day (3.1% of distribution input), with the majority of the savings arising from the repair and replacement of supply pipes, where arguably the driver was to meet leakage targets rather than to comply with the water efficiency duty.

OFWAT has stated that it expects a 'basic, minimum level of activity from all companies, unless supplies are under pressure where a more active approach is necessary'. A 'more active approach' is not defined. Most companies are complying with the requirement to carry out this basic, minimum level of activity and doing little more (with one or two notable exceptions).

11.3.2.2 The Water Supply (Water Fittings) Regulations 1999

The primary purpose of the Water Supply (Water Fittings) Regulations 1999 (see DETR, 1999) is to give the water undertaker powers to prevent the waste, undue consumption, misuse or contamination of the public water supply. Prior to 1999 the water companies had the power to make and enforce water byelaws. The Water Supply (Water Fittings) Regulations are now the property of the Secretary of State (for the Environment, Food and Rural Affairs) and water companies have a duty to enforce them.

The main thrust of the regulations has always been to prevent contamination of the public water supply but Table 11.1 shows that the water use of standard appliances is also a consideration. The regulations are enforceable at the point of installation rather than at the point of sale or manufacture.

Table 11.1 Appliances and Water Regulations.

Appliance	Water Byelaws, pre 1999	Water Supply (Water Fittings) Regulations 1999
Toilets	7.5 litre single flush	6 litre and re-introduction of dual flush
Clothes washers[1]	180 litres/cycle	120 litres/cycle
Dish washers	7 litres/place setting	4.5 litres/place setting
Showers	None	None

The regulations (SI 3442: The Water Industry (Prescribed Conditions) Regulations 1999) also created a notification procedure for high water using appliances defined as:

* Unattended garden watering equipment

---

[1] Quantities for clothes washers are based on a 5kg load, a typical washing machine capacity – the regulation states no more than 24 litres of water per kg of washload capacity

- Automatically replenishing a pond or swimming pool, with a capacity of greater than 10,000 litres
- Large bath (>230 litres to overflow)
- A unit that incorporates reverse osmosis (e.g. water softener)
- A power shower (undefined, and subsequently removed from the list)
- A pump or booster drawing more than 12 litres/min.

In advance of the installation the person proposing to undertake the work is required to notify the water company that can then exercise its right to install a meter at the customer's premises. Early indications are that the number of notifications are few and where they are made water companies have not taken advantage of the metering opportunity.

11.3.2.3 Building Regulations

The references to water management are found in Parts G (Hygiene, 1991) and H (Drainage and Waste Disposal, 2000). The requirements in Part G relate to provision of suitable numbers of sanitation and wash basin installations and their ease of cleaning. There is no mention of specific design or water volumes. The developer/builder is responsible for complying with the Building Regulations, which are enforced by local authorities. Recent revisions to Buildings Regulations aimed to clarify specific regulations and at the same time become less prescriptive. This does not facilitate introducing more specific definitions of suitable measures for water efficiency. Neither the Building Regulations nor Water Regulations make specific references to rainwater harvesting or greywater recycling. Following the Deputy Prime Minister's initiative *Sustainable Communities: Building for the Future*, launched in February 2003 there was concern in relation to the environmental impacts of housing growth in the south-east of England. He further announced (Better Buildings Summit, 21 October 2003) that *"new standards for water conservation would be introduced by 2005. They will cover new and refurbished buildings"*. A revision of Part G of the Building Regulations, to incorporate improved standards of water efficiency, is expected in 2005.

11.3.2.4 Development planning

The current development planning system is multi-tiered, from Central Government guidance through to County Structure Plans, Unitary Development Plans and District Local Plans. Both Government (Planning Policy Guidance or PPGs) and Regional Planning Bodies (Regional Planning Guidance or RPGs) have historically advised on how regional and local plans should be developed. The guidance is usually issue led and has no structured development or regular review. The Planning and Compulsory Purchase Act 2004 will result in PPGs and RPGs being replaced with more concise guidance that clearly separates policy from guidance. The only specific PPG for water management relates to the issue of development within the flood plain (PPG 25).

Additionally there are supplementary planning guidance notes and development briefs for specific aspects and areas of development. These are potentially a mechanism to provide very detailed advice to Planning Authorities relating to suitable water efficient options for a particular site.

There has been some speculation that Section 106 Agreements could be used as a means of ensuring water efficient developments. Section 106 Agreements are part of the Town and Country Planning Act 1990, and allow planning obligation agreements between the local authority and developers. The agreements can be used for a variety of purposes:

Government advice (PPG 1, see (ODPM, 1997)) states that: "Planning obligations are useful instruments, where they are necessary to the development and fairly and reasonably related in scale and kind, as they can enable a property owner to overcome obstacles which would otherwise prevent planning permission for being granted. Planning obligations should be directly related to the proposed development" (Para 36).

Circular 1/97 'Planning Obligations' gives more detailed advice on when a Section 106 agreement is or is not acceptable: 'In granting planning permission, or in negotiating with developers, a local planning authority may seek to secure modifications or improvements to the proposals…where appropriate they may seek to enter into planning obligations with a developer." (para 5).

"Planning obligations have a positive role to play in the planning system. Used properly, they can remedy genuine planning problems and enhance the quality of development. They can provide a means of reconciling the aims and interests of developers with the need to safeguard the local environment or to meet the costs imposed as a result of development." (para B7).

There is uncertainty over whether a Section 106 agreement could be used to stipulate a (high) level of water efficiency, for example 4/2 litre dual flush toilets in all households.

With respect to planning consultation the Environment Agency and English Nature are statutory consultees for development plans (structure and minerals). Water companies are not although they are often consulted. However, the Planning Inspectorate does not have to act on advice given. Questions asked of water companies are generally restricted to their ability to provide infrastructure ahead of development, in accordance with their statutory duty to do so.

11.3.2.5 Charging for water

Household metering remains a contentious and emotive issue. Since the Water Industry Act 1999 (SI 3441, the Water Industry (Charges) Vulnerable Groups Regulations 1999) compulsory metering of households has been prohibited but metering is allowed in the following circumstances:

- Free option scheme – water companies must provide a free meter to any customer who requests one. The customer has a 'right to revert' up to one year after the installation.
- New homes – metering is the normal charging method.
- Households that use water for 'non-essential' purposes, e.g. garden sprinkler.
- Water stressed areas – a water company can appeal to the Secretary of State for designation, if granted the company can then meter all households compulsorily (there have been no applications to date).
- At change of occupancy (since customers only have a right to remain unmeasured only in their *present* home).

The difference in the proportions of metered households across the water companies is largely due to the varying degree with which the above policies are pursued, with some companies showing considerably more enthusiasm than others.

Data in the public domain are limited that demonstrate the effect on demand of moving from unmeasured to measured charging. The National Metering Trials Group (1993) concluded that the introduction of a volumetric charge reduced average demand by 10% and peaks by 30%. Recent work undertaken by UKWIR (EA, 2003) has shown that on average optants reduce their use by 9%, with an additional reduction of 0.2% per month although it is not known for how long this effect lasts. In the last price setting process (1999) OFWAT allowed for a 5% reduction in demand by meter optants (those electing to be metered, rather than it being imposed), unless the company could demonstrate otherwise. Very little is known about the price elasticity of water supplied to households and although Government, OFWAT and the Environment Agency wish to see companies experiment with tariffs (e.g. rising block, seasonal) few companies have shown themselves willing to do so.

11.3.2.6 Enhanced capital allowance

Government has stated that it would like to see environmental considerations playing a greater role in business decisions. In 2001 the Government proposed, subject to EU State Aids approval, further tax incentives in the form of enhanced capital allowances for environmentally friendly investments. The incentive scheme is called the Green Technology Challenge and aims to tackle climate change, air quality, and water use and water quality. The scheme is not yet operational but a product list approach is being adopted for water conservation technologies. Short-listed technologies include: sub meters, flow restrictors, smart metering, leakage detection, water efficient toilets and taps (faucets), recycling and re-use systems. Where a business purchases technologies on the product list they will qualify for an enhanced capital allowance.

Capital allowances are given to businesses when they invest in plant and machinery. The allowance reduces the amount on which the company has to pay income or corporation tax. The allowance is 25% of the purchase cost to be offset against tax per year (on a declining balance basis). An enhanced capital allowance allows companies to offset the whole of the capital purchase costs of qualifying equipment against their taxable profits in the year of purchase. This improves the cash flow for the company involved making it (more) worthwhile to invest in environmentally friendly goods.

The scheme, previously in place for energy saving technologies was introduced in 2003.

11.3.2.7 Competition

The Government has stated that the forthcoming Water Bill will include proposals for new entrants (to the market) to be licensed to compete with the existing water companies by using their own source of water (accessing the existing water company's network to transport the water to a customer, i.e. "common carriage") or by buying water from the statutory undertaker to supply a customer (i.e. providing retail services only). The Government is adopting a cautious approach by proposing that competition should be subject to industrial consumers using over 50Ml/year, although this could be reduced in future if the Government considers that this would be in the consumer interest. Competition is likely to affect demand in two ways. Firstly the price will fall leading to a rise in demand (price elasticity has been estimated to be approximately –0.3% in the non-household sector). But in a competitive market it has been suggested that one way those new entrants would compete for customers would be by offering water conservation expertise as part of the arrangement. The Government believes that competition will make industrial consumers more aware of costs. It remains to be seen what the overall effect on demand will be.

11.4 RECENT LEGISLATION

11.4.1 The European Water Framework Directive

The European Water Framework Directive, a common framework across Europe to address water issues, came into force on 22 December 2000. Its overall aim is to protect and enhance the 'ecological status' of surface waters and groundwaters in the European Union through a framework with a common approach based on the river basin as a management unit (not administrative or political boundaries). Each river basin district will have a river basin management plan that will contain a 'programme of measures' to resolve any problems identified in the river basin characterisation process. The objective is to achieve 'good ecological status' for all waters by 2015.

Previous European policies have been fragmented in both objectives and means (e.g. Urban WasteWater Treatment Directive, Nitrates Directive, Drinking Water Directive). These will effectively be replaced by a single piece of legislation. Key timetable milestones are as follows:

2003   Member states will have modified their current legislation to comply with the Directive
2004   Complete river basin characterisation process
2005   Start public consultation
2009   Finalise River Basin Management Plans (including the programme of measures)
2010   Introduce pricing policies
2015   Meet environmental objectives (good ecological status)

The areas where the Directive may influence demand management policy are as follows (Official Journal of the European Communities, 2000):

Article 5 – The characterisation process for each River Basin District includes an economic analysis of water use. Annex II specifies the *'estimation and identification of significant water abstraction for urban, industrial, agricultural and other uses, including seasonal variations and total annual demand, and loss of water in distribution systems'*.

Article 9 – Recovery of costs of water services. This states that *'member states shall take account of the principle of the costs of water services, including environmental and resource costs, having regard to the economic analysis'*……and Member States shall ensure by 2010 that *water pricing policies provide adequate incentives for users to use water resources efficiently and thereby contribute to the environmental objectives of this Directive…*

Article 11 – Programme of measures. 'Basic measures' are the minimum requirements to be complied with and shall consist of…..(c) *measures to promote an efficient and sustainable water use to avoid compromising the achievement of objectives specified in Article 4* (The Environmental Objective) and (e) *controls over the abstraction of fresh surface water and groundwater, and impoundment of fresh surface water, including a register or registers of water abstractions and a requirement of prior authorisation for abstraction and impoundment.*

Article 14 – Public Information and Consultation. *1. Member states shall encourage the active involvement of all interested parties in the implementation of this Directive, in particular in the production, review and updating of the river basin management plans.*

Article 14 is important in the context of demand management because water conservation/efficiency programmes are unlikely to succeed in the absence of public involvement and support. It could stimulate a greater dialogue with the public where there is an opportunity to demonstrate that people cannot have increasing quantities of water and an improved water environment. The emphasis on *active involvement,* described in the supporting documentation as '*a higher level of participation than consultation'* has the potential to initiate a culture change where the public become part of the solution, as opposed to their water use being the problem.

11.4.2 The Water Act 2003

In November 2003, the Water Bill received Royal Assent. The Act is intended to provide a more robust legislative framework to facilitate both sustainable water resources management and economic growth through new opportunities in the water sector The following pertain to the efficient use of water:

Section 72: Efficient use of water resources

In section 6 of the Environment Act 1995 (c. 25) (general provisions with respect to water), in subsection (2)(b), after "Wales" there is inserted "(including the efficient use of those resources)".

Section 81: Duty to encourage water conservation

(1) The relevant authority must, where appropriate, take steps to encourage the conservation of water.

(2) The relevant authority is-

   (a)   the Secretary of State, in relation to England,

   (b)   the Assembly, in relation to Wales.

(3) After the period of three years beginning with the date on which this section comes into force, and after each succeeding period of three years, the Secretary of State must prepare a report about the steps taken by him under this section, and about any such steps which he proposes to take.

(4) The Assembly may make an order requiring the preparation by it of corresponding reports, and such an order may make provision about when, or in relation to what periods, they are to be prepared.

(5) Each such report must-

   (a)   if prepared by the Secretary of State, be laid before Parliament,

   (b)   if prepared by the Assembly, be laid before, and published by, the Assembly.

Section 82: Water conservation: requirements on relevant undertakers

In section 3(2)(a) of the WIA (environmental duties in relation to proposals relating to the functions of a relevant undertaker), after "special interest" there is inserted "and, in the case of the exercise of such a power by a company holding an appointment as a relevant undertaker, as to further water conservation".

Section 83: Water conservation by public authorities

(1) In exercising its functions and conducting its affairs, each public authority shall take into account, where relevant, the desirability of conserving water supplied or to be supplied to premises.

(2) In subsection (1), "public authority" means any of the following-

    (a) a Minister of the Crown (within the meaning of the Ministers of the Crown Act 1975 (c. 26)),

    (b) a Government department,

    (c) the Assembly,

    (d) a local authority (within the meaning of section 270(1) of the Local Government Act 1972 (c. 70)),

    (e) a person holding an office-

        (i)   under the Crown,

        (ii) created or continued in existence by a public general Act, or

        (iii) the remuneration in respect of which is paid out of money provided by Parliament,

    (f) a statutory undertaker (being any person who, by virtue of section 262 of the Town and Country Planning Act 1990 (c. 8) is or is deemed to be a statutory undertaker for any purpose), and

    (g) any other public body of any description.

The water conservation sections, 81,82 and 83 were included in the first commencement order, and came into force on 8 March 2004. In response to a question by Norman Baker MP, when the Water Bill was being debated in the House of Commons, the Minister for the Environment, Eliot Morley committed the Environment Agency to an investigation into the feasibility of a Water Saving Trust.

In relation to section 81, in three years time the Secretary of State and the National Assembly for Wales are due to present reports on progress on water conservation. For sections 82 and 83 DEFRA is planning to issue guidance for water companies and public authorities during 2004.

In relation to abstraction licensing the Water Act 2003 makes a number of changes to the abstraction licensing regime.

- All new licences will be time-limited.
- New powers for the Agency to require information from abstractors on how they use the water.

- The Agency gains powers to enter into enforceable water management arrangements with licence holders.
- Allows the Agency to propose bulk supplies between companies where it thinks this would promote the efficient use of resources.
- Reducing the period after which the Agency can revoke a licence for non-use without compensation.

The Water Act 2003 is not being used specifically to transpose the Water Framework Directive. However provisions in the Act will help deliver much of what is required in respect of controlling water abstractions. The new duty on water undertakers to further water conservation is also thought to be helpful in relation to implementation of the Directive.

11.5 SUMMARY AND PERSPECTIVES

A summary of regulatory policies and how they control or influence different components of water use is shown in Table 11.2.

Table 11.2 Summary of regulatory policies for the efficient use of water.

<table>
<tr><td colspan="6">Total water abstracted (and discharged)</td></tr>
<tr><td colspan="6">Environment Agency abstraction licensing policy
Environment Agency discharge consenting policy</td></tr>
<tr><td colspan="2">Public water supply</td><td colspan="4">Direct abstraction</td></tr>
<tr><td colspan="2"><ul><li>Water resource plans and supply demand balance</li><li>Drought plans</li></ul></td><td colspan="4">Enhanced capital allowance
Development Planning</td></tr>
<tr><td>Leakage</td><td>Supply to buildings</td><td>Power generation</td><td>Industry</td><td colspan="2">Agriculture</td></tr>
<tr><td>Leakage targets</td><td><ul><li>Water Supply (Water Fittings) Regulations 1999</li><li>Development planning policies</li><li>Duty to promote the efficient use of water</li></ul></td><td>IPPC</td><td>IPPC</td><td colspan="2">IPPC (Pigs & Poultry)</td></tr>
<tr><td></td><td>Households | Non-households</td><td></td><td></td><td colspan="2" rowspan="2"></td></tr>
<tr><td></td><td>Metering policy | <ul><li>IPPC</li><li>Enhanced capital allowance</li><li>Competition</li></ul></td><td></td><td></td></tr>
</table>

It is apparent that for direct abstraction there is a high dependence on abstraction licensing compared to the public water supply where there is a multitude of regulatory instruments. Although development planning policies also apply to direct abstractors, the planning process is currently unlikely to add anything in terms of expectation of the efficient use of water, compared to the public water supply where direct controls on abstraction are more difficult because the supply to the development is likely to be only a small proportion of a larger abstraction. Table 1.3 explores whether the regulatory instruments detailed in this paper by considering the positive and negative incentives from the perspectives of those stakeholders who are in a position to control or influence demand.

Table 11.3 demonstrates the greater role that sticks have played in introducing water efficiency.

Table 11.3 Sticks and carrots.

From the perspective of:	Sticks	Carrots
Water Company	• Efficient use duty • Leakage targets • Water Resource Plans/PR04	• Economic level of leakage (2004 on) • Good Public Relations (PR)
Direct abstractor	• Abstraction licensing • Discharge consenting • IPPC • Restrictions	• Good PR • Enhanced Capital Allowance (ECA)
Non-household supplied by water company	• IPPC • Water Fittings Regulations	• ECA • Free audit by some water companies • Reduced bills
Household	• Water Fittings Regulations • Restrictions	• Reduced bills (for the 20% metered)
Building designer/installer	• Water fittings regulations • Development Planning conditions?	• Good PR

11.6 DISCUSSION

11.6.1 Abstraction licensing

The Environment Agency must ensure that the needs of the applicant are reasonable before it can grant an abstraction licence. A key test is whether the purpose for which water will be used is efficient. The growing body of knowledge on water use by sector means that this assessment is becoming progressively more rigorous and widespread across the range of purposes for which licences are granted.

The aim is to ensure a sustainable balance between the needs of society and those of the environment. Historically, some licences were granted which are now known to be unsustainable. In catchments where this is the case the Agency will seek to reduce the volumes abstracted. One potentially cost-effective, low impact way of doing this is by the active adoption of water efficiency measures by abstractors within the catchment.

11.6.2 Water resources plans and the periodic review process

The fourth Periodic Review marks the second round of Water Resources Plan submissions to the Agency. The process is still maturing. We have yet to see water companies making progressive choices, in the form of demand management options other than leakage control and metering, in their submissions. In OFWAT (2002) it states: '*Companies will need to provide robust estimates of the costs and savings of efficient use of water initiatives based on actual (measured rather than estimated) changes in consumption. We also expect companies to state their assumptions about the sustainability of savings and to demonstrate that* these are consistent with available evidence'.

The Periodic Review process is currently a barrier to a progressive approach to water resource planning by the water companies. Placing the 'burden of proof' firmly on the water companies to demonstrate that a particular demand management option is cost effective means that the companies are in a 'Catch 22' position. They lack the capital (water efficiency studies are expensive because of the monitoring involved) to produce the robust estimate and because they don't have the robust estimate they cannot obtain the necessary capital at price setting. At present this is a major barrier to the progress of water efficiency programmes involving the public. Despite the 'principal guidance' (DEFRA 2004) stating '*The Government believes that such investigations are necessary before further commitment is made to any new reservoirs, and that as part of the investigative work there should be a sound and evidence based understanding of the potential benefits of water efficiency and demand management measures*' 2004 Water Resources Plan submissions largely propose new resource development as the supply-demand balance solution.

11.6.3 Water Supply (Water Fittings) Regulations

The major area of progress has been reducing maximum toilet flush volume to 6 litres. Elsewhere the regulations are so far behind the market of water efficient appliances no claim could be made that they were following it. A decision needs to be made to what extent the Regulations should be leading the market and driving manufacturers to produce ever more efficient goods. Other areas where progress needs to be made, for effective regulations, is enforcement at the point of sale and/or manufacture rather than at the point of installation, and an effective and publicised notification procedure allowing the water companies to install meters so the purchasers of high water using goods face the economic consequences of their purchasing decision. Incorporation of water efficiency measures into the Building Regulations should improve compliance as a result of a more efficient enforcement process.

11.6.4 Leakage targets

Leakage targets have undoubtedly been a success. They have reduced leakage from its all time high of 5112 Ml/day in 1994/95 to a low of 3240 Ml/day in 2000/01. The previous reliance on water companies to act in their own economic interests failed to produce the necessary reductions. With most companies now claiming to be at or close to their economic level there seems to be acceptance that only minor reductions will be achieved in future. The development of leakage performance indicators, recommended by the tripartite study (EA, OFWAT and DEFRA 2002), that will explain differences between company targets and performance is a logical next step.

11.6.5 Water efficiency/water conservation duty

The existing duty on water companies to promote water efficiency by their customers has encouraged cheap (as opposed to cost-effective) options which have delivered only very small and transient savings. It is not clear at present what Section 82 of the Water Act 2003 (an additional duty for water companies to further water conservation) will mean in practice. Hopefully it will provide the impetus for more concerted approaches.

11.6.6 Water charging policy

Payment by volume as well as being a water conservation measure in itself, underpins all other measures targeted at the household. Metering is an essential prerequisite for other options. Only when large proportions of households are metered within a company area will it be possible for the company to consider introducing tariffs that distinguish between essential and non-essential uses of water.

The current water charging policy, although something of a compromise between pro- and anti-metering views provides many opportunities for water companies to progress metering. In addition to metering optants companies could establish a procedure for metering houses at change of ownership since the right to remain unmeasured is only applicable to householders in their *present* home. They could also seek water-scarcity designation and increase efforts to use the notification procedure in order to meter households with high water using appliances.

11.6.7 Development planning

The current absence of strategic policy in this area provides little incentive for planners to include water efficiency as an objective in their local plans and thus apply the principles to individual planning applications. There remains the issue that water efficient design is not a material consideration for planners and unless they are tasked with some responsibility to progress this aspect there seems little likelihood, at present of using the planning process to help manage water demand.

However, with the development of Regional Spatial Strategies by Regional Assemblies in England, it seems likely that in southern England at least there will be policies setting out expectations of building new houses to high environmental standards. The key documents here are likely to be the Supplementary Planning Guidance (SPG) which should contain the detail of how the policies in the Regional Spatial Strategies will be delivered. To prevent planners getting embroiled in technical specifications of water using appliances, a reference point is needed, whether this is one level of the BRE's EcoHomes standard, or a level in the proposed code for Sustainable Buildings (Sustainable Buildings Task Group, 2004) or even a water efficiency labelling scheme that has been proposed by DEFRA.

Water companies (and others) should also be able to advise on options for water efficiency as part of any development design brief given their duty to promote water efficiency to their customers and the need to balance demand with available supply.

Section 83 of the Water Act specifies that 'each public authority shall take into account, where relevant, the desirability of conserving water supplied or to be supplied to premises'. This has been deliberately written to capture the role of planners.

It would appear that the avenues for making water efficiency appliances a condition of planning have as yet, not been fully explored, but the new Regional Spatial Strategies and associated documents will provide an opportunity to progress in this area.

11.6.8 Enhanced capital allowance

Enhanced Capital Allowances represent a change in Government thinking – from sticks to carrots. Hopefully this will be the first of many initiatives that reward good performance as opposed to penalising poor performance. One area that would benefit from such a change would be a financial incentive for water companies to introduce water efficiency measures.

11.6.9 Recent legislation

The proposals in the Water Act 2003 are to be welcomed principally because the Agency will have greater powers to protect and enhance the environment through the abstraction licensing process. Any additional amendments that place a water conservation duty on public bodies and water companies are to be welcomed.

The Water Framework Directive potentially offers 'hooks' for further development of policy to ensure the efficient use of water.

11.7 CONCLUSIONS

- The regulatory system in England and Wales, with clear separation of roles between the regulators and regulated has brought about a degree of transparency that did not exist previously; for example leakage and per capita consumption figures are now in the public domain.
- The regulatory approach in England and Wales has in many respects been successful; total demand in 2002/03 is less than it was in 1992/93 largely due to reductions in water company leakage.
- This success has largely been achieved by the 'stick' variety of regulation. But stick or 'negative incentive' regulation may have run its course; if future demand reductions are to be achieved and there is evidence that they are necessary more 'positive incentive' (carrot) regulation may be required. The Government's Enhanced Capital Allowance points the way.
- The 'Catch 22' situation that the water companies found themselves in over funding for water efficiency at the Periodic Review was not resolved for PRO4. Detailed investigations establishing robust costs and savings data are essential for planning at PRO9.
- Despite the need for more incentive based regulation there are many areas where improvements could make significant contributions to delivering water efficiency. For example, the Water Supply (Water Fittings) Regulations could be more stringent, the 'minimum level of activity' for a water company to fulfil it's duty to promote water efficiency could be higher and water companies could take greater advantage of the opportunities to install meters afforded by current legislation.
- The Water Act 2003 could make a significant contribution to water efficiency, depending on what is contained within DEFRA's guidance.
- The Water Framework Directive presents opportunities for promotion of water efficiency measures and further development of regulatory policy.
- The Regional Spatial strategies currently being developed offer opportunities to enshrine the efficient use of water in planning policies and guidance.

The views expressed are those of the author and not necessarily those of the Environment Agency.

11.8 REFERENCES

DETR (1999) Water Supply (Water Fittings) Regulations 1999. Department of the Environment, Transport and the Regions.

DETR (2000) *Economic instruments in relation to water abstraction; A consultation paper.* Department of the Environment, Transport and the Regions.

DEFRA (2002) *Extending opportunities for competition in the water industry in England and Wales.* Department of the Environment, Food and Rural Affairs, July.

DEFRA (2004) *Principal guidance from the Secretary of State to the Director General of Water Services. 2004 periodic review of water price limits.* Department of the Environment, Food and Rural Affairs, March.

DoE (1992) *Using water wisely.* Department of the Environment.

DoE (1997) *Circular 1/97, Planning obligations.* Department of the Environment.

EA (2001) *Water resources for the future: A strategy for England and Wales.* Environment Agency.

EA (2003) Response to the National Consumer Council consultation paper *Towards a sustainable water charging policy*, Environment Agency, Worthing.

EA (2003) *Water Resources Planning Guidelines.* Environment Agency, www.environment-agency.gov.uk.

The National Metering Trials Group (1993) *Water metering trials – final report.* Water Services Association, Water Companies' Association, Office of Water Services, WRc and Department of the Environment.

ODPM (1997) *Planning Policy Guidance Note 1, General Policy and Principles.* Office of the Deputy Prime Minister, February.

OFWAT (2002) *Periodic Review 2004, Setting Price Limits for 2005-10: Framework and Consultation Paper.* Office of Water Services.

OFWAT (2003*) Security of supply, leakage and efficient use of water 2002-03 report.* Office of Water Services.

Official Journal of the European Communities (2000), OJ Ref. L327, Vol 43, pp 1-73.

Sustainable Buildings Task Group (2004) *Better buildings, better lives.* http://www.dti.gov.uk/construction/sustain/EA_Sustainable_Report_41564_2.pdf

UKWIR, OFWAT, DEFRA and the EA (2002) *Economics of balancing supply and demand.* UK Water Industry Research.

12

Consumer reactions to water conservation policy instruments

Paul Jeffrey and Mary Gearey

12.1 INTRODUCTION

The central objective of this contribution is to critically review current knowledge about the (largely behavioural) response of water consumers to different water conservation instruments. The text covers a range of issues relating to water use practices, responses to water policy instruments, and emerging paradigms which provide new opportunities for designing conservation strategies. The text focuses primarily on domestic contexts but does refer to industrial and agricultural examples where relevant. We first provide a comprehensive picture of the motivations for water use behaviour within different settings (Section 12.2) and indicate the relative performance of different conservation measures (Section 12.3). We also draw attention, where relevant, to the benefits and problems associated with specific conservation instruments. However, water conservation (including reuse & recycling) is a field in transition as we move from supply, through demand, to 'structural' approaches to resource management (see Section 12.4). The material presented here suggests that water conservation cannot be dissociated from wider considerations concerning socio-cultural context, regional development, technology choice, and legitimacy. Conservation is but one tool in the arsenal of water resource management; its effective deployment is dependent on its relationship not only with our communities, but also with other tools and strategies being deployed by government, regulators, and commercial enterprises.

12.1.1 Supply, demand, and social expectations

One of the key state functions in any country is to make available reliable and safe water resources to its populace. Water resources are the freshwater supplies that a society depends on for a range of essentials: For survival - drinking, raising crops and livestock, maintaining fauna and flora, disposing of effluent and dispersal of pollutants; for economic development – industry, agriculture and public health; for social stability – for improved life quality, recreational activity and so much more. Water resources come in multiple forms – rivers, springs, lakes, reservoirs, underground streams, aquifers and water in soil, all of which vary in distribution both spatially and temporally. All states need a minimum quota of accessible water to function; without water even the greatest of civilisations such as those of the Mesopotamians and the Akkadians have fallen (Schama, 1995) and water stress has evidenced itself in water wars (Bulloch, 1993) and political fragmentation.

Water supply has been the dominant paradigm governing water resources planning in developed countries in the modern era. Consequently water management systems have historically been shaped by the need to plan around meeting consumer demands, both in terms of water quantity and assured quality. As a result, water has been privileged as a social 'right' and attempts to restrain or qualify water use, either through fiscal, legislative or technological means, have played a role only in times of drought events, with social acceptance of curtailments accomplished only through a perceived crisis (Lawson, 2002).

However, both the quality and availability of water are now changing at a faster rate than before. The rate and nature of change is not uniform across space and time – with improvements in some areas, for instance pollution control, being mirrored by deterioration in overall quality, for instance a growth in nitrogen levels - either between developed countries or within national catchments. This is due to a number of factors; demographic trends in the density and spatial distribution of population, climate change, improved pollution controls and a broader understanding of the need to provide water in support of environmental functions. Juxtaposed against these changing conditions which contemporary consumers will experience within their lifetime are the existing expectations of consumers. The current supply orientated paradigm, prevalent in almost all developed countries, is focused on meeting consumer demand in terms of quantity and quality. For theorists such as Allan (2001) the epoch of the hydraulic society represents the apex of modernism. The sustainability of this paradigm is now being questioned. The privileging of hydraulics within society has then created a scenario of excess supply and excess quality for many routine uses, such as toilet flushing and garden watering. We therefore have a delivery system with few restraints to quantity or quality. For amelioration in water resource management the two very different tensions of resource depletion and cultural stasis will need to be reconciled.

Our water management systems (in both a technological and institutional sense) are functions of history and as such they are functions of our culture. At the level of the nation state, different countries have different institutional and technological arrangements for water provision and wastewater disposal, each of which has been influenced by dominant political and socio-economic processes. Hence, the current configuration of a water management system is complementary with a set of social and institutional expectations, norms and capacities which determine the terms of reference for system operation and future development. Public acceptance of individual institutional actor practices and policies will, to a large extent, be determined by how well the policies conform to perceived norms and also by how well the actor is identified as part of the community and is thereby acting on the community's behalf.

12.1.2 Water use and policy instruments

Before considering those policy instruments that are used to influence water use behaviour, it might be useful to briefly look at how individuals and communities relate to and interact with water as both a commodity and a natural resource. Influences on the patterns of water use in domestic contexts are now more fully understood than they were even a few years ago. Chapter 1 of this volume presented data on consumption trends and a selection of the major empirical evidence for modifiers in such trends is presented in Table 12.1.

It is worth noting that most of the studies reported in Table 12.1 were conducted in either Europe or the USA. The determinants of water consumption in developing world contexts can be very different. For example, Arimah and Ekeng (1993) report on the determinants of water consumption in Nigeria (where residential water consumption accounts for 80-93 per cent of total water consumption) using data from the city of Calabar. The key factors affecting domestic water consumption were identified as income, price per unit of water, household size, average age of household, educational status of the head of the household, number of taps in the housing unit, number of water using habits, number of water supply sources, and distance from source of supply and the frequency in the supply of pipe-borne water. Household demand for water was found to be largely inelastic to a variety of influences.

Table 12.1 Sources of evidence for specific influences on water consumption.

Influence on consumption	Reference	Comments
Number of residents in a household	Höglund (1999)	Increase in water use is less than proportional to increase in household size
Number of residents in a household	Arbués *et al.* (2000)	Optimum household size, above which the economies of scale identified in Höglund (1999) do not operate.
Children in household	Nauges and Thomas (2000)	Households with children and young people use more water than other households.
Tariff structure	Stevens *et al.* (1992)	Price elasticities are not significantly affected by the choice of a uniform, increasing, or block rate tariff.
Price	Pint (1999)	Study conducted during a drought event in California – demonstrated price elasticity of -0.04 to -1.24.
Property value	Aitken *et al.* (1994)	Positive relationship between property value and water consumption.
Property value	Detwyler and Marcus, (1972)	Statistical analysis of water use in 23 U.S communities revealed a linear relationship between the average market value of a property and the domestic per capita use of water
Household income	Zhou *et al.* (2000)	Identified a positive correlation between income and water consumption.
Household income	Dziegelewski *et al.* (1990)	Study carried out in California, USA. Found that an increase of between 8-28% in average personal income results in an increase of 3% in the total per capita water demand.
Household income	Twort *et al.* (1993)	Study of U.K and European households. Found that in properties characteristic of the highest income groups about 80-100 litres of water more per capita are consumed daily than in houses of the low income groups

Influence on consumption	Reference	Comments
Housing type mix	Baumann *et al.* (1998)	Multifamily housing units tend to share common water uses such as landscaping irrigation, swimming pools, etc and generally have fewer water using appliances and therefore consume less water per capita than single family housing structures
Climate and precipitation	Gundermann (1986)	Pan European study. Also looked at differences between town and country, and the importance of consumer habits and lifestyles.
Age of residents	Lyman (1992)	Retired citizens use more water for gardening
Age of residents	Mayer *et al.* (1999)	Study carried out between 1996 and 1999 in 14 American cities. Data loggers fitted in 1,188 homes. 6,000 households responded to a survey about their water use, and 12,000 households had their billing records incorporated into the study. Study found that the presence of teenagers tended to increase a household's water usage. The presence of adults working full-time decreased water usage
Bathroom number	Mukhopadhyay *et al.* (2001)	Study conducted in Kuwait. Number of bathrooms positively correlated to higher water consumption.
Garden size	Mukhopadhyay *et al.* (2001)	Study conducted in Kuwait. Garden size correlates positively to increased water consumption.
Precipitation	OFWAT (2000)	Precipitation events positively correlate to reduced demand for water.
Temperature	Stern & Gardner (1996)	Temperature rises are directly correlated with increase in water use.
Number of residents in household	Butler (1993)	Study based on wastewater discharges. More household members generate more water consumption per head.

Having briefly reviewed the influences on domestic water consumption, the following sections include a consideration of how individuals, communities, and institutions respond to attempts by water management agencies to influence water use behaviour. Section 12.2 delivers a comprehensive literature review on how water consumers respond to economic, educational, technological and regulatory instruments. Section 12.3 considers the particular case of responses to water recycling systems whilst the final section discusses some of the theoretical contributions recently made to the water policy debate that move beyond supply and demand side strategies.

12.2 ATTITUDES AND RESPONSES TO CONSERVATION INITIATIVES

Encouraging water conservation is perhaps the most obvious policy instrument available to water supply professionals in their efforts to balance supply and demand. National and pan-national bodies tend to have favoured approaches or mixes of instruments which they promote (often informed by local cultural and development considerations), although the overall effect of many strategies is that of a 'carrot and stick' approach. Such demand side approaches rely upon a range of instruments and techniques that can be divided into four broad categories; economic, regulatory, technological, and educational. It is worth noting that economic and educational policy instruments are primarily designed to encourage, whilst regulatory and technological instruments are largely obligatory in nature.

Before moving on to a consideration of these instruments, there are some general points relating to the formulation of policy instruments which have been raised in the literature and which are worthy of note. For example, Latorre *et al.* (2000) have noted the importance of creating a policy dialogue of some kind in relation to water management. They posit that a 'no policy' message will simply encourage the erosion of a water resource. Although we will be commenting on the cultural status of water later in the text, Veronica Strang's contribution on the cultural landscapes of water are relevant in the context of policy formulation. Drawing on studies carried out in the UK and elsewhere, she concludes that policy instruments which conflict with the local cultural meaning of water will not be successful (Strang, 2001).

12.2.1 Economic instruments

Economic instruments to promote water conservation comprise a range of monetary incentives (e.g., time of use, rebates, tax credits) and disincentives (e.g., real cost, penalties, fines) which relay to users accurate and clear information about the value of water. These incentives and costs can be employed as temporary or permanent measures and can either be targeted at specific types of user or be applied universally. On the incentives side, financial instruments such as low interest or forgivable loans,[1] tax credits and rebates can be used to promote installation of water efficient devices. Inefficient units can be bought back and larger sums can be used to jump-start neighbourhood or wider scale conservation initiatives. Providing grants and subsidies for research into water efficient technologies and distribution system management practices can also yield benefits. The use of economic instruments as a deterrent can include fines for non-compliance with regulations or emergency legislation and adjustments to price tariffs. In a broader context, pricing structures can be used as both incentive and disincentive to help utilities manage load profiles (e.g. levelling out peaks in demand). Block or marginal cost rate structures can be used with variations which reflect the need to manage demand across daily, weekly and seasonal timeframes.

Design and implementation of economic policy instruments in the water sector requires awareness of the implications of such instruments and the impacts they may have on particular groups of users. The essential role of water in the lives of humans and its cultural status in many societies needs to be recognised and valued respectively. Ability to pay for water, either as a commodity, a social good, or an environmental resource varies across communities and through time. This fact, when combined with the nature of water as a primary good,[2] raises issues of equity and fairness in water allocation (Collinge, 1992; Herbert and Kempson, 1995); particularly as the volumes consumed by some types of water use, such as drinking, are relatively inelastic to price. Availability of a reasonably priced supply of water has also been linked to regional and national economic growth, particularly in the agricultural and primary industrial sectors (Schama, 1995).

Evidence to support the view that economic instruments are effective in modifying water consumption behaviour is (as suggested in chapter 9) variable. Although the application of simple pricing instruments such as block rates has generated expected gross responses from domestic consumers,[3] more detailed pictures of response envelopes have been difficult to construct. Different groups of water users clearly respond to economic instruments in different ways and at

---

[1] A forgivable loan has no repayment schedule or interest charges. If criteria concerning the use of the loan are met (e.g. a low flush toilet is installed and used for a 5 year period), the loan is written off.

[2] Those resources which humans require for survival are often referred to as 'primary goods.' (ref: The United Nations document on Human Rights, Article 25/1)

[3] For example Nieswiadomy and Molina (1989) found that a decreasing block rate scenario encouraged greater water use, whereas an increasing block rate scenario resulted in a reaction to the price increase and a corresponding decrease in water use.

different times. Although many studies have demonstrated a link between water price and consumption (see Table 12.1) results from a study carried out in The Netherlands reported by Achttienribbe (1998) recently raised serious doubts about the price elasticity of water consumption in different sectors. The study suggested that price rises were likely to influence industrial water consumption to a greater degree than domestic consumption, but that it was difficult to calculate elasticity values for either sector as the effects might not become apparent immediately.

The motivation to save money (a financial incentive) rather than to save water (an environmental incentive) has been demonstrated to be the dominant driver for reactions to many conservation initiatives. For industrial actors this suggests that they will follow a water saving policy if total savings are likely to exceed the costs of implementing that policy (Holt *et al.*, 2000). It has also been noted that larger organisations more dependent on water resources, are more likely to be favourable to more efficient pricing systems (Rees, 1969). Although there is limited scope for the application of price mechanisms in developing countries, Arntzen (1995) reported that in Botswana the scope of economic instruments is limited because of the large non-market water sector, ambiguity about property rights and low incomes.

12.2.2 Regulatory instruments

As described in the previous Chapter, regulatory instruments can include both mandatory and enabling legislation, regulations, policies, standards and guidelines. These can be used to reduce institutional, legal or economic barriers to more efficient water use or to create barriers against unnecessary or wasteful water consumption. Although central governments often play an active role, and indeed are typically the ultimate source of executive power, responsibility for regulation definition, implementation and enforcement is increasingly being devolved to local government, regulatory agencies or, in some cases, commercial institutions. The range and format of regulatory instruments available for influencing water use will reflect the legal and administrative structure of a nation state or pan-national body. Some of the more common tools adopted include; guidelines on planning, building and plumbing (e.g. standards & codes of practice), provision of enabling legislation which enhances planning powers, setting targets for specific conservation techniques, bestowing emergency powers on regulatory bodies, and implementing framework legislation which requires consideration of water efficiency. A particular case of this latter option was reported by Ward & King (1998) who found that legislation that removes institutional barriers to conservation can promote greater total economic benefit from scarce water. Regulatory instruments are often used in times of emergency and instruments such as hosepipe bans have been shown to have varying effects dependent on the credibility associated with the problem being addressed. For example, one such study suggests that consumers will only accept hosepipe bans and standpipes if the issue is identified as a regional rather than water company problem (Lawson, 2002).

Whilst the use of regulatory measures can generate a more predictable and immediate effect on consumption patterns, there are a number of considerations to be taken into account. Firstly, the perceived legitimacy of a regulatory measure can significantly influence its impact. Communities will ask questions about whether the regulation is based on a sound and broadly accepted understanding of the problem, and the credibility/competence of the regulating body in setting the measure. They will also be concerned that any price increases are not being used to take advantage of the situation to increase profits. Secondly, many regulatory measures rely on effective monitoring and enforcement, activities which in themselves are resource consuming. Finally, and as is the case with any regulatory measure, evasion, deception, and abuse will both adversely impact the effectiveness of the instrument and challenge its credibility as an effective policy instrument. Actors will struggle to ensure the right outcomes for themselves despite policy changes, with low level, incremental resistance rather than through conflict or demonstrations (Long, 1992).

12.2.3 Technological instruments

Technological instruments include structural or physical improvements to water supply and use systems and installation of water efficient devices or processes (such as those described in Chapter 4). Such actions can, of course, be carried out by many different stakeholders, sometimes in response to the other types of policy instrument noted in these pages (regulatory, educational and economic). What we are interested in here are those technological initiatives which are directly driven by governmental or private institutions rather than by individuals. For example, water supply utilities can use pressure reduction programmes to reduce both potable consumption and leakage, invest in leakage repair programmes, and provide water use reduction fittings to customers. Governmental and regulatory bodies can set up technology demonstration projects to exhibit the benefits of particular pieces of equipment or whole systems. Finally, commercial companies have made major advances recently in the design and development of water efficient domestic appliances, high-efficiency irrigation scheduling systems, low water requirement crops, and other water saving technologies. Difficulties associated with technology based policy instruments typically concern the availability of complimentary knowledge and skills required for effective deployment. In addition, new technologies cannot simply be located in our houses, streets and utility infrastructures without some understanding of how they impact existing system performance.

The impacts of water metering on demand has attracted a lot of attention over the past decade. Documented case studies generally suggest that meter adoption motivates water savings (at least in the short term) in terms of total volume (Herrington, 1998) and can also help reduce peak demands (EA, 2000). Table 12.2 shows the summarised results of a number of studies which demonstrate water savings as a result of meter adoption in domestic premises.

Table 12.2 Estimated Water Savings Due to Metering and Charging by Volume (Primarily in Single-family Houses) Data from OECD (1999).

Location	Study period	Comparison	Savings due to metering	Reference
Collingwood, Ontario, Canada	1986-90	Summer peak	37%	Anon (1992)
Nine metering trial sites in England	1988-92	7000 houses in 9 trial groups and 9 control groups	Average 11.9%	Herrington (1997)
Isle of Wight, England	1988-92	Metered population rose from 1% to 97% in 1992 (50 000 homes)	Annual: 21.3%	DoE (1993)
Barcelona, Spain	early 1990s	2,927 connections switched UM to M	Annual: 12.8%	Sanclemente (undated)
East Anglia, UK	1990s		Annual: 15-20% Summer peak: 25-35%	Edwards (1996)
Oaks Park, Kent, UK	1993-96	61 houses	Annual: 27.5% Summer peak: up to 50%	Mid-Kent (1997)
Abu Dhabi City	2000	Looked at the effect of introducing metered pricing, comparing it with a previous flat rate tariff.	Consumption reduced by an average of 29%	Abu Qdais and Al Nassay (2001)

Public responses to retrofit programmes (e.g. supply and fitting of low volume cisterns) has been shown to be positive if (a) the equipment is offered free and (b) if the programme is high-profile and aggressively managed (Sarac *et al.*, 2002). However, initiatives may be rejected on aesthetic grounds, particularly if bathrooms or kitchens have recently been refurbished. Scheduling fitting activities at weekends has also been shown to promote a positive response. (EA, 2000). The success of retrofit programmes to promote water conservation can be linked to other, seemingly unrelated, opportunities. For example, Cameron & Wright (1990) found that household decisions to install shower retrofit devices were influenced by the potential to save money on water heating bills.

More generally, authors such as Lam (1999) and Shiva (1991) have pointed out that cultural change is often required as a pre-requisite for technology adoption. External constraints may dictate take up rate of new technologies – either through availability of the product or cost of the product, even if attitudes and behaviour correspond positively. Consequently, cultural issues must be addressed at the same time as technology deployment to prevent unforeseen problems and social upheaval. Efforts have been made to identify sections of the population which might be more receptive to information on conservation actions. Although the correlations from these studies are not statistically significant, the trends do show that higher income and experience of higher education are positively correlated with a positive response to water conservation programmes (de Oliver, 1999).

12.2.4 Education instruments

Education of water users through different contact routes and media is largely utilised to modify water use behaviour and encourage voluntary water conservation actions. Often seen as the core instrument for use in long-term conservation strategies (Grisham and Fleming, 1989), educational programmes make use of printed, video, and audio media as well as face-to-face methods. Developments in the fields of participative planning (House, 1999) and social learning (Parson & Clark, 1995) have influenced the design and execution of this type of water policy instrument as more consensual and community informed approaches to water management have been developed. Indeed, although the term 'education' has traditionally been used to characterise this form of water policy instrument, there is increasing impetus to use a term which better reflects the collaborative nature of the process (e.g. 'communication', or 'dialogue'; the latter of which is the preferred term here). Examples of dialogue based water conservation instruments include:

- Competitions, awards and recognition programs.
- Demonstration sites and information centres;

- Face to face meetings with major water users;
- Social marketing campaigns such as public broadcasting announcements, brochures and handouts, public displays, slogans, bill inserts, advertising and news bulletins, special public events, internet sites, door-to-door campaigns, newspaper articles and radio/television programs;
- Published materials such as "how to" manuals, case studies, technical reports, resource libraries;
- School programs and materials including activity books, games, videos and CDs, poster contests, in-class visits and demonstrations, "teach the teacher" guides, curriculum guides;
- Special project committees, seminars and workshops with specific water users;
- Irrigation audits and water use audits.

Dialogue, as noted above, is an instrument which encourages behavioural change, and consequently its effectiveness is posited on the assumption that beliefs determine values, values determine attitudes, and attitudes determine behaviour. However, the ability of attitudes to predict behavioural intentions and overt behaviour continues to be a major focus of theory and research in psychology. Ajzen (2001), in a review of attitude research, concluded that although it is now generally recognised that attitudes are relevant for understanding and predicting social behaviour, many important questions remain unanswered. Indeed, many studies, such as that conducted with specific reference to the water sector by de Oliver (1999) tell us that none of these links can be taken for granted, and that measuring the causal process is itself a non-trivial activity. For example, Zelezny (1999) criticised the great majority of environmental education efforts with claims of changing behaviour when most relied on self-reported behavioural change measures noting that self–reporting of behavioural change is often inconsistent with actual behaviour.

These limitations to managing water use behaviour through dialogue have led to calls for more targeted campaigns (Aitken *et al.*, 1994), greater public participation during the early stages of programme design (House & Fordham, 1997; Chambers, 1992; Garin *et al.*, 2001), best practice exemplars to demonstrate the benefits of conservation (Holt *et al.*, 2000) and programmes which generate a commitment to act (Stern and Gardener, 1996).

Another strong theme of the research on dialogue-based approaches to influencing water consumption is communication. Aitken *et al.* (1994) found both that informing a consumer of their water use led to positive short-term changes, and that the difference between attitude and behaviour (dissonance) could be reduced if consumers are confronted with that dissonance. Other authors writing on the public understanding of science have commented on the way in which different sources of communication are valued and interpreted by consumers, noting that an uneducated populace is potentially less harmful than an ill-educated one (Stott & Sullivan, 2000).

Other problems relating to the use of dialogue as a water policy instrument include difficulties in evaluating the impact of programmes (EA, 2000),[4] the need for reinforcement of messages to maintain the level of impact (Wang *et al.*, 1999), and the need to design dialogue processes capable of engaging with a wide variety of social actors. It has also been noted that information and education based conservation programmes do not appear to be effective by themselves in achieving a conservation goal without at the same time imposing significant price increases to provide a financial incentive to conserve water (Martin and Kulakowski, 1991).

12.3 ATTITUDES AND RESPONSES TO WATER RECYCLING INITIATIVES

The use of treated and recycled wastewater in agricultural, municipal, or domestic applications (several types of which are described in Chapter 3) is quite properly a source of concern for a variety of consumer groups. Irrespective of what conclusions the scientific evidence lead to, the impressions and attitudes which the public hold can speedily and effectively bring a halt to implemented schemes. The issues here are both complex and complicated, having to do with beliefs, attitudes and trust. Although in some contexts there appears to be no intrinsic cultural barrier to grey water re-use (Karpiscak *et al.*, 1990), it is important to expose the public's own agenda for discussing and debating water reuse topics.

Public perception is viewed as one of the major obstacles to effective water conservation and reuse projects (DeSena, 1999). The development of sustainable water recycling schemes needs to include an understanding of the social and cultural aspects of water use. In order to do this we need to enlarge traditional design and management activities to include:

- Development of a portfolio of technologies appropriate to different scales and circumstances.
- Descriptions of these in terms of their risk, performance and cost functions and their contribution to catchment scale water conservation.

---

[4] However, Michelsen *et al.* (1999) studied the effectiveness of nonprice conservation programmes in reducing water demand across Los Angeles, San Diego, Broomfield, Denver, Santa Fe, Albuquerque and Las Cruces during 1984-1995 Nonprice conservation programmes resulted in reductions in demand of between 1.1 and 4.0%.

- Methods for understanding the way in which stakeholders and institutional structures will respond to various types of initiative (pricing, technology application etc.)
- Methods for understanding how the final user will view and respond to the application of existing and novel technologies.

Whilst the development of suitable technologies and policies which provide opportunities for water conservation and recycling has moved on apace over the past decade, their practical application will not depend solely on effective and reliable administration or engineering performance. Successful employment of preferred strategies and technologies will require an understanding of the social environment in which they are to be applied. The drivers which promote conservation and recycling behaviour may vary between households and cultures, will often be a function of the context within which a debate about reuse is being conducted (Simpson, 1999), and will certainly be different for domestic, commercial and industrial users (see Chapter 10 for a more detailed discussion of drivers and barriers).

Studies of public attitudes to water reuse have been carried out since the late 1950's (originally in the USA, but latterly in Europe, Central America and Africa). Bruvold & Crook (1981) provide a valuable summary of research during these early years, highlighting that individuals who consider their potable supplies to be under threat (in terms of either quality of quantity) or perceive an economic benefit are generally more positive towards the idea of recycling water. Other work has demonstrated that acceptance of water recycling schemes in general is influenced by the degree of human contact associated with the reuse application (WPCF, 1989) although occupation could be a modifying factor here (Adams & Templer, 1980). Uses such as garden irrigation and toilet flushing are consistently preferred to uses such as food preparation and cooking (Bruvold, 1985).

Knowledge and previous experience of water recycling has also been suggested as having an influence on the acceptance of recycling systems, although there is some suggestion that this influence has been overestimated (Olson & Bruvold, 1982). Trust in the technology being used to treat the source water has also been found to positively influence acceptance of recycled water (Johnson, 1971). Other factors which have been identified as promoting acceptance of reuse projects include focusing on financial savings (Marks *et al.*, 2002) and using demonstration projects effectively (Gibson & Apostolidis, 2001).

Age, gender and level of education are other factors that have been shown to influence water recycling acceptability. For example, Olson & Bruvold (1982) showed that age is correlated negatively with acceptance of recycled water and that generally speaking, women appear to be less accepting than men (although this is specific to the populations studied). Also, those with a higher level of education appear to be more accepting than those with a lower level.

Risk perception has played a key role in characterising the acceptability of water recycling schemes. Various contributions in other fields of research (e.g. Slovic, 1993) have found that people are more willing to accept risks if they are voluntary, of low catastrophic potential, familiar, and emanate from a trustworthy source. Other identified psychological actors include 'Disgust' and 'Sensitivity' (Bixler and Floyd, 1997), aversion to the unclean, over-concern with health, and aversion to human waste, all of which have been postulated as having a negative association with the acceptability of reusing water (Olson & Bruvold, 1982). Cultural factors such as religion may also play a large part in determining attitudes (see for example Warner, 1999).

Although extensive research has been carried out on the relative acceptability of different recycled water applications (i.e. the use to which the recycled water is to be put), less is known about individual responses to variations in the scale or context of the recycling system. (i.e. the 'source water – treatment process– water use' configuration). There is also a lack of knowledge concerning indicators of willingness to use water recycling systems; knowledge which is required if niche markets are to be exploited and educational/promotional material effectively designed.

The source of recycled water as well as the environment in which it is to be used are likely to influence attitudes towards the system as a whole. Source waters which are perceived as clean or safe will be more willingly accepted than those that appear dirty or dangerous. However, the 'use history' of the water is likely to be a moderating factor in willingness to use. Ask yourself the question … would you prefer to use your own bathwater to flush your toilet or your next-door-neighbour's bathwater? Likewise, the setting within which a recycling system is located is also anticipated to be a significant moderator of willingness to use. The economics of many recycling technologies favour larger scale systems for applications such as public or institutional buildings (Surendran & Wheatley, 1998). Whilst such large scale uses serve to optimise the economic (and perhaps the technological) performance of the system, there is less evidence that the public is as willing to consider using water recycling systems within these settings as they would be in their own homes.

Clearly, the consequences of adopting a particular technology are not always obvious at the design and planning stage. Risk analysis is able to focus on the positive and negative outcomes associated with investing in and using recycling technologies. However, in terms of generating an understanding of technology adoption issues, there is a requirement for complementary activities which focus more generally on the attitudes and potential behaviour of stake holders to new technology i.e. on the attributes that potential users consider to be relevant. Such attributes may not be those which are perceived to be significant by the technology developers in the design and development stage. Features of the technology which are of importance to users could, for instance, involve such issues as size and location or ease of use. The aesthetic characteristics of the water produced will also impact willingness to use, but may provide unreliable cues to perceptions of quality (as highlighted by McDaniels *et al.*, 2000) Therefore, the identification of user concerns needs to be achieved before the technology is brought to market, even to the extent of providing feedback into the engineering design process.

In studying some of these issues amongst a sample of over 300 households in England & Wales, Jeffrey (2002) has shown that:

- Use of a water recycling system where the source and application are located within their own household is acceptable to the vast majority of the population as long as they have trust in the organisation which sets standards for water reuse. Using recycled water from second party or public sources is less acceptable, although half the population show no concern, irrespective of the water source.
- Water recycling is generally more acceptable in non-urban areas than in urban areas. This disparity is most pronounced for systems where the source and use are not within the respondent's own residence.
- Willingness to use recycled water, particularly from communal sources, is higher amongst metered households than amongst non-metered households, and higher amongst those households which take water conservation measures than amongst those who do not.

12.4 BEYOND DEMAND MANAGEMENT?

In this final section some of the developments in other areas of research which are increasingly impacting thinking on the design of water policy instruments and their deployment are considered. Specifically, the potential contribution of current thinking on socio-cultural adaptability, complexity & coevolution, and water as a cultural identifier are discussed. Our intention is not to provide a comprehensive overview of these areas but merely to introduce them as a provocation to new thinking on water resources management, and in particular water conservation.

12.4.1 Gauging socio-cultural adaptability

Before a solution can be offered to the problems of increasing water stress we need to begin to define what is individually and communally acceptable as response options and what the barriers are to adaptation. One barrier may be convenience. Given that water supply in many parts of the world is universal and that there are few barriers to delivery, access couldn't be easier. Low levels of water metering and relatively low pricing, signal to the market that the product is cheap and abundant. Asking people to change their consumption patterns needs to be correlated with an explanation as to why change is needed. Another barrier to adaptation may be awareness: although people are aware of global warming, the uncertainty of predictions means that a guaranteed prognosis cannot be delivered; asking people to reduce water use when flooding experiences are still fresh sends conflicting messages. A third problem might be the cultural significance of water; hygiene, health and prosperity are all linked to access to water – for some it represents modernity at its highest apex. The goal is to identify what triggers need to be put in place before individuals and communities accept their responsibility in cutting water demand. Without this shift in attitude policy targeted at individual and community consumption will face legitimacy problems. The inability of organisations and the state to enforce water use quotas has been demonstrated both by the negative outcomes of debt disconnection programmes (Herbert and Kempson, 1995) and by the negative feedback to regionally enforced hosepipe bans and standpipe measures (Lawson, 2002). One key aspect will be to identify the opportunity spaces available to individuals and communities (Lemon, 1999) and how we can tailor these so that they produce the 'right' outcome without being in conflict with attitudes towards and perceptions of water. The key to tackling individual and community consumption will be to recognise that consumers are not homogenous groups: in the same way that market consumption is heterogeneous, so is water use.

But what about institutions or organisations: What prevents them from adapting? In the same way that drivers of behaviour can be identified for individuals, so too for organisations. Given the profit orientated nature of industry and agriculture some financial benefit must be factored in to induce change, again without conflict. Given the sectoral nature of water use, adaptation will be required on a macro scale; this inevitably requires long term planning strategies which may be at odds with short term profit making. The success of waste minimization clubs (Holt *et al.*, 2000) has shown that organisations are willing to change – but not in isolation. Again, finding the triggers behind sectoral change will be key.

12.4.2 Complexity and co-evolution

As noted in the introductory section to this chapter, the primary focus of water policy has moved from supply to demand side approaches over the past 50 years. However, new thinking in the fields of complex systems, coevolutionary processes, and the roles of culture, power structures and legitimacy in framing water policy, are providing opportunities for the development of novel approaches to water resources management.

Throughout the 1990s the work of Norgaard (1994), Tainter (1995) and Allen (1994) can be seen as being exemplary in leading the field in the 'humanising' of theory which address more than one aspect of our world (e.g. socio-natural or socio-technical systems). The crux of the argument shared by Allen and Norgaard is the desire to move away from mechanistic explanations of socio-economic processes. Whilst Norgaard focuses on Newtonian physics as the reason behind the need for theory to drop into 'cause and effect' categorisations, Allen cites economic theory as the paradigm which tries to create closed systems of knowledge. For both, such closed systems generate theoretical models which do not reflect real world experience and they promote the creation of more fluid or 'plastic' models which accept the inherent chaotic features of socioeconomic and natural resource interactions. The broad aim of these 'complex systems' representations is to provide frameworks of understanding which create a platform on which different disciplines and actors can share knowledge to aid the promotion of resource management in the context of sustainable development. Norgaard's work is explicitly a critique of development practise – the attempts to promote industrialisation and economic growth modelled on the developed countries' experience. His analysis denigrates past development practise and forwards 'coevolutionary' theory as a means to develop an alternative model. Allen's work is concerned with developing a decision making instrument, or set of criteria, that will enable policy makers and planners to quantify the 'fuzziness' aspect of real life that gets channelled out of socio-economic modelling. For both the struggle is to make people aware of the opportunity space they inhabit, together with all its inherent contradictions, messy knowledge, and difficult choices.

Within the context of water resources management, it is worth noting that these ideas have been taken up by several researchers in the field; perhaps most fluently by the Water Issues Group at the University of Pretoria, South Africa, headed by Tony Turton which has considered the social adaptive capacities of communities facing water scarcity in developing countries. Turton's work provides an insight into the coevolutionary dynamic at play in communities which have few resources and who are mostly dependent on water for their livelihoods (Turton, 1999). What the co-evolutionary perspective brings to the problem of socio-cultural changes to water availability is the recognition that societies have been able to adapt throughout the course of history. Adaptation is not the preserve of the environment but is a two way process. How does this inform our possible envelopes of response and the planning instruments we need ? What co-evolutionary perspectives provide is the setting of boundary space – that change can happen; what policy makers need to assess is the manner and speed with which adaptivity is introduced.

A co-evolutionary perspective also suggests that appropriate policy responses to water stress might be focused on structural changes which are not *per se* concerned with water. Such structural responses might impact demand for water by reconfiguring industrial or agricultural practices (e.g. by promoting low water requirement crops in water stressed regions), promoting land use patterns which better conform to long term water availability distribution, or by influencing regional development trends towards low water requirement futures.

Lattorre *et al.*'s (2000) study of water use in Semi-arid Spain has proved insightful in terms of challenging the apolitical approach of the co-evolutionary theorists. The work mapped the changing land use and hydraulic systems of a semi arid environment and matched these with population flows and the dynamics of the local economy to try to understand the different social processes and developments of the region. Their research showed that land use systems were not a simple response to the environment, but rather that species and technology were transferred inwards after regional conquests. Rather than simply adapting the landscape to fit the new expectations of the populace, what occurred was an iterative process whereby imported technology was adapted to the environment. These adaptations were for political and social reasons rather than for any pressing need to fit in with the environment – i.e. it was a totally artificial process. Hence, technological adaptation is as much connected with iterative socio-cultural processes as political necessity.

Mazmanian and Kraft (1999) explore the obstacles to, and potential for, reshaping our communities towards the fulfilment of multiple sustainable civic goals. Again, they focus on the need to develop strong grass roots collectives to increase public participation and hence stakeholder investment in new developments. This work's ethos of motivating and visioning a more sustainable collectivised way of operating is also developed in a contribution by Schrama (2000) who focuses on tackling organisational behavioural change.

12.4.3 Water as a cultural signifier

Any discourse on water consumption is predicated on cultural understandings of water and its institutional framework. In many countries the institutional framework is based on private property rights – access to water 'belongs' to a person or institution, whether private or public. There is not universal access to water and the rights to water are not on a needs basis but a legal entitlement basis. This is not the place to review water rights but this brief outline helps us to select the type of literature that could inform further research. Research on water scarcity in Western USA (Davis 2001) is a good basic guide to the types of issues and institutional structures concerning the allocation of water rights, linking it with a historical frontier mentality of man vs. nature. However, it is the work conducted by Aguilera-Klink *et al.* (2000) which is exemplary in this area: deconstructing concepts of water 'scarcity' they are able to build a strong argument about what explains the development of a society's water structure, what shapes attitudes towards water and examine how consumption patterns become engrained within an institutional framework. This covers not just the way water is accessed and priced but highlights that perceptions of scarcity can create 'panic consumption' leading to more acute conditions of scarcity. By linking progress with water, consumption creates its own dynamism cemented within power structures in society. Again this paper is one of the few to emphasise how power relations and water have a direct effect on consumption levels. What also is made clear is how those with a direct dependency on access to water have specific local knowledge (seasonal water flow, depth of aquifer) but limited understanding of the holistic hydraulic process.

Water use, perceptions and attitudes to water and water governance adaptivity must include a perspective on how water acts as a conductor of power relations, how it becomes representative of forms of knowledge and means of operating power/knowledge discourses. We cannot talk of human behaviour without recognising that we also need to talk about power. Behaviour is learnt and is socio cultural – we learn to adapt with our environment. Part of that process is gaining knowledge and as a consequence of that our actions help us move through the various networks of power that exist in our society. Using power as a theoretical underpinning enables us to analyse water as a vehicle of control rather than just as a social or economic good.

What links a lack of knowledge and a lack of power is an understanding as to how consumers engage with water. Power and knowledge are formally linked through Foucault's work (e.g. Foucault, 1970) – public participation will always be hampered by the existence of the 'expert' with a wider knowledge base and in whom power then resides. An environmental psychology approach allows us to unravel the way that the human psyche is shaped and determined by how we perceive ourselves in the environment and as part of the natural cycle. Within the UK the work of Strang (2001) and House and Fordham (1997) have both focused on water's immersion into the human psyche. For Strang, rural communities have strong cognitive bonds with water courses; local water bodies' health and status reflect the condition of the wider environment (Strang, 2001). House and Fordham's work highlighted the role of the river corridor and responses by the public to changes within it (House & Fordham, 1997). Both contributions note how strongly communities respond to changes in their environment and the frustration of their non participation in decision making processes.

12.5 CONCLUSION: SOME COMMENTS ON CAPACITY DEVELOPMENT

The foregoing sections have provided an overview of how individuals and communities respond to different water conservation initiatives and has explored some of the theoretical frameworks that might inform future thinking. The emphasis here is on understanding human and institutional behaviour. Although research into such processes has increased (and its quality improved) over the past 10 years, knowledge exploitation is beset by three problems.

Firstly, the knowledge base itself is dispersed and typically located within the confines of a disciplinary community such as sociology or anthropology rather than with water management per se. This makes it difficult for water sector professionals to locate relevant knowledge (in terms of both research findings and knowledgeable individuals). Possible responses to this issue are difficult to envisage although dedicated publications or events which provide an opportunity for commercial concerns to access contributions on the human dimensions of water management would be of benefit.

A second problem is that many organizations in the water sector are poorly equipped to recognise and exploit the potential contributions of the 'softer' sciences. Decades of emphasis on engineering, technology and infrastructures has left its imprint on water supply and management institutions to the extent that the only incentives to understanding any human association with water is in terms of marketing (selling people the products of engineering) and public relations (convincing people of the benefits of engineering). However, the issues here go deeper than the educational background of individuals. Studies of human behaviour or attitudes typically produce results of low predictive power. However, this does not mean that they are of no value. Research contractors need to identify the contribution of such studies before they are executed and accept that they are more likely to 'inform' than 'resolve' a particular problem.

The final problem is the selection of appropriate research competencies and perspectives. The range of potential contributions available to the water industry from the human and social sciences is immense. The strengths of traditional paradigms such as history, psychology, sociology, anthropology etc. are today being supplemented by new theoretical frameworks, methods and perspectives which bridge the intellectual gaps between disciplines. It is important to recognise that these perspectives are (normally) not in competition. Useful insights are usually only realisable via complimentary multi-perspective studies, creating a problem in terms of how studies are to be structured, and suitable research skills engaged. Resource constraints (who is available and at what price) will play a large part in determining what mix of skills is considered 'appropriate'. However, this guarantees neither the rigour nor the value of the findings. Again, there is no simple solution to these problems. Water management institutions should learn by doing and be prepared to disseminate experiences of structuring human focused research and exploiting the findings.

The economic, environmental and social development of our communities co-evolves with the availability and quality of water and we need to enrich and deepen our understanding of these relationships. Sustainable development is fundamentally about the adaptive capacity of the human race. In relation to water, the broad objective should be to enhance adaptive potential in the context of safeguarding water supplies; not only for human consumption but also in support of viable ecosystems. People adapt and change at a faster rate than policies, technologies and infrastructures. The challenge is to understand this potential as it impacts on water supply, and exploit it as a beneficial tool for adaptive response.

12.6 REFERENCES

Abu Qdais, H. A. and Al Nassay, H.I. (2001) Effect of pricing policy on water conservation: a case study. *Water Policy* **3**(3), 207-214

Achttienribbe, G.E. (1998) Water price, price elasticity and the demand for drinking water. *Aqua* **47** (4) , 196-198.

Adams, R.A. & Templer, D.I. (1998) Body elimination attitude and occupation. *Psychological Reports* **82**, 465-466.

Aguilera-Klink, F., Perez-Moriana, E., Sanchez-Garcia, J., (2000) The social construction of scarcity. The case of water in Tenerife (Canary Islands). *Ecological Economics* **34**, 233 – 245.

Aitken, C.K., McMahon, T., Wearing, A., Finlayson, B.L. (1994) Residential water use: predicting and reducing consumption. *Journal of Applied Social Psychology* **24** (2), 136 – 158.

Ajzen, I. (2001) Nature and operation of attitudes. *Annual Review of Psychology* **52**, 27-58.

Allan, J.A. (2000) *The Middle East Water Question: Hydropolitics and the Global.* I B Tauris, London.

Allen, P. M. (1994) Coherence, chaos and evolution in the social context. *Futures* **26** (6), 583-597

Anon (1992) Canadian Water Utility Makes Successful Switch to Metering. *Water Engineering and Management* **139**, 20-26.

Arbués, F., Barberán, R., Villanúa, I. (2000) Water price impact on residential water demand in the city of Zaragoza. A dynamic panel data approach. Paper presented at the *40th European Congress of the European Regional Studies Association* (ERSA), Barcelona, Spain, August.

Arimah, B.C. and Ekeng, B. E. (1993) Some factors explaining residential water consumption in a Third World city - the case of Calabar, Nigeria. *Aqua* **42** (5), 289-294.

Arntzen, J. (1995) Economic instruments for sustainable resources management: The case of Botswana's water resource. *AMBIO* **24**, 335-342.

Baumann, D.D., Boland, J.J. and Hanemann, W. M. (1998) *Urban Water Demand Management and Planning*, McGraw-Hill, Inc., New York.

Bixler, R.D. and Floyd, M.F. (1997) Nature is Scary, Disgusting, and Uncomfortable. *Environment and Behaviour*, **29** (4), 443-467.

Bruvold, W. H. (1985) Obtaining public support for reuse water. *Journal of the American Waterworks Association* **77**(7), 72.

Bruvold, W.H. and Crook, J. (1981) What the public thinks: Reclaiming and Reusing Wastewater. *Water Engineering & Management*, April, 65.

Bulloch, J. (1993). *Water Wars: Coming Conflicts in the Middle East.* Victor Gollancz, London.

Butler, D. (1993) The influence of dwelling occupancy and day of the week on domestic appliance wastewater discharge. *Building and Environment* **28** (1), 73 – 79.

Cameron, T. A. and Wright, M.B. (1990) Determinants of Household Water Conservation Retrofit Activity: A Discrete Choice Model Using Survey Data. *Water Resources Research* **26** (2), 179-188.

Chambers, R. (1992) *Rural appraisal: rapid, relaxed and participatory.* IDS Publications, Brighton.

Collinge, R. A. (1992) Revenue neutral water conservation: Marginal cost pricing with discount coupons. *Water Resources Research* **28** (3), 617-622.

Davis, S. K. (2001) The politics of water scarcity in the Western states. *The Social Science Journal* **38**, 527 – 542.

De Oliver, M. (1999) Attitudes and inaction: A case study of the manifest demographics of urban water conservation. *Environment and Behaviour* **31**(3), 371 – 394.

De Sena, M. (1999) Public opposition sidelines indirect potable reuse projects. *Water Environment & Technology*, May.

Detwyler, T.R. and Marcus, M.G. (1972) *Urbanization and Environment: the physical geography of the city.* Duxbury Press, California.

DoE (1993) *Water Metering Trials: Final Report.* Department of the Environment, London.

Dziegielewski, B. Rodrigo, D.M. and Opitz, E. M. (1990) *Commercial and industrial water use in Southern California (Final Report).* Planning & Management Consultants Ltd. Report for the Metropolitan Water District of Southern California, Los Angeles.

Edwards, K. (1996) The role of leakage control and metering in effective demand management. In *Conference Proc. Water '96: Investing in the Future*, London.

EA (2000). *On the right track: A summary of current water conservation initiatives in the UK.* Environment Agency.

Foucault, M. (1970) *The order of things.* Tavistock., London.

Garin, P. Rinaudo, J. D. and Ruhlman, J. (2001) Linking expert evaluations with public consultation to design water policy at the watershed level. In *Proc. IWA World Water Congress*, Berlin, October 2001.

Gibson, H. E. and Apostolidis, N. (2001) Demonstration, the solution to successful community acceptance of water recycling. *Water Science & Technology* **43** (10), 259-266

Grisham, A. and Fleming, W. M. (1989) Long-term options for municipal water conservation. *Journal of the American Water Works Association* **81** (3), 33.

Gundermann, H. (1986) Primary influence factors on domestic water demand in Europe. *Aqua* **2**, 81-85.

Herbert, A. Kempson, E. (1995) *Water debt and disconnection.* Policy Studies Institute, London.

Herrington, P.R. (1997) Pricing Water Properly, ch. 13 in (ed.) O'Riordan, T., *Ecotaxation.* Earthscan: London.

Herrington, P. R. (1998) Analysing and forecasting peak demands on the public water supply. *Journal of the Chartered Institution of Water & Environmental Management.* **12**, 139-143.

Höglund, L. (1999) Household demand for water in Sweden with implications of a potential tax on water use. *Water Resources Research* **35** (12), 3853–3863.

Holt, C. P., Phillips, P. S. and Bates, M. P. (2000). Analysis of the role of waste minimalisation clubs in reducing industrial waste demand in the UK. *Resources, Conservation and Recycling* **30**, 315 – 331.

House, M.A. (1999) Citizen Participation in Water Management. *Water Science and Technology* **40** (10), 125-130

House, M. A. and Fordham, M. H. (1997) Public perception of river corridors and river works. *Landscape Research* **22** (1), 25-44.

Jeffrey, P. (2002) Influence of technology scale and location on public attitudes to in-house water recycling in England & Wales. *Journal of the Instiitution of Water & Environmental Management* **16** (3), 214-217.

Johnson, J.F. (1971) *Renovated Wastewater: An Alternative Source of Municipal Supply in the U.S.* University of Chicago, Department of Geography Research, Chicago, IL.

Karpisck, M., Foster, K. and Schmid, N. (1990) Residential water conservation: Casa del agua. *Water Resources Bulletin* **26** (6), 939-948

Lam. S.P. (1999) Predicting intentions to conserve water from the theory of planned behaviour, perceived moral obligation and perceived water right. *Journal of Applied Social Psychology* **29**, 1058 – 1071.

Latorre. Juan.G., Picon. A.S., Latorre. Jesus.G. (2000). Water, irrigation systems and society in semi-arid Mediterranean environments. In *Proc. Third Biennial Conference of the European society for Ecological Economics*, Vienna Austria, May.

Lawson, R. (2002) Demand management during a drought. In *Proc. CIWEM conference on Planning & Managing Drought.* May, London.

Lemon, M. (1999). Social enquiry and natural phenomena. In Lemon.M. (Ed.). *Exploring environmental change using an integrated method.* Gordon and Breach, Australia.

Long, N. (1992) From paradigm lost to paradigm regained? The case for an actor oriented sociology of development. In Long N. and A. Long (eds.), *Battlefields of Knowledge: The Interlocking of Theory and Practice in Social Research and Development,* Routledge, London.

Lyman, R.A., 1992. Peak and off-peak residential water demand. *Water Resources Research* **28** (9), 2159–2167.

Marks, J., Cromar, N., Fallowfield, H., Oemcke, D. and Zadoroznyj, M. (2002) Community experience and perceptions of water reuse. In *Proc. IWA 3rd World Water Congress,* Melbourne, Australia, April.

Martin, W. E., and Kulakowski, S. (1991) Water price as a policy variable in managing urban water uses: Tuscon, Arizona. *Water Resources Research* **27** (2), 157-166.

Mayer, P., DeOreo, W., Opitz, E., Kiefer, J., Davis, W., Dziegielewski, B., and Nelson, J. (1999) *Residential End Uses of Water.* American Water Works Research Foundation, Denver, Colorado.

Mazmanian, D. A. and Kraft, M. E. (1999) *Towards Sustainable Communities.* MIT Press, Cambridge, Massachusetts, USA.

McCully, P. (1996) *Silenced Rivers: The political ecology and politics of large dams.* Atlantic Highlands, London.

Mid-Kent (1997) *Meter Pilot Project Report 1*. Mid-Kent Water Plc., Snodland, UK.

McDaniels, T. L., Axelrod, L. J. and Cavanagh, N. (2000) Public perceptions regarding water quality and attitudes towards water conservation in the lower Fraser Basin. *Water Resources Research* **34** (5), 1299-1310.

Michelsen, A. M., McGuckin, J.T. and Stumpf, D. (1999) Nonprice water conservation programmes as a demand management instrument. *Journal of American Water Resources Association* **35** (3), 593-602.

Mukhopadhyay, A. Akber, A. and Al-Awadi, E. (2001) Analysis of freshwater consumption patterns in the private residences of Kuwait. *Urban Water* **3**, 53 -62.

Nauges, C. and Thomas, A. (2000) Privately-operated water utilities, municipal price negotiation, and estimation of residential water demand: the case of France. *Land Economics* **76** (1), 68–85.

Nieswiadomy, M. L., and Molina, D. J. (1989) Comparing residential water demand estimates under decreasing and increasing block rates using household data. *Land Economics* **65** (3), 280-289

Norgaard, R. B. (1994) *Development Betrayed: The end of progress and a coevolutionary revisioning of the future*. Routledge, London.

OECD (1999) *Household water pricing in OECD countries*. Final report by Working Party on Economic and Environmental Policy Integration, OECD, Paris.

OFWAT (2000) *Patterns of demand for Water in England and Wales 1989 – 1999*. Office of Water Services. Birmingham, UK.

Olson, B.H. & Bruvold, W. (1982), Influence of social factors on public acceptance of renovated wastewater. In *Water Re-Use*, MiddleBrookes.

Parson, E. A. and Clark, W. C. (1995) Sustainable development as social learning: theoretical perspectives and practical challenges for the design of a research program. In L. H. Gunderson & C. S. Holling (eds.) *Barriers and Bridges to the Renewal of Ecosystems and Institutions*, Cambridge University Press, New York, 428-460.

Pint, E. (1999) Household responses to increased water rates during the California drought. *Land Economics* **75** (2), 246–266.

Reynard, N. (2002) Climate change and drought. Paper presented at the CIWEM Conference on Planning & Managing Drought, 16th May 2002, London.

Rees, J. (1969) *Industrial demand for water: a study of South East England*. London School of Economics, London.

Sarac, K., Day, D. and White, S. (2002) What are we saving anyway? The results of three water demand management programs in Melbourne. In *Proc. IWA 3rd World Water Congress*, Melbourne, Australia, April.

Schama, S. (1995) *Landscape and memory*. HarperCollins, London.

Schrama, G. (2000) *Stimulating environmental innovations by optimising the organisation's scope of choice*. CSTM-report presented at the 16th EGOS Colloquium 'Organisational Praxis', Helsinki School of Economics and Business Administration, Finland, ISSN 1381-6357, July.

Shiva, V. (1991) *The violence of the green revolution: Third world agriculture, ecology and politic.*, Zed Books, London.

Simpson, J. M. (1999) Changing community attitudes to potable re-use in South-East Queensland. *Water Science and Technology* **40** (4-5), 59-66

Slovic, P. (1993) Perceived Risk, Trust and Democracy. *Risk Analysis* **13** (6), 675-682.

Stern, P. C. and Gardner, G. (1996) *Environmental problems and human behaviour.* Allyn and Bacon, Boston.

Stevens, T.H., Miller, J., Willis, C., (1992) Effect of price structure on residential water demand. *Water Resources Bulletin* **28** (4), 681–685.

Stott, P. and Sullivan, S. (2000) *Political ecology: science, myth and power*. Arnold. London.

Strang, V. (2001) *Evaluating water: cultural beliefs and values about water quality, use and conservation*. Water UK, Suffolk.

Surendran, S. and Wheatley, A. D. (1998) Grey-water reclamation for non-potable reuse. *Journal of the Chartered Institution of Water & Environmental Management* **12** (6), 406-413.

Tainter, J. (1995) Sustainability of complex societies. *Futures* **27** (4), 397 – 407.

Turton, A. R., (1999) *Water scarcity and social adaptive capacity: Towards an understanding of the social dynamics of water demand management in developing countries.* MEWREW Occasional paper No.9. Water issues study group, SOAS.

Twort, A., Law, F., Crowley, F. and Ratnayaka D. (1993) *Water Supply* (Fourth edition). Arnold, London.

Wang, Y. D. Song, J.S. Byrne, J. and Yun, S.J. (1999) Evaluating the Persistence of Residential *Water Conservation Journal of the American Water Resources Association* **35** (5), 1269-1276.

WPCF (1989) *Water Reuse: Manual of Practice* (Second Edition). Water Pollution Control Federation, Alexandria, USA.

Ward, F.A. and King, J.P. (1998) Reducing institutional barriers to water conservation. *Water Policy* **1** (4), 411-420.

Warner, W.S. (1999) The influence of religion on blackwater treatment. In *Proceedings of the 4th International Conference on Managing the Wastewater Resource, Ecological Engineering for Waste Water Treatment*, Aas, Norway, June.

Zelezny, L.C. (1999) Educational interventions that improve environmental behaviours: A meta–analysis. *Journal of Environmental Education* **31**, 15-18

Zhou, S.L., McMahon, T.A., Walton, A. and Lewis, J. (2000) Forecasting daily urban water demand: A case study of Melbourne. *Journal of Hydrology* **236**, 153 – 164.

13

Decision support tools for water demand management

Christos K. Makropoulos

13.1 INTRODUCTION

Previous chapters have discussed issues of water demand forecast, identified technologies and systems for reducing demand (including, *inter alia*, water recycling, low water use devices and leakage reduction in networks) and have presented legal, economic and customer perception issues related to water demand management in the developed and developing world. This chapter concentrates on decision support tools, in the software sense of the word, designed to assist decision-making in some of these areas. Clearly, as identified in the discourse throughout this book, demand management entails much more than simply tools. Following this rationale, the chapter, provides an overview of current research into the development and use of "intelligent" tools, supporting, but not substituting, decision-makers in managing urban water and water demand in particular, including but not restricted to decision support systems based on genetic algorithms, fuzzy logic, knowledge engineering, distributed intelligence, neural networks and system dynamics.

13.2 DECISION SUPPORT SYSTEMS

The core concept of decision support systems (DSS) was introduced in 1960 by Simon (Simon, 1960) through his work on structured versus unstructured systems. According to Simon, every problem falls within the continuum ranging from completely structured to unstructured decision problems. Possible stages of a (dynamic) decision process based on Simon's ideas can be seen in Figure 13.1. In the case of *structured* problems, the decision maker knows all relevant information on the system and its rules in advance. The solution to the problem is then scaled-down to a series of programmable tasks, which can be essentially solved by a computer (Malczewski, 1999).

In contrast, ill-defined problems, with no programmable/repetitive solution are termed *unstructured*. The experience of the decision maker is the key factor determining the solution process. Computers can be of no significant help to this type of problem. In real life situations however, problems are seldom within the domain of these two extremes. The intermediate type of problem is defined as *semi-structured* (Malczewski, 1999). This is the domain of application of *decision support systems*. Model-based decision support systems for environmental and resource management have been discussed and advocated for some time (e.g. Fedra, 1996), although success stories of actual use in the public debate and policy-making process are somewhat rarer, in particular at the societal rather than commercial end of the spectrum of possible applications.

The decision-maker in the urban environment, in particular, is in need of tools that would be able to address the following issues (Seder *et al.*, 2000):

- Integrate and coordinate information on a domain-oriented scale
- Support analysis, observation, valuation and forecast of systems and their conditions
- Support decisions as a balance between economic, social and environmental objectives based on expert knowledge
- Utilise an information system which is natural to the user and which offers transparency without requiring knowledge of some computer language.

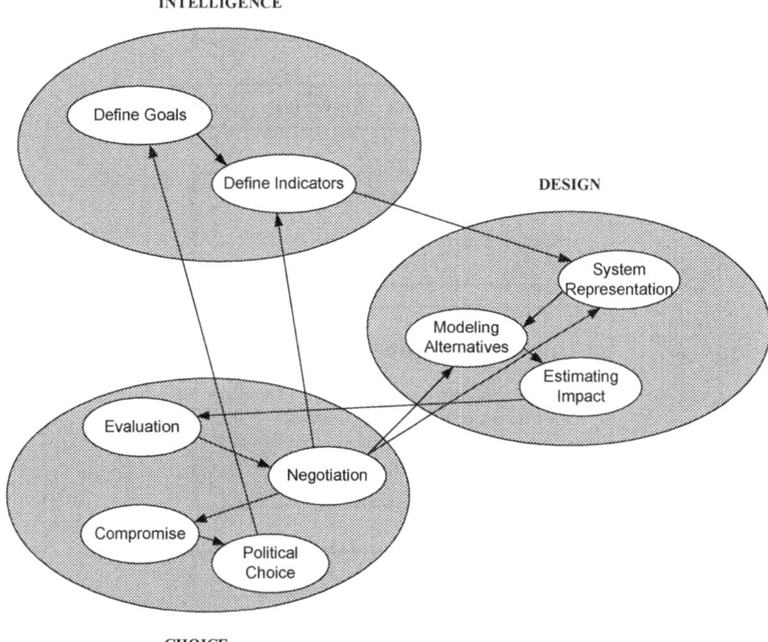

Figure 13.1. Decision Process based on Simon's intelligence-design-choice concept.

The stated requirements imply that the tools (or decision support systems) should in principle integrate knowledge and reasoning as an essential part of the system's functionality while dealing with the inherent ambiguities and uncertainties of any reasoning/decision making process. Water demand management in particular is a complex decisional environment where optimal planning presupposes a synthesis of heterogeneous information of variable spatial and temporal resolution and thus requires intelligent support. A general schematic of a DSS architecture can be seen in Figure 13.2.

The following paragraphs will present research into decision support systems developed for: (a) forecasting what the demand will be, (b) supporting the (site-specific) implementation of strategies to reduce this demand (c) managing demand as part of the overall urban water system and (d) negotiating between decision makers and communicating with the public. The chapter will conclude by discussing state-of-the-art in decision support systems and present thoughts on the usefulness and proper function of these tools relevant to decision making within the water demand management domain.

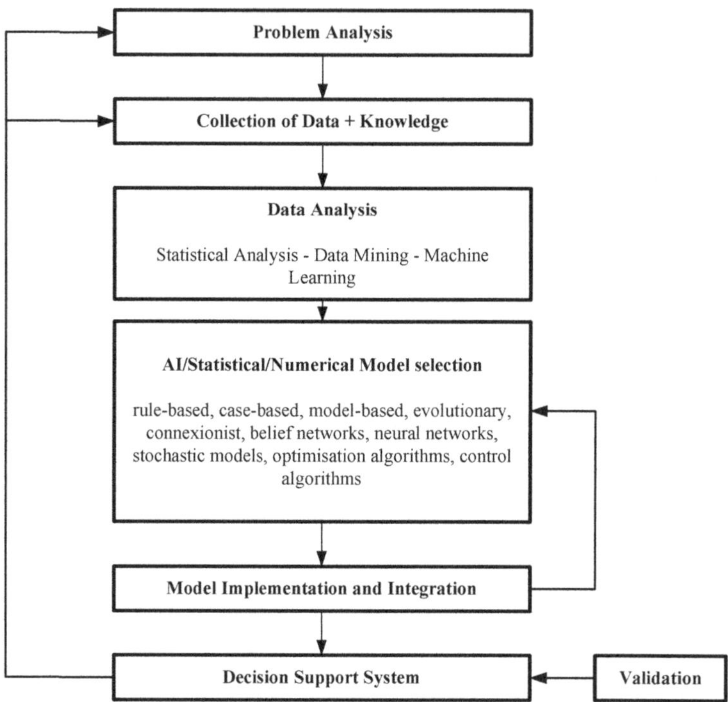

Figure 13.2. A flow diagram of a Decision Support System (adapted from Poch *et al.*, 2004).

13.3 TOOLS FOR FORECASTING DEMAND

As indicated in Chapter 1, demand forecasts can be developed for both long-term trend assessment (e.g. time-extrapolation, disaggregate end-uses, single-coefficient methods and multiple-coefficient methods) and for short-term, supply and distribution system operation purposes (e.g. probabilistic methods, memory -based learning technique, time-series models such as Box Jenkins and Arima models and Neural Networks (NN)).

Customer demand forecasts (for utilities in general) are based on analysis of relationships between climatic conditions, behavioural patterns, special events and previous customer demand in an attempt to estimate future demand (Lertpalangsunti *et al.*, 1999). In practice, adopting a single technique tends to produce high prediction errors (Mohammed *et al.*, 1995) especially when the problem domain has different classes of patterns. Using a combination of prediction techniques and tools could improve prediction performance (Lertpalangsunti *et al.*, 1999). The following paragraphs will therefore attempt to present some of these tools separately followed by applications of their integration.

Liu *et al*. (2003), present the development of a simple Neural Network-based tool for forecasting domestic water demand in the city of Weinan in China. Neural Networks (NN) are programs which try to emulate the way the human brain processes information, by using an interconnected network of simple processing elements, called neurons (Figure 13.3). Neural networks can *learn* to approximate an unknown function by being trained on a series of input-output data points. They can thus derive process generalizations from complex and incomplete information. Jain and Ormsbee (2002) compare conventional approaches to short-term demand forecasting (including regression and time series analysis) to tools based on NN and rule-based systems and discuss their respective performance when dealing with a common problem. By comparing 8 different demand-forecast models (four conventional and four based on NN or rule-based systems) they conclude that the latter outperform the former in all cases although their actual performance depends on the nature of the data. This a generally valid point, worth bearing in mind when using NN. NN is a "black-box" type of tool that has the ability to undergo a learning process. The learning process is not, however, knowledge based but data driven and relies solely on locally provided data sets, for which the system is trained. This is considered a limitation of the technique as, for example, redundant data points or conflicts in the dataset will result in less than perfect system training (see Figure 13.4, based on Shi and Muzimoto, 2001).

Expert knowledge can be better captured through rule based systems (employing a number of IF-THEN type of rules) and efforts to couple the two (through, for example, neuro-fuzzy systems) are becoming increasingly common in literature, from Wang and Mendel (1992) to Shi and Muzimoto, (2001). These tools, attempt to use the learning processes of NN to set up appropriate (IF-THEN) rules for rule-based systems. The rule base provides a starting point for the NN algorithm that then fine-tunes the rules on the basis of available data.

Another approach to the development of appropriate rules for rule-based systems dealing specifically with demand management can be found in An *et al*. (1996), where data mining, in the form of classification rules based on a variable precision rough set model (Pawlak, 1991) is employed to develop probabilistic rules. Rough sets are based on the assumption that for every object within a given universe of discourse we associate some information (data, knowledge). Objects characterized by the same information are *similar* in view of the available information about them. Due to this similarity, some objects of interest cannot be differentiated and appear as the same. As, a consequence vague concepts cannot be (fully) characterized in terms of information about their elements. Rough set theory proposes that any vague concept can be replaced by a pair of precise concepts - called the lower and the upper approximation of the vague concept. The lower approximation consists of all objects which surely belong to the concept and the upper approximation contains all

 Water Demand Management

objects which possibly belong to the concept. The difference between the two constitutes the boundary region of the vague concept. Rough set theory is used to discover patterns in data. The importance of the approach adopted by An *et al.* (1996) stems from the fact that their tool develops and uses simple to understand "IF-THEN" type of rules, easily understood by the users, based on incomplete and imprecise datasets, which is most often the case in water demand management.

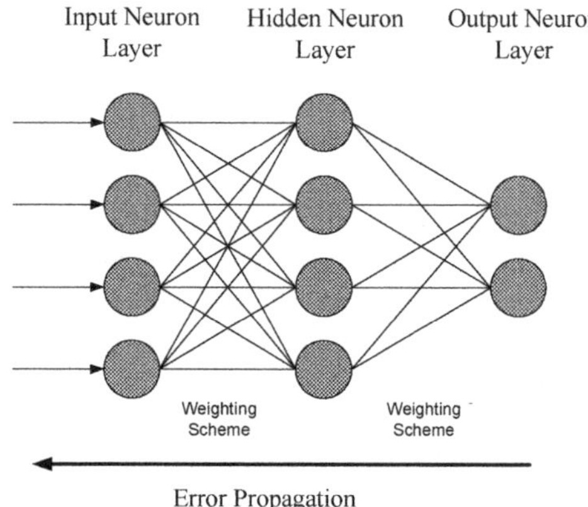

Figure 13.3. A general schematic of a NN. The error minimisation between input and output training pairs drives a weighting tuning scheme until the prediction error reaches a predefined minimum.

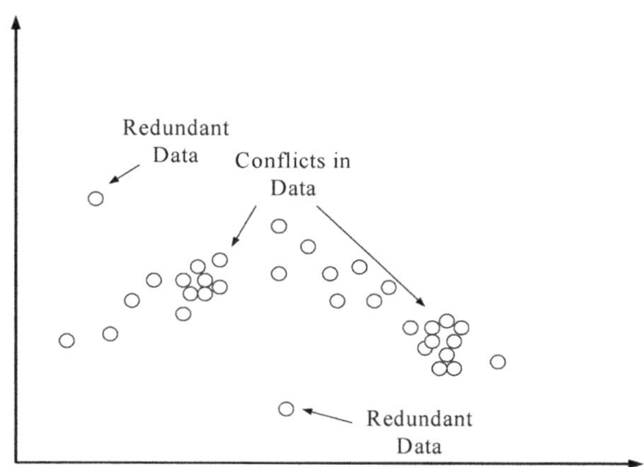

Figure 13.4. Hypothetical training data for a neural network (adapted from Shi and Muzimoto, 2001).

The (logical) next step in R&D has been to integrate tools of this nature into coherent and user friendly decision support systems. Such efforts include Lertpalangsunti *et al.* (1999), who present a toolset (named IFCS) which, further to NN (discussed above) also supports fuzzy logic (FL) and case-based reasoning (CBR). FL provides a formal mathematical framework for expressing linguistic variables and rules. In classic Set Theory, all elements either belong to a set or not (i.e. their membership function is either 1 or 0). In fuzzy sets, however, elements belong to a (fuzzy) set to "a certain degree". This provides a mathematical tool to capture linguistic ambiguity (e.g. to model an expert judgment suggesting that a specific demand management technique is *not very suitable* for a given situation). IF-THEN type of linguistic rules can be modelled through FL). In CBR, the system's expertise is embodied in a library of past cases, rather than being encoded in IF-THEN rules. Each case contains a description of the problem, plus a solution and its outcome. In IFCS the developer can construct a forecasting application using rules, procedures and flow diagrams. The IFCS was implemented on a commercial real-time expert system shell (which is software that allows construction of expert systems). The authors suggest that the NN approach tends to over-fit the training data when irrelevant parameters such as humidity and wind speed are used. This is because the network cannot generalize from the training data due to the presence of irrelevant parameters, and tries to "remember" them case by case. This reinforces the point that to correctly apply "intelligent techniques" both an accurate dataset and a good knowledge of what drives demand in the particular situation is needed to identify irrelevant parameters that would cause over fitting of the tools (from NN to CBR to rule-based FL).

Froukh (2001) provides a further level of tools integration, as the system he proposes has the capability of predicting (long-term) domestic-water demand by various methods and can also compute (long-term) conservation effectiveness due to the implementation of various demand-management measures, forecasting the number of customers for different consumption units (person, household and water connection) and facilitating the development of demand-scenarios for evaluating various options. The tool eventually forms part of a decision-support system for river-basin management (WaterWare, see Jamieson and Fedra, 1996) and as such is one of the rare attempts to integrate water demand management tools into catchment management. The decision support system includes a database, a GIS, an expert system, predictive models, a multi-objective decision component, hypertext files and a user-interface component.

Further to forecasting demand, tools have been developed to support actual implementation of demand management strategies and techniques. The following paragraph discusses some of the relevant research.

13.4 TOOLS FOR SUPPORTING THE IMPLEMENTATION OF DEMAND REDUCING STRATEGIES

Each location within the city boundaries has its own properties and its own set of constraints (social, economic and engineering). Taking into account these site-specific characteristics rather than drawing a black box around cities results in more realistic and therefore more applicable planning. Four types of strategies for WDM have been identified in decision support literature: technical measures, financial implementation incentives, legal instruments and institutional arrangements (Mohamed and Savenije, 2000). Most of the actual development of supporting tools has been centred upon the first of these strategies.

The strategy that has received the most attention in terms of tools development is leakage reduction, as this has been linked to pipe replacement prioritisation (e.g. Cooper *et al.*, 2000; Goulter and Kettler, 1985), which in turn is linked to assets management (Babovic *et al.*, 2002; Fenner and Sweeting, 1999). Sinske and Zietsman (2004) present a spatial decision support system assisting pipe breakage susceptibility analysis based on fuzzy logic and pipe-break theory. The system takes into account pipe age (through a simple cut-off rule, where pipes are considered old if they are older than 25 years), air pocket formation (through a rule based fuzzy inference system taking into account pipe slopes, network form, low and high point of the network) and damages to pipes by tree routes (through a fuzzy inference that takes into account distance from trees and tree size). The information needed as input to the fuzzy inference system is obtained by querying a GIS database. The overall susceptibility results are obtained by a multi-factor weighting process, calibrated against pipe-break occurrences.

Due to the complexity and multi-objective nature of optimisation support required for network rehabilitation, there has been considerable focus on the use of Genetic Algorithms for over a decade in response to this problem. Genetic algorithms (GAs) are a class of optimisation techniques used for complex large-scale combinatorial optimisation problems. They are stochastic algorithms whose search methods model some natural phenomena: genetic inheritance and Darwinian mechanisms of survival (Michalewicz, 1996). Darwinian Evolution suggests that natural species evolve by developing characteristics allowing them to adapt to a particular environment. These characteristics are incorporated in their genes and the best ones survive through the process of natural selection, from one generation to the next. The metaphor is conceptually a powerful one: Darwinian natural selection and survival can be considered as one of the most complex and demanding "optimisation" problems ever encountered. Evolution is a natural process and is only "natural" that a mathematical equivalent would be used for some of our own complex optimisation problems. GAs maintain a population of individuals P(*t*) for iteration *t*. Each individual in the population is a potential solution of the problem and is evaluated against a measure of *fitness*. Typically, fitness is related to the objective function to be optimised. In the next iteration *t+1*, a new population is formed by selecting the fitter individuals from the initial population and allowing them to undergo transformations (*mating*) through a set of genetic *operators*. There sare broadly two types of operator: unary (*mutation* type) operators which affect a single individual by changing one or more of its genes through probabilistic procedure and higher order (*crossover* type) which affect a number of individuals by combining parts of two or more initial individuals to create a new one. As the fitter individuals get selected more often for mating they have more *offsprings*, therefore the better solutions they represent become more dominant in successive iterations (*generations*). After a number of generations (subject to a convergence criterion) it is hoped[1] that the best individual represents a near optimum solution.

Kim and Mays (1994) presented a methodology that makes a replacement/rehabilitation decision for each pipe in the network based on the costs of replacing/relining pipes and increasing pumping to satisfy demands at all nodes. Halhal *et al.* (1997) proposed a multiobjective rehabilitation formulation that also utilized GAs. The two objectives considered were the benefits to the system and the rehabilitation cost, with the latter constrained to a predetermined budget limit. The benefits include the improvements to the system in hydraulic performance, the increase in flexibility, savings in maintenance and operating costs, and improvements in water quality. These were combined using weights, set by the decision maker, resulting in a trade off (pareto) curve between the two objectives. Dandy and Engelhardt (2001) provide an overview of water distribution pipe rehabilitation tools based on GAs and present their problem formulation, which is a single-objective GA minimising the present value of capital, repair, and damage costs. The tool is demonstrated in a case study in Adelaide, Australia. What is interesting in this work is that the GA tool was able to handle the scheduling of the works (i.e. when would each individual pipe need to be replaced) within budget constraints as well as the diameters of the replaced pipes as decision variables. It is suggested that the ability of GA tools to support future works scheduling can be used to investigate the sensitivity of the results to the various values used throughout any rehabilitation strategy. This is especially true of the indirect

---

1 For a detailed explanation of the theoretical background of GA convergence properties through natural selection (based on the *Schema* Theorem) the reader is referred to various introductory textbooks on GAs (for example Michalewicz (1996)).

repair costs, whose role in a rehabilitation policy remains a subject of dispute (Dandy and Engelhardt, 2001).

Makropoulos *et al.*, (2003) and Makropoulos and Butler (2004a; 2004b), present a spatial decision support system, based on rule-based fuzzy inference and GAs supporting three demand management strategies: In addition to leakage reduction (through pipe replacement prioritisation), the tool supports the site-specific implementation of water metering and grey-water recycling. The analysis undertaken provides a tool to: (a) Compare different water demand reducing scenarios, (b) Plan the site-specific implementation of these scenarios in the form of GIS maps and (c) Assess the investment needed to achieve a desired reduction in demand or (d) Identify the best composite scenario for a given investment scheme. The system operates in 4 steps: STEP 1: The user chooses which WDM strategies he/she wishes to take into account. STEP 2: Each strategy is broken down into a number of attributes (including pipe diameter, pipe class, pipe age, pipe material, system pressure, road types, soil ph, age of buildings, type of use for the various buildings units, income per building unit and education level per building unit), which influence its applicability in a given location. STEP 3: Each selected attribute is imported into the system as a raster GIS layer and is then used to develop a suitability map for the relevant strategy insofar as it is related to the attribute. The link between attribute and suitability for application of a water demand strategy is performed through a rule-based fuzzy inference system. STEP 4: The suitability maps for each attribute of a given strategy are then aggregated to provide one all-inclusive suitability map for the strategy's application, via a series of aggregation rules able to take into account spatially variable risk indexes (e.g. flood risk, environmental and health risk etc.) as well as the decision maker's attitude towards risk (Makropoulos and Butler (in press)). STEP 5: A multi-objective GA is used to develop a composite strategy to achieve the desired water reduction objective while minimising investment cost and maximising spatially variable socio-economic benefits. As with all multi-objective techniques the result is not a unique solution but rather a pareto front which can provide a basis for further negotiation and political choice. An example of a composite water demand strategy map developed within this tool can be seen in Figure 13.5.

Figure 13.5 Example of a composite water demand management intervention scheme (result from the tool presented in Makropoulos *et al.*, 2003).

13.5 TOOLS FOR MANAGING WATER DEMAND AT A SYSTEMS LEVEL

In systems terms, water demand management goes beyond predicting demand and implementing demand reducing strategies. It implies the management of the entire water supply/distribution system from source to tap, a system which includes from water resources to consumers and everything in between. This clearly requires comprehensive planning, taking account of numerous interdependencies between subsystems. There is of course an issue of complexity related to such large systems (Poch *et al.*, 2004). If the degree of this complexity is represented as a function of uncertainty and the magnitude or importance of the decisions to be taken, then Funtowicz and Ravetz (1990) distinguish three levels of complexity:

(1) The first level of complexity would correspond to simple, low uncertainty systems where the issue at hand has limited scope. A single perspective and simple models would suffice to provide satisfactory descriptions of the system.

(2) The second level would correspond to systems with enough uncertainty that simple models can no longer provide satisfactory descriptions. Knowledge becomes then more important, and the need to involve experts in problem solving becomes advisable.

(3) The third level would correspond to truly complex systems, where much epistemological or ethical uncertainty exists, where uncertainty is not necessarily associated with a higher number of elements or relationships within the system, and where the issues at stake reflect conflicting goals. It is then crucial to consider the need to account for a plurality of views or perspectives.

Despite the issues of complexity identified above (particularly 2nd and 3rd level complexity), there is a growing awareness (see for example WHO, 2004) that the management of water supply and demand should be viewed as one all-encompassing system and that tools should be developed to support this analysis. Even though this will be certainly be an area of R&D development within the next few years, current state of the art tools, address the two parts of the overall system (source-to-treatment and treatment-to-tap) to a large extent independently.

The development of decision support tools for the first subsystem is well documented (for example Westphal *et al.*, (2003), Koutsoyiannis *et al.*, (2003), Mysiak *et al.* (2005) and Holmes *et al* (2005) to name a few of the most recent publications) and to the extent the two sub-systems are viewed separately further discussion on these falls outside the scope of this chapter.

Work in the second subsystem focuses on the urban environment in an attempt to support decisions related to the interactions between the various parts of the system resulting from the application of demand management strategies and techniques. Within this concept, research has attempted, not so much to develop tools to *identify* "optimal solutions", since there has been growing recognition of the complexity and uncertainty of such an assertion, but to *explore* the possibilities and interaction between subsystems, technologies and strategies.

Balkema (2003) developed a mass-balance decision support tool, which is a combination of existing tools, such as life cycle assessment, cost-benefit analysis, and social inventories. The main features that make the combination more than a sum of the parts are: the explicit representation of interactions between the systems, a set of sustainability indicators that are used to measure the system's performance, the process design oriented approach for modelling and the selection of optimal technical solutions by multi-objective integer optimisation. The tool's core is constructed by superimposing a large number of alternative (technical) options managing different water sources (drinking water, household water, and rainwater), in-house water disinfection, water conservation, and wastewater treatment. The optimisation algorithms are then allowed to select combinations of these technologies and rate them against sustainability indicators. A schematic of the core mass balance tool, developed in Matlab/Simulink (by MATHWORKS) can be seen in Figure 13.6.

At a more integrated, and less detailed level, Foxon *et al.* (2000), developed a tool (known as the Reference Sustainability System (RSS)) that adopts a systems approach to model the entire water cycle in order to assess a range of water management measures including leakage reduction, water metering, low flow WCs and greywater systems. The methodology is based on representations of material and resource flows through the city and includes a representation of energy and water flow (resources) as well as wastepaper and bottled water (materials). The tool enables the simulation of a number of scenarios for future development of the system relating to future water demand. The introduction of demand management measures is compared with a base line scenario of rising per capita consumption through a number of "sustainability indicators". The three indicators used in this study were: water saved from going into supply, wastewater generation avoided and cost per unit water saved. An important aspect of this work is the fact that it has enabled the analysis of a wider range of impacts throughout the water and wastewater distribution system resulting from demand management measures.

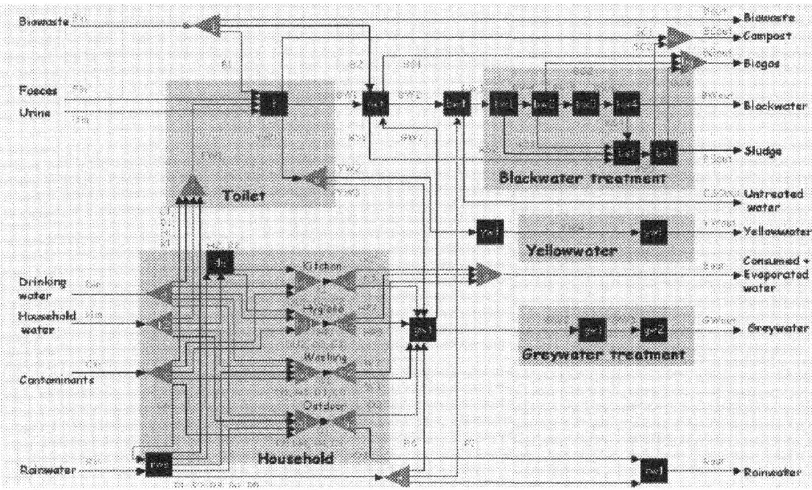

Figure 13.6 A schematic of a Simulink-based mass balance water management tool (Balkema, 2003).

Even more generic is the approach taken by Morley *et al.* (2004) who attempt to support the development of decision support systems for urban water management, including water demand management, by producing a decision support development *workbench*. They describe an architecture for a system that facilitates the rapid development of DSS applications. Custom DSS implementations supporting decision-making for a wide range of problems in urban water management may be generated through an extensible, interactive environment, featuring "drag and drop" object-oriented components and dynamic connections between them. They also present a number of components for the workbench derived from existing tools for integrated modelling, spatial visualization and advanced decision support.

13.6 TOOLS FOR NEGOTIATIONS

Developing or even exploring "appropriate" solutions based on methodologies and tools is one thing; reaching "appropriate" decisions is very much another. Brunner and Starkl (2004), in their review of methods and tools for assessing sustainable urban water management, suggest that the art of decision making consists in selecting an efficient and feasible solution on the boundaries of what the involved stakeholders consider tolerable and then explaining to the stakeholders, in what sense the selected solution was the best and why they can be satisfied with it. There are literally hundreds of decision support methodologies and techniques (see for example Brunner and Starkl, (2004) for an account of some of the most important ones and their limitations, including PROMETHEE, ELECTRE, NAIADE etc.). Apart from these general purpose techniques for collaborative decision making, there also exist computer based group decision support systems which are designed to assist the decision makers in handling complex decision making processes. These tools allow for the participation of numerous stakeholders by the use of e.g. communication and database technology, expert systems and information reporting systems. The following discussion will focus on some of the latter tools, developed for water demand management, under the assumption that communication between stakeholders can provide a basis for consensual outcomes, especially if the process is based on underlying stakeholder interests rather than explicit stakeholder positions (Brown *et al.*, 1995; Crowfoot and Wondolleck, 1990).

Becu *et al.*, (2003) present a multi-agent system (MAS) that enables the simulation of a whole catchment as well as stakeholder's individual decisions. *Agents* are computational systems that inhabit some complex dynamic environment, sense and act autonomously in this environment, and by doing so realize a set of goals or tasks for which they are designed (Maes, 1995). The term agent systems refers to systems containing agents which are: autonomous—they act independently of any controlling intelligence; social— they interact with other agents; communicative—they can communicate with other agents explicitly via some language; reactive—they observe and respond to changes in the environment and pro-active—they are goal-driven (Ferber, 1999). A review and a taxonomy of agent systems modelling and its application in environmental and water management can be found in Hare and Deadman (2004). Becu *et al* (2003) have applied their tool in water demand management for irrigation, but the work is relevant to negotiations support for water demand management in its broader sense. The general advantage of MAS comes from their agent based, systemic and highly dynamic approach. As far as human behaviour is concerned, this kind of model is better described as prospective and exploratory rather than purely predictive (Bousquet *et al.*, 1999). Expectations and beliefs of the various stakeholders play an important role in reaching the final equilibrium with a view of providing greater interaction between socio-economic and biophysical dynamics and therefore assist the development of what-if scenarios related to socio-economic as well as physical and engineering changes (Feuillette *et al.*, 2003).

Moss *et al.*, (2001), present a tool that investigates how social structure and learning affects the effectiveness of measures designed to urge consumers to save water as part of a drought management policy. Interacting agents, in this case, include households and policy makers.

Another approach to negotiations modelling is games theory, as is the case in the tools developed by Adams *et al.* (1996) based on non-cooperative models of multilateral negotiations, applied in the case of water policy development in California. The outcome of such negotiations depend on the "constitutional" structure of the game: the input each group has in the decision making process, the coalitions between groups, the scope of the negotiations and, the outcome if the parties fail to reach agreement.

Stave (2003) presents yet another approach to facilitate public understanding of water management options (in the case of Nevada) based on system dynamics. System dynamics is a problem evaluation approach based on the premise that the way essential system components are interrelated generates the system's behaviour (Sterman, 2000; Stave, 2003). If dynamic behaviour arises from feedback within the system, finding effective policy interventions requires understanding system structure. Stave (2003) suggests that system dynamics is well suited to the analysis of problems where feedback relationships and long-term time horizons are of the essence – as is the case in water demand management. They have used their dynamic models in a number of workshops to test the effectiveness of using a system dynamics model for engaging stakeholders in discussions about water demand policy. The stakeholders could suggest policy changes and measures and then observe the system response their interventions caused. Their results and experiences show considerable potential of using such tools to involve the general public and facilitate its understanding of impacts of alternative water demand policy options in an interactive and intuitive way.

13.7 CONCLUSIONS AND FUTURE TRENDS

The review of DSS and tools presented in these paragraphs, does not of course cover all relevant research initiatives related to DSS development, especially since the rate of appearance of scientific literature relevant to DSS has been steadily increasing over the past decade. It is meant to provide a flavour of options and approaches towards DSS development in the water demand management domain and act as a first portal of information and references for interested parties. Since the rate of IT evolution is such that forbids any realistic prediction about its future, this paragraph will now try to answer some key questions summarising what is considered state-of-the-art (and slightly beyond) based on the current trends in decision support tools R&D:

(1) **Who will be using these tools in the future?** There is a clear trend for more collaborative use of the tools, or development of tools explicitly for collaborative decision making, including web-based application (e.g. Salewicz and Nakayama, 2004).

(2) **How easy will these tools be developed?** Object Oriented Programming, GIS Objects as well as the emergence of the concept of tool development *workbenches* (e.g. Morley *et al.*, 2004) suggests that developing customised tools tailored to specific needs from a wide variety of pre-made components, in a drag-and-drop fashion will soon become easier and more affordable.

(3) **Where will the data come from?** Web-based Data Warehouses coupled with data mining techniques (e.g. Poch *et al.*, 2004) are becoming more accessible and cost-effective means of storing, accessing, sharing and analysing information through the web.

(4) **What kind of mathematical techniques will there be able to incorporate into these tools?** This discussion has tried to touch on a number of state of the art techniques used within the water demand management domain (including fuzzy logic, system dynamics, genetic algorithms, agents etc). Open platform, mathematically oriented, decision support system development environments (including for example MATLAB (by MATHWORKS)) coupled with development workbenches mean greater flexibility to choose appropriate techniques for each problem.

(5) **What kind of knowledge will the tools include?** There seems to be growing interest towards more intelligence and linguistic representation of knowledge to be incorporated in decision support. Interacting Agents

(Ferber, 1999) or Modelling with Words (Zadeh, 1999) are just a few examples of efforts to encode knowledge, preference, perception and other characteristics associated with human decision making into the tools that assist it.

(6) **How fast will the analysis be performed?** Parallel computing and more recently Grid computing (Berman *et al.*, 2003) has drastically improved the potential to handle highly computationally demanding problems and it now seems that the limiting factor is not the computing power but our understanding of the problems at hand, including our own perceptions about them.

Although the power of IT technology has drastically increased in recent decades, and the tools and methods have become "more intelligent", one could argue that the promise of improved management resulting from decisions based on an increased understanding of the issues and facts, currently possible through the use of decision support tools and systems, has not been fulfilled. Indeed, there is potential for quite the opposite: on occasions poor decisions may have increased due to unrealistic decision support system results generated by users without appropriate training in the use of the tools and little understanding of the fundamental operation of the system or indeed, of the decision making process. Parker *et al.* (1995) have identified this early on, in their work related to misuse of the water demand management tools. Having said this, the effort towards integrated water management solutions, including water demand management under its various interrelated aspects (from technical solutions, to legal incentives and public perception) is bound to increase the complexity of the decision space and with it the conflicts of interest arising from the (necessary) participation of multiple actors. As Brunner and Starkl, (2004) suggest, in the context of sustainable water management, these conflicts will no longer be resolvable by the elementary means of intuitive decision making. Rather they will ask for computer-aided decision making in many dimensions. As the chapter tried to illustrate in brief, there are numerous tools and decision support systems, able to assist parts of the assessment, planning and decision making process. These tools have different underlying principles and approaches and as such may generate different outputs for similar decision problems (Brunner and Starkl, (2004)). In that respect they cannot be thought of as decision *making* tools, but rather decision *support* aids, to be used to increase understanding of the issues related to demand management, including stating decision makers' attitudes and priorities explicitly – in itself not a minor achievement. The issue of the use of these tools within a decision making process and the issue of tailoring the tools to the specific needs of each process and interpreting the results within the context of each decision rather than using blindly off-the-shelf products to justify decisions already made, is an important one, but goes beyond the scope of this chapter. To pursue this issue further the reader is referred to a wide range of academic literature ranging from generic scientific debates (e.g. Ravetz (1999)) to debates on sustainable water management (Brunner and Starkl, (2004)). Concluding this discussion it may be worth keeping in mind that that the period of euphoria about tools that can do *everything* in decision making (particularly in the 70s-80s) has been followed by an equally militant view that tools can do *nothing* (mostly in the 90s), in response to the growing understanding of the importance of social and political parameters which come into play in any decision making process. It is hoped that this period of discussion about extremes is coming to an end, and that the resulting maturity in the field of both decision making and decision support will mean that tools will stop being eulogised or demonised, resuming their appropriate position as just that: tools.

13.8 REFERENCES

Adams, G., Rausser, G., and Simon, L. (1996) Modelling multilateral negotiations: An application to California water policy. *Journal of Economic Behavior & Organisation* **30**, 97-111.

An, A., Shan, N., Chan, C., Cercone, N. and Ziarko, W. (1996) Discovering rules for water demand prediction: An enhanced rough-set approach. *Engineering Applications in Artificial Intelligence* **9** (6), 645-653.

Babovic, V., Drécourt, J.P., Keijzer, M. and Hansen, P. F. (2002) A data mining approach to modelling of water supply assets. *Urban Water* **4** (4), 401-414.

Balkema, A. J. (2003) *Sustainable Wastewater Treatment: Developing a methodology and selecting promising systems*. PhD Thesis, University of Eindhoven.

Becu, N., Perez, P., Walker, A., Barreteau, O. and Le Page, C.(2003) Agent based simulation of a small catchment water management in northern Thailand Description of the CATCHSCAPE model. *Ecological Modelling* **170**, 319–331.

Berman, F., Fox, G. and Hey, T. (2003) *Grid Computing: Making the Global Infrastructure a Reality*, Wiley, New York.

Bousquet, F., Barreteau, O., Le Page, C., Mullon, C. and Weber, J., (1999) An environmental modelling approach. The use of multi-agent simulations. In: *Advances in Environmental and Ecological Modelling*, (Ed. F. Blasco, and A. Weill), pp. 113–122, Elsevier, Amsterdam.

Brown V., Smith, D.I., Wiseman, R. and Handmer, J., (1995) *Risks and Opportunities: Managing Environmental Conflict and Change*. Earthscan Publications Ltd., London.

Brunner, N., and Starkl, M. (2004) Decision aid systems for evaluating sustainability: a critical survey. *Environmental Impact Assessment Review* **24**, 441–469.

Cooper, N.R., G. Blakey, C. Sherwin, T. Ta, J.T., Whiter and Woodward, C.A. (2000) The use of GIS to develop probability-based trunk main burst risk model. *Urban Water* **2**, 97-103.

Crowfoot, J.E. and Wondolleck, J.M. (1990) *Environmental Disputes: Community Involvement in Conflict Resolution*. Island Press, Washington.

Dandy, G. C. and Engelhardt, M. (2001) Optimal Scheduling Of Water Pipe Replacement Using Genetic Algorithms. *Journal of Water Resources Planning and Management* **127** (4), 214-223.

Fedra, K., (1996) Distributed models and embedded GIS: integration strategies and case studies. In: *GIS and environmental modelling: progress and research issues*, (Eds. M. Goodchild, L.T. Steyaert, B.O. Parks), pp. 413-417, GIS-World, Fort Collins.

Fenner, R. A. and Sweeting, L. (1999) A decision support model for the rehabilitation of "non-critical" sewers. *Water Science and Technology* **39** (9), 193-200.

Ferber, J. (1999) *Multi-Agent Systems: An Introduction to Distributed Artificial Intelligence*, Addison-Wesley, New York.

Feuillette, S., Bousquet, F., and Le Goulven, P. (2003) SINUSE: a multi-agent model to negotiate water demand management on a free access water table. *Environmental Modelling & Software* **18**, 413–427.

Foxon, T., Butler, D., Dawes, J., Hutchinson, D., Leach, M., Pearson, P. and Rose, D. (2000). An assessment of water demand management options from a systems approach. *Journal of Chartered Institution of Water and Environmental Management* **14**, 171-178.

Froukh, M.L. (2001) Decision-Support System for Domestic Water Demand Forecasting and Management. *Water Resources Management* **15**, 363–382.

Funtowicz, S. and Ravetz, J. (1990) *Uncertainty and quality in science for policy*. Kluwer Academic Publisher, Dordrecht.

Goulter I. and Kettler, A. (1985) An analysis of pipe breakage in urban water distribution networks. *Canadian Journal of Civil Engineering* **12**, 286-293.

Halhal, D., Walters, G. A., Ouzar, D., and Savic, D. (1997) Water network rehabilitation with a structured messy genetic algorithm. *Journal of Water Resources Planning and Management*, ASCE, **123** (3), 137–146.

Hare, M. and Deadman, P (2004) Further towards a taxonomy of agent-based simulation models in environmental management. *Mathematics and Computers in Simulation* **64** (1), 25- 40.

Holmes, M. G. R., Young, A. R., Goodwin, T. H. and Grew, R. (2005) A catchment-based water resource decision-support tool for the United Kingdom. *Environmental Modelling & Software* **20** (2), 197-202.

Jain, A and Ormsbee, L. (2002) Short-Term water demand forecast modelling techniques: conventional methods versus AI. *Journal American Water Works Association* **94** (7), 64-72.

Jamieson, D.G. and Fedra, K. (1996) The WaterWare decision-support system for river basin planning: I. Conceptual Design. *Journal of Hydrology* **177** (3-4), 163-175.

Kim, J. H., and Mays, L. W. (1994) Optimal rehabilitation model forwater-distribution systems. *Journal of Water Resources Planning and Management* ASCE, **120** (5), 674–692.

Koutsoyiannis, D., Karavokiros, G., Efstratiadis, A., Mamassis, N., Koukouvinos, A. and Christofides, A. (2003) A decision support system for the management of the water resource system of Athens. *Physics and Chemistry of the Earth*, Parts A/B/C, **28**(14-15), 599-609.

Lertpalangsunti, N, Chan, C.W., Mason, R., and Tontiwachwuthikul, P. (1999) A toolset for construction of hybrid intelligent forecasting systems: application for water demand prediction. *Artificial Intelligence in Engineering* **13**, 21-42.

Liu, J., Savenije, H., Xu, J., (2003) Forecast of water demand in Weinan City in China using WDF-ANN model. *Physics and Chemistry of the Earth* **28**, 219–224.

Maes, P (1995) Artificial Life Meets Entertainment: Life like Autonomous Agents *Communications of the ACM* **38** (11), 108-114.

Makropoulos C. and Butler, D. (2004a) Spatial Decisions under Uncertainty: Fuzzy Inference in Urban Water Management. *Journal of Hydroinformatics* **6** (1), 3-18.

Makropoulos C. and Butler, D. (2004b) Planning site-specific Water Demand Management Strategies. *Journal of the Chartered Institution of Water and Environmental Management* **18** (1), 29-35.

Makropoulos C., Butler, D. and Maksimovic, C. (2003) Fuzzy Logic Spatial Decision Support System for Urban Water Management. *Journal of Water Resources Planning and Management*, ASCE, **129** (1) 69-78.

Makropoulos, C. and Butler, D. (in press) Spatial Ordered Weighted Averaging: Incorporating spatially variable attitude towards risk in spatial multicriteria decision-making, *Environmental Modelling & Software.*

Maltczewski, J., (1999) *GIS and multicriteria decision analysis.* John Wiley & Sons, New York.

Michalewicz, Z., (1996) *Genetic Algorithms + Data Structures = Evolution Programs.* 3rd edition, Springer- Verlag, New York.

Mohamed, A.S. and Savenije H.H.G. (2000) Water Demand Management: Positive Incentives, Negative Incentives or Quota Regulations? *Physics and Chemistry of the Earth* **25** (3), 251-258.

Mohammed O, Park D, Merchant R, Dinh T, Tong C, and Azeem A. (1995) Practical experiences with an adaptive neural network short-term load forecasting system. *IEEE Transactions on Power Systems* **10** (1), 254-265.

Morley, M., Makropoulos, C., Savic, D. and Butler, D. (2004) Decision-Support System Workbench for Sustainable Water Management Problems. In *Proc. International Environmental Modelling and Software Society Conference,* University of Osnabrück, Germany.

Moss, T. Downing, J. Rouchier, (2001) *Demonstrating the Role of Stakeholder Participation: An Agent Based Social Simulation Model of Water Demand Policy and Response,* CPM Report 00-76, Centre for Policy Modelling, Manchester Metropolitan University, Manchester, UK.

Mysiak, J., Giupponi, C. and Rosato, P., (2005) Towards the development of a decision support system for water resource management. *Environmental Modelling & Software* **20** (2), 203-214.

Parker, M., Thompson, J. G., Reynolds, R. R. and Smith, M. D., (1995) Use and Misuse of Complex-Models - Examples from Water Demand Management. *Water Resources Bulletin* **31**(2), 257-263.

Pawlak, Z. (1991). *Rough Sets: Theoretical Aspects of Reasoning about Data.* Kluwer Academic Publishers, Dordrecht.

Poch, M., Comas, J., Rodrıguez-Roda, I., Sanchez-Marre, M., and Cortes, U. (2004). Designing and building real environmental decision support systems. *Environmental Modelling & Software* **19**, 857–873.

Ravetz, J. (1999) What is Post-Normal Science. *Futures,* **31**, 647-653.

Salewicz, K.A. and Nakayama, M. (2004) Development of a web-based decision support system (DSS) for managing large international rivers. *Global Environmental Change,* 14, Supplement 1 , 25-37.

Seder, I., R. Weinkauf and T. Neumann (2000) Knowledge–based databases and intelligent decision support for environmental management in urban systems. Computers. *Environment and Urban Systems* **24**, 233-250.

Shi, Y and Mizumoto, M. (2001) An improvement of neuro-fuzzy learning algorithm for tuning fuzzy rules. *Fuzzy Sets and Systems* **118**, 339-350.

Simon, H. A., (1960) *The new science of management decision.* Harper & Row, New York.

Sinske, S. A. and Zietsman, H. L. (2004) A spatial decision support system for pipe-break susceptibility analysis of municipal water distribution systems. *Water SA* **30** (1), 71-79.

Stave, K.A (2003) A system dynamics model to facilitate public understanding of water management options in Las Vegas, Nevada. *Journal of Environmental Management* **67** (4), 303-313.

Sterman (2000) *Business Dynamics: Systems Thinking and Modelling for a Complex World.* McGraw-Hill, Boston.

Wang, L.X. and Mendel, J. (1992) Back-propagation fuzzy system as nonlinear dynamic system identifiers. In *Proc. IEEE Int. Conf. on Fuzzy Systems*, San Diego, 1409-1416.

Westphal, K.S, Vogel RM, Kirshen P and Chapra, S. (2003) Decision support system for adaptive water supply management. *Journal of Water Resources Planning and Management*, ASCE **129** (3), 165-177.

WHO (2004) *Guidelines for Drinking-water Quality*, vol. 1, 3rd Edition. Geneva.

Zadeh, L.A. (1999) From computing with numbers to computing with words – From manipulation of measurement to manipulation of perceptions. *IEEE Transactions on Circuits and Systems* **45** (1), 105-119.

Index

IWA Publishing's authorised EU representative for General Product Safety Regulations is Diane D'Arras, 15 rue Duret, 75116 Paris, France, e-mail: safety@iwap.co.uk.

Printed and bound by CPI Group (UK) Ltd, Croydon, CR0 4YY

27/03/2026

02079974-0003